BIODIVERSITY AND LANDSCAPES

BIODIVERSITY AND LANDSCAPES

A paradox of humanity

Edited by

KE CHUNG KIM
The Pennsylvania State University

ROBERT D. WEAVER
The Pennsylvania State University

CAMBRIDGE UNIVERSITY PRESS
Cambridge, New York, Melbourne, Madrid, Cape Town, Singapore, São Paulo, Delhi

Cambridge University Press
The Edinburgh Building, Cambridge CB2 8RU, UK

Published in the United States of America by Cambridge University Press, New York

www.cambridge.org
Information on this title: www.cambridge.org/9780521119337

First published 1994
Reprinted 1997
This digitally printed version 2009

A catalogue record for this publication is available from the British Library

ISBN 978-0-521-41789-1 hardback
ISBN 978-0-521-11933-7 paperback

Contents

Contributors

Paulo Barreto
Instituto do Homeme do Meio Ambiente da Amazonia, Caixa Postal 1015, Belém, Pará 66.000, Brazil

Michael J. Bean
Environmental Defense Fund, 11616 P. Street, NW, Suite 150, Washington, DC 20036, USA

John Cairns, Jr.
Center for Environmental and Hazardous Materials Studies, Virginia Polytechnic Institute and State University, Blacksburg, VA 24061-0415 USA

M. Rupert Cutler
Virginia Explore Park, 101 South Jefferson St., 6th Floor, Roanoke, VA 24011 USA

Eugene Hargrove
Dept. of Philosophy, University of North Texas, P.O. Box 13496, Denton, TX 76203-3496 USA

James Karr
Institute for Environmental Studies, Engineering Annex FM-12, University of Washington, Seattle, WA 98195 USA

Eric Katz
Department of Social Science and Policy Studies, New Jersey Institute of Technology, Newark, NJ 07102 USA

Ke Chung Kim
Center for BioDiversity Research, The Pennsylvania State University, Land and Water Research Building, University Park, PA 16802

Marli Maria Mattos
Instituto do Homeme do Meio Ambiente da Amazonia, Caixa Postal 1015, Belém, Pará 66.000, Brazil

Norman Myers
Consultant in Environment & Development, Upper Meadow, Old Road, Headington, Oxford, OX3 8SZ, England

Zev Naveh
Faculty of Agricultural Engineering, Technion, I.I.T., Technion City, Haifa 32000, Israel

Bryan G. Norton
School of Public Policy, Georgia Institute of Technology, Atlanta, GA 30332 USA

Howard T. Odum
Center for Wetlands, University of Florida, Gainesville, FL 32611-2013 USA

Alan Randall
Department of Agricultural Economics and Rural Sociology, Ohio State University, 2120 Fyffe Road, Columbus, OH 43210-1099 USA

Holmes Rolston, III
Department of Philosophy, Colorado State University, Fort Collins, CO 80523 USA

William T. Sanders
Department of Anthropology, The Pennsylvania State University, 323 Carpenter Building, University Park, PA 16802 USA

Leslie Sauer
Andropogon Associates, Ltd., 374 Shurs Lane, Philadelphia, PA 19128 USA

Ricardo Tarifa
Instituto do Homeme do Meio Ambiente da Amazônia, Caixa Postal 1015, Belém, Pará 66.000, Brazil

Harold Tukey
Center for Urban Horticulture, University of Washington, Box GF-15, Seattle, WA 98195 USA

Christopher F. Uhl
Department of Biology, The Pennsylvania State University, University Park, PA 16802 USA

Adalberto Verissimo
Instituto do Homeme do Meio Ambiente da Amazônia, Caixa Postal 1015, Belém, Pará 66.000, Brazil

Robert D. Weaver
Department of Agricultural Economics and Rural Sociology, The Pennsylvania State University, University Park, PA 16802 USA

David Webster
Department of Anthropology, The Pennsylvania State University, 316 Carpenter Building, University Park, PA 16802 USA

Garrison Wilkes
Department of Biology, Harbor Campus, University of Massachusetts at Boston, Boston, MA 02125-3393 USA

Preface

Can civilization be sustained, and for how long, without fundamental changes that ensure the conservation and restoration of natural landscapes and biological diversity? What role will science and technology play in strategies for human civilization? What fundamental changes must we make for the sustained evolution of human civilization?

The issues raised by these questions represent some of the most difficult challenges we face today. Biodiversity is the natural resource that is basic for human life. The preservation of plant, animal, and microbial diversity and of our landscapes is essential for the well-being of humans and for all other organisms. The destruction of natural and biological resources by human enterprise, for over two decades, has reached critical proportions. Current human processes conflict with and impair the earth's ability to properly respond to changing environmental conditions, threatening the capacity of natural ecosystems to meet our future needs. The control of these processes requires interdisciplinary approaches encompassing all scientific and humanistic disciplines, with a wide range of ideas and expertise.

To explore these issues in detail, Biodiversity and Landscapes: Human Challenges for Conservation in the Changing World, an international event, was held October 22–25, 1990 at University Park, Pennsylvania, U.S.A. This first-of-its-kind event, organized by the Penn State Center for BioDiversity Research, included the symposium, a "Discourse on Environmental Art," art exhibitions, films, videos, broadcasts, and a public forum. The event brought together artists, biologists, conservationists, economists, sociologists, technologists, and philosophers. Participants discussed and explored the essence of biodiversity and its relationships with culture, socioeconomics, and values and ethics; assessed historical and contemporary impacts of human civilization on biodiversity; considered the consequent effect of rapidly declining biodiversity on human existence; and formulated a consensus concerning options and tactics that could be adopted as a strategy to resolve the current human dilemma.

This book represents a culmination of this international event as well as subsequent related efforts. This volume is not the proceedings of the event's symposium. Instead of creating a compilation of presentations, chapters were specifically prepared for the book following the original ideals of the event on biodiversity and landscapes. All manuscripts were reviewed by two or more referees and revised before final editing. This volume has seven parts and 22 chapters contributed by distinguished scholars and scientists who participated in the event. In the first part, Ke Chung Kim and Robert D. Weaver, editors of this volume, set the basic theme by introducing the essence and values of biodiversity and assess the current state of human impacts on biodiversity. Part II deals with aspects of human values of biodiversity: ethical imperatives by Bryan G. Norton, moralistic and biblical perspectives by Holmes Rolston, and the moral dimension of the biodiversity crisis by Eric Katz. In Part III, both historical and contemporary human processes and related impacts on biodiversity are discussed. William T. Sanders and David Webster confront the relationships between preindustrial human impacts and environmental degradation, Robert L. Peters explores the impacts of contemporary global climate changes on biodiversity, and Norman Myers details the detriments of anthropogenic extinction on humans. Garrison Wilkes completes Part III of the volume by discussing germplasm conservation and agriculture. Four contributions in Part IV address the management of biodiversity and landscapes: Eugene Hargrove provides biological and humanistic perspectives on humans and their relation to biodiversity, Zev Naveh and Leslie Sauer present an ecologist's view and landscape architect's thoughts on landscape management, James Karr elaborates the concept of landscapes and management of ecological integrity, respectively. In Part V, Robert D. Weaver and Alan Randall discuss the values and valuation of biodiversity in the economic and political contexts. This is followed by a contribution by Christopher F. Uhl and his colleagues Adalberto Verissimo, Paulo Barreto, Marli M. Mattos, and Recardo Tarifa, presenting the case history of conservation efforts in the aging Amazonian frontier. Part VI includes six chapters dealing with different approaches to biodiversity conservation: economic perspectives are discussed by Robert D. Weaver. John Cairns and Howard T. Odum consider the role of technology and its implications on biodiversity and Howard Tukey presents a conservation approach to urban horticulture. M. Rupert Cutler discusses the role of nongovernment environmental organizations and Michael J. Bean describes legislative and public agency initiatives in ecosystem and biodiversity conservation. Robert D. Weaver and Ke Chung Kim conclude the book with a summary of the issues on biodiversity, landscapes, diverse conservation strategies, and their significance for human survival.

This book could not have been possible without the patience and commitment of all the contributing authors. We are deeply grateful for their efforts and cooperation. Special appreciation is due to Judy Cranage, who provided excellent administrative and clerical assistance throughout the book's preparation. We also wish to thank Robin Smith, Editor at Cambridge University Press, New York, for his support and encouragement, which greatly helped us get through difficult times. The success of the event, which culminated in the preparation of this book, was possible due to the talents and commitment of many other people. We take this opportunity to acknowledge these people who served in various capacities for the event and subsequent activities related to the book's preparation: The Organizing Committee—Stephen Beckerman, Charles R. Garoian, K. C. Kim, Neil Korostoff, Carl Mitcham, J. R. Pratt, J. R. Stauffer, Gerald L. Storm, James Ross Sweeney, Richard H. Yahner, and Laksham Yapa; The Advisory Board—Ralph W. Abele, Camille George, Maurice K. Goddard, John Heissenbuttel, Benjamin A. Jayne, Charles R. Krueger, Marlane Liddell, and Paul Wiegman. Special thanks are also extended to Rae D. Chambers for graphic design, Judy Cranage for administrative assistance, Deb Hager and Shawna Reppert for coordination of the event activities, Gary Petersen for conference coordination, and Lanny Sommese for the event poster. Finally, the numerous scientific and private organizations and academic units of the Pennsylvania State University, which provided support and endorsements for the event, are duly acknowledged for their contributions to the event that provided the basis for this book.

University Park, Pennsylvania Ke Chung Kim
April 25, 1993 Robert D. Weaver

Part one

Introduction

1

Biodiversity and humanity: paradox and challenge

KE CHUNG KIM and ROBERT D. WEAVER

Introduction

Biodiversity – the biotic basis of plants, animals, and microbes on the earth – is being reduced at an alarming rate, just at the time when we need it most for sustaining human life. The human destruction of this essential resource has become a major global issue and the 1992 Earth Summit at Rio de Janeiro (United Nations Conference on Environmental and Development) heightened this concern at the global level. The issue is a manifestation of a paradox that is central to technological society. Technological humanity, through both the magnitude of its populations and its actions, has caused an increasingly devastating effect on the stability of the earth system, which in turn affects all organisms and entire ecosystems, including humans themselves. This process also undermines the biosphere's capability to respond to changing environmental conditions and impairs options for human sustenance. Rapid destruction of biodiversity as a result of technological and other human activities threatens the integrity of ecosystems and landscapes; it also affects human enterprises, such as agriculture, development, and recreation. In other words, the loss of biodiversity actually constrains and counteracts economic development, which is the immediate goal of all nations of the world.

The Earth is no longer a world of distant lands and unfamiliar cultures. Our small blue planet has become crowded with 5.4 billion human inhabitants, increasingly homogenized by technological culture. No country is now insulated from global environmental and socioeconomic changes, and local/regional development activities cannot remain independent of each other. Exponential growth of the human population has occurred only for the last hundred years, reaching the 5.4 billion mark by 1992 – more than three times the barely 1.6 billion of 1900. Economic and environmental activities are now closely interrelated. Industrial production and per capita income have grown even faster than population, allowing a continual increase in our material standard of living, food production, and life expectancy (Meadows et al., 1992; World Resources Institute, 1992). The Gross World Product has increased almost five times since 1950 to $18.7 trillion, with an average per capita of $3,477 (Brown et al., 1992).

In the midst of the remarkable technological advances of the 20th century, our world has also undergone dramatic changes in the natural and sociopolitical landscapes, many of which are irreparable or irreversible (e.g., Turner et al., 1990). Nothing now looks like it did at the end of World War II. Natural resources, the basis for contemporary economic development, are increasingly scarce. Environmental problems are looming at a global level, no longer limited to local or national concerns. With the collapse of Soviet communism, the threat of nuclear war between two ideological giants has been lifted from our shoulders. The unification of Germany and the collapse of the Eastern European communist states are most recent evidence of the political change that has affected dozens of nations over the past several decades. People of all nations aspire to the material prosperity and humane economic/political systems of democracy enjoyed by the western industrial nations. Toward that end, nations rapidly accelerate economic development, with consequent high environmental costs, including the destruction of habitats for living organisms. The result has been a precipitous loss of biodiversity.

As a biological species in the biosphere, *Homo sapiens* is a part of the Earth's life-support system. Humans are not exempted from the natural functioning of ecosystem processes by science and technology. As a recent arrival in the history of life, humans must be supportive partners in sustaining the natural dynamics of ecosystem processes without which civilization cannot persist. Yet, western civilization and modern technological society have encompassed many basic beliefs premised on the dominance of human needs and the belief that humankind is above nature (e.g., White, 1967; Weiskel, 1990). Current mass extinction of species is a manifestation of the rapid global economic expansion of the last quarter of the 20th century and is now beginning to show signs of threatening the sustainability of human life-support systems. The demands of rapidly expanding human population require continued economic development to meet basic needs and to satisfy expanding economic aspirations. If current economic practices persist, this development will likely further erode stocks of already scarce natural resources, seriously deteriorating environmental quality and invariably resulting in continued impoverishment of the remaining biodiversity (Western, 1989).

The ultimate human challenge at the dawn of the 21st century is to protect the Earth's life-support system from collapse, while meeting the basic economic needs of a rapidly expanding human population. More simply put, the challenge will be one of conserving biodiversity and the dynamics of the biosphere, while limiting anthropogenic impacts and altering human roles within those systems. To achieve this goal will require new tactics and strategies drawn from a deep and thorough understanding of the implications of the current, and possible alternative, human roles in the biosphere. Thus, in this chapter we will hereafter refer to the goal of resolution of the paradox of biodiversity versus humanity as one of *conserving biodiversity*. In this context, *biodiversity conservation* will be interpreted as a tactic of systematic preservation and utilization of biological resources through means which do not fundamentally alter the general relationship between humanity and the biosphere.

The loss of biodiversity is a human paradox and a crisis of technological culture. The solution to this paradox will require no less than a rapid and major transformation of anthropocentric cultural values and the technological value system (Weiskel, 1990). The reformation of human values in the technological society must be framed within the context of the functioning parameters of the earth system, but not within the current anthropocentric, technological context. This is the ultimate challenge of biodiversity conservation. Toward resolving the dilemma we face today, biodiversity conservation becomes a most serious challenge for humankind because it is a matter of human survival. In this chapter we will discuss the underlying concept of biodiversity in the context of human ecosystems and establish the link between biodiversity conservation and sustaining the life-support system for human life.

Humanity in the biosphere

The biosphere

Life on the planet Earth has existed without humans but humanity could not sustain itself without the life of other organisms. The planet is a complex, dynamic system in which all biotic and physical factors are interconnected with subsystems such as oceans, land, atmosphere, and biomes. The system has cyclic dynamics over time and maintains a relatively steady state within which humankind has evolved. As the environmental malaise produced by the technological society for the past 100 years demonstrates, humans are not outside of nature. On the contrary, we are directly under the influence of changes in the biosphere, which are generated by our own activities (Hargrove, 1994).

The environment of this planet has supported life forms for the past 3.5 billion years (Schopf, 1993). The planet Earth, formed some 4.6 billion years ago, has been inhabited by unicellular and bacterial organisms for over 2.1 billion years (most of its formative years). Multicellular, eucaryotic organisms first appeared 1.4 billion years ago, and the first animals evolved about 570 million years ago, at the beginning of the Cambrian period (Cloud, 1983). Throughout geological time the global environment has been changing and life has adapted continuously to these changes. Natural landscapes have constantly changed and new habitats been created. This process has resulted in the formation of new species and the extinction of many others, forming the unique interactive system called the *biosphere.* With the lithosphere, hydrosphere, and atmosphere all living things – animals, plants and microbes – are integrated to constitute the biosphere (Cloud, 1983).

For the next 100 million years, the oceans, which contained about 10% of the current supply of oxygen, were occupied by a worldwide succession of ancient animals (Cloud, 1983), and the land began to be colonized by diverse plants and animals (Siever, 1983; Gray and Shear, 1992). Toward the end of the early Silurian period, about 420 million years ago, the landscape began to change dramatically with a rapid evolution of vascular plants (Gray and Shear, 1992). The terrestrial environment continued to change and complex ecological systems were developed on land, with a wide diversity of plants and animals by the Devonian and Carboniferous periods (300–400 million years ago). By the Permian–Triassic periods, about 200 million years ago, large mammal-like reptiles dominated, and for the next 150 million years dinosaurs flourished and dominated much of the planet. They disappeared suddenly at the end of Cretaceous period, 65 million years ago. Following the massive extinction of dinosaurs the Cenozoic era brought the age of mammals. With the relatively warm, moist climates diverse mammals evolved and flourished on the planet. The evolution of hominoids began during the Pliocene epoch bringing the first hominid *Australopithicus,* which evolved in Africa.

The evolution of humankind

Two important steps in hominoid evolution, the evolution of bipedalism and the enlargement of the brain, developed during the late Miocene epoch. The evolution of *Australopithecus,* which appeared in Africa 5–5.5 million years ago and disappeared about 1.3 million years ago, set the path for human evolution. The first tool-making hominid, *Homo habilis,* appeared in a wide area of eastern and southern Africa about 2 million years ago and survived for about 500,000 years. Following the evolution of *H. habilis,* the first ancestor of modern humans, *Homo erectus,* with a strong similarity to our own species, was widespread in Africa and later through Europe and Asia from 1.7 million to 250,000 years ago. Although the evolution of *Homo sapiens* has been debated, modern humans most likely appeared about 500,000 years ago (e.g., Dobzhansky, 1962; Mayr, 1978; Pilbeam, 1984).

Human ancestors lived by gathering, herding, and hunting, and they had to have a knowledge of their natural environment and of how to utilize animals and plants to sustain their lives. As with all species, their survival depended on this consumption and use of other species. As natural environments evolved worldwide, and as the human population expanded, humans had to change their way of life drastically by adopting land-intensive systems of food production such as cropping and animal breeding (Ponting, 1991). The large human brain allowed the evolution of intelligence and the capacity for learning and invention. This led to the development of the use of symbols and language, the emergence of culture, including aesthetics and the arts (Dobzhansky, 1962; Birch and Cobb, 1990), and the development of cultures which placed high levels of demand on their environment (Malde, 1966; Meggers, 1966). The culmination of this process came as diverse human cultures became ecologically dominant and self-centered, losing their consciousness and appreciation of ecological interactions and interconnectedness and the human role as a component of the biosphere.

Nevertheless, as the capability of intensive food production evolved, stable societies developed worldwide and human population growth accelerated rapidly. The world population gradually increased from 4 million about 10,000 years ago to 5 million in 5000 BC, doubling every millennium to 50 million by 1000 BC. This doubled to 100 million in the next 500 years. At the time of the Han Dynasty and the Roman Empire in 200 AD, the world population reached 200 million and the pace of human population expansion continued to accelerate (Ponting, 1991). The world was inhabited by 910 million people during the Industrial Revolution (circa 1800), but by 1900, a mere 100 years later, the number of human inhabitants had increased to 1.6 billion. During the twentieth century, human population has been growing exponentially, reaching the 5.4 billion mark by 1992.

Intelligent species and culture

In terms of geological history *Homo sapiens* is a recent arrival in the succession of life forms. Yet the development of a brain conferred the capacity of high intelligence, with which the capacity for abstract thought and communication by language evolved, thus developing human cultures and technology. Modern humans have the largest cranial capacity of the hominoids, 1,200–1,500 cc, whereas *Australopithecus* had barely 450–550 cc and *Homo erectus erectus* measured barely 700–1,000 cc (Dobzhansky, 1962).

Homo sapiens is the most intelligent species, with unique genetic and biological traits. This level of intelligence, unprecedented in the history of life, has provided the human species with a unique capacity to consume, utilize, and manipulate other species and the processes of ecosystems. As early as 30,000 years ago, ancestral humans had adapted to different geographic areas and climatic conditions and already had developed a variety of cultural modes in tool making, shelter, diet and food production, social arrangements, and religious expressions (Ponting, 1991). The Neolithic Revolution, beginning perhaps 10,000 years ago, brought significant technological and social changes such as agriculture, domestication of animals, and manufacturing of pottery and textiles (e.g., Gregg, 1992). The advancement of cultures has influenced the biological and physiological capacity of humans (Mayr, 1982). Between genetic and cultural inheritances there has developed a positive feedback loop through which the human species has increasingly dominated the Earth's ecosystem. Through this process a modern technological society developed. These revolutionary changes promoted a rapid increase in human population and increased the scope of its control over its environmental surroundings (e.g., Tudge, 1989). Great civilizations arose in many parts of the world and many phonetic scripts were invented before 1000 BC. Beginning with the religious awakening of the first millenium, human civilization went through a tumultuous history – development of nation states, the Renaissance, and the Reformation (Ponting, 1991). Finally, the Industrial Revolution transformed agrarian cultures into technological societies and this process allowed us to substantially intensify the productivity of human labor. In turn, the intensity of human utilization of and impact on the ecosystem processes and the dynamics of the biosphere has expanded exponentially.

Humans are a biological species

The life and destiny of humankind has been shaped by the biosphere or "nature" through its interactions with all other organisms and their environment. Now, however, our dominance has expanded to encompass the dynamics of the Earth's ecosystem, yet we neither clearly understand how it functions nor can we control or predict the direction of the human selection process. Our science cannot yet define the full extent of human genetic capacity necessary to meet the biological and environmental challenges we have created. Humans, like other species, do not have unlimited genetic and biological capacity to continuously adapt to rapidly changing environments. Within this context, human control of the dynamics of the biosphere is *not* a reasonable expectation.

Since the Industrial Revolution, we have transformed the planet into a web of human ecosystems in which the cultural force has begun to overtake our biological capacity. We have changed our environments and landscapes (e.g.,

Jacobsen and Firor, 1992; Sanders and Webster, 1994). Many of these changes are permanent and irreversible. In other words, humans no longer live in a natural environment but rather inhabit human ecosystems which are increasingly devoid of nature. Urban environments such as city slums, where there is no nature, no trees, and no wildlife, are infested with poverty, drugs, and crime. The continuous advancement of technology and economic development with the acceptance of human dominance over the biosphere has made humans increasingly detached from nature. Technological humans have become a major factor in global environmental problems such as climate change, biodiversity destruction, and the stress on the genetic and biological capacity (Peters and Lovejoy, 1992; Stern et al., 1992; Myers, this volume). This reality may ultimately determine the destiny of the human species.

Biodiversity and human life

Biodiversity and species

Biodiversity is not a simple collection of species but a reference to the diversity of life (Wilson, 1992). No species is endowed with all the genetic and phenotypic attributes needed to fit into every ecological niche, and no single species can survive alone without interaction with other species (Noss, 1990). The human species is no exception. Conversely, no healthy ecosystem can function without its primary component species (Odum, 1989; Robinson et al., 1992). Therefore, biodiversity is the integration of the varieties and variation of all living organisms in relation to their habitats and ecological complexes (Norse et al., 1986; Noss, 1990). It has composition, structures, and processes, all organized into a nested hierarchy (Franklin, 1988; Walker, 1992). It involves all biological and environmental levels from genes through populations and species to ecosystems and landscapes.

Today's biodiversity represents the manifestation of a long evolutionary process through which many species became extinct and new species have arisen. The composition and structure of biodiversity has changed through time as organic evolution has persisted. In this process, genetically unique species were molded over time by their own genetic and environmental forces interacting with other species. In addition to an ethically based right to exist (Rolston, 1994), all the species have evolutionary and ecological reasons for existence that should be respected and preserved to sustain the life-support system shared by both humans and all other organisms (Wilson, 1984).

Every species occupies a specific niche with a specific set of habitat requirements and has a definite range of distribution. Each species has definite and specific roles to play in sustaining the dynamics of ecosystem processes as producers, consumers, decomposers, parasites, and predators (Olembo, 1991; Pimm, 1991; Walker, 1992; Holt and Gaines, 1992). Biodiversity, in a practical sense, is the basic resource for ecological services and for sustaining Earth's life-support system. Thus, the value of biodiversity to humanity lies not only in its economic utility but also in its capacity to sustain human life and the systems that support it (Hargrove, 1994; Norton, 1994; Katz, 1994; Randall, 1994; Weaver, 1994). Furthermore, the value of biodiversity to humanity must also be viewed as involving intrinsic aspects (Katz, 1994; Norton, 1994) that go beyond the current utilization value of its resources (e.g., Bormann et al., 1991).

Species richness

Approximately 1.5 to 1.7 million species of plants, animals, and microbes have been named and described (Wilson, 1985a,b; Groombridge, 1992). Since Erwin (1982, 1983a,b, 1991) provided an unexpectedly high estimate of insect diversity, global species richness has been estimated differently (May, 1986, 1988, 1990a; Williams et al., 1991; Gaston, 1991a,b; Groombridge, 1992). Regardless of which estimate may turn out to be accurate, the actual total number of existing species is much higher than those discovered so far. The total number of species on earth may be somewhere between 10 to 30 million including 8 million of insects. In numbers, arthropods alone make up 91% of all the living animals and 79% of the global diversity (Groombridge, 1992). If we consider current global biodiversity to be 10 million species, our present knowledge of 1.7 million species amounts to a mere 17% of the total. That our present biological knowledge is based on such a small fraction of total biodiversity is unsettling.

Biodiversity as a source of natural resources

Biodiversity is the nested composite of plants, animals, and microbes and is the basis for ecosystem processes and the fountain of humankind's life-support system. Thus it may be considered as a unique, irreplaceable source of natural resources (Weaver, 1994), most of which are as yet undiscovered or undocumented for their biological roles in natural ecosystems (Myers, 1983, 1994). Furthermore, biodiversity provides a highly valuable source of knowledge of the function and evolutionary history of organisms, as variation and interactions of species that constitute the basis for ecosystem processes are the manifestation of long evolutionary processes. They include potential sources of food, fiber, fuel, ornamentals, antibiotics, pharmaceuticals, and other products that satisfy human needs (Myers, 1979a,b; Kim and Knutson, 1986; Wilson, 1988, 1992; Weaver, 1990). Unfortunately, the consumption benefit of the species being utilized for food, fiber, and pharmaceuticals is at present only a minute fraction of the total benefits biodiversity could provide for humanity. Much of the unexplored biodiversity includes many thousands of species of plants that may prove to be superior to those already cultivated. For example, only 5–15% of the extant plant species have been surveyed for their biologically active natural products (Balandrin et al., 1985). Similarly, there are vast numbers of insects and arachnids that are superior to known pollinators or parasitoids and predators of pests. Only 20 crop species provide 90% of the food supply, of which just three species, namely wheat, maize, and rice, supply more than half of the global food supply (Myers, 1979a,b).

The potential of today's biotechnology is limited because of scarce information on the biological properties of potentially useful species and because of the limited number of species available for genetic engineering (Oldfield, 1989; Weaver, 1990). Further limiting the advancement of biotechnology is the highly homogenized genetic resources upon which present biotechnology is based, representing a narrow range of genetic variation developed by human genetic manipulations over the last century. By expanding our knowledge of biodiversity we will expand the material basis for future biotechnology and food production (Markle and Robin, 1985; Boussienguet, 1991; Bull, 1991). Future advances of agriculture and biotechnology will be linked closely to new wild germplasms including many species that are currently unknown or poorly studied (Weaver, 1990).

Nature for humanity: biophilia

Trees, flowers, gardens, wild animals, and landscapes are not merely decorative or aesthetic entities. They are aspects of nature in which humanity exists. Living organisms exist only because of their interactions with other things, both living and inanimate, which constitute their environments. Our life support system is a network of constituent elements, with which we have "internal relations" despite our current domination (Birch and Cobb, 1990). In other words, humans have *biophilia,* defined here as "the human's innate need of nature," an essential aspect of humanity and an element of human conscience. Satisfaction of biophilia is fundamental to maintaining the natural course of human evolution and sustaining our cultural life.

The term *biophilia* was first used by E. O. Wilson in 1979 (*New York Times Book Review,* 4 January 1979, p. 43), and subsequently became the title of his book (Wilson, 1984). Wilson defined the term as "the innate tendency to focus on life and lifelike processes," considering that "to explore and affiliate with life is a deep and complicated process in mental development." Here, we dare to expound the Wilsonian concept of biophilia a step further as an aspect of humanity that is essential to human evolution – humanity cannot be sustained without it. Over the last one hundred years, the development of technological society has increasingly abstracted humanity from contact with nature, thus diminishing the satisfaction of biophilia. This process will continue with further evolution of the demands of our technological society for increasingly high rates of consumption, utilization, and manipulation of biological resources.

The rise and evolution of human species has been shaped by nature for about 500,000 years, interacting with other organisms and their environment. Being a part of nature, humans require a continued interaction with plants, animals, and microbes (i.e., the "biophilia"). Biodiversity is a source of the human spirit, inspiration, and intelligence (Orr, 1992). We draw our human spirit from nature, which consists of biodiversity and landscapes, that is formed by synergistic interactions of plant, animal, and microbial species, as Wilson (1984) eloquently describes in his book *Biophilia.* Simply put, humans need to interact with wildlife in a natural environment to sustain their biological and cultural inheritance. Biodiversity conservation, therefore, can be viewed as necessary to ensure the preservation of opportunities for biophilia.

For the past century humans have been gradually abstracted from nature through technological advancement. In primitive cultures humans were clearly a part of their ecosystems. As their subsistence completely depended on plants and animals in their environment, humans had to maximally utilize biodiversity in a sustainable way. However, as technological process has been made, human life has increasingly become preoccupied with its man-made environment, thus failing to sustain biophilia. In many developing countries, people who survive at a subsistence level presently are doubly threatened by the low level of their technology. They neither enjoy material comfort nor have the opportunity to nurture their biophilia as their efforts are directed to improving their standard of living. In city slums, there is virtually no natural environment and no opportunity to fulfill biophilia. As urbanization intensifies [in 1985, 41% of the human population resided in cities (Ponting, 1991)], this trend is likely to be accelerated.

Biodiversity conservation in a broad sense constitutes a strategy for restoring the biophilia of humans. This strategy includes nature reserves, zoos, botanical gardens, urban horticulture (Tukey, this volume) and in situ germplasm conservation (Wilkes, 1989, 1994). Similarly, civic and social programs related to plants and animals may contribute to the satisfaction of biophilia. A recent upsurge of these activities, such as The People–Plant Council in Virginia, garden clubs associated with botanical gardens (Aldrich, 1993), and a recent doubling of membership in the Sierra Club (now reaching the million mark), is witness to human efforts to satisfy biophilia.

Extinction and technological society

Extinction as an evolutionary process

Extinctions have been occurring since life began 3.5 billion years ago (Raup, 1988, 1991; Margulis and Olendzenski, 1992). Extinction is a natural process of terminating an evolutionary lineage and is brought about by the disruption of ecological processes. The process of extinction may be initiated by biological or environmental factors operating at the habitat and ecosystem levels. Geologic mass extinctions were caused by natural environmental disturbances, whereas the current mass extinction is caused by the activities of a single species, *Homo sapiens*.

Extinction provides an opportunity for speciation and ecological reorganization in a selectional ecosystem, but this process takes a long evolutionary time. In retrospect, although past geological extinctions appear to have occurred relatively rapidly, they were gradual in evolutionary time, taking millions of years before species permanently disappeared and the ecosystem-level impact became apparent (Patrusky, 1986; Kaufman and Malloy, 1986; Raup, 1988, 1991; Jablonski, 1991). The Cretaceous–Tertiary extinction, for example, killed 60–80% of the biodiversity, including dinosaurs and clam species, over a span of more than 50 million years (Raup, 1991; Raup and Jablonski, 1993). However, the current mass extinction is particularly alarming because it is, unlike past geological mass extinctions, caused by the activities of a single species over a very short period of time (less than 100 years).

Human role in the extinction process

The Industrial Revolution ignited the process of rapid technological innovations. The onset of rapid economic development after World War II resulted in further expansion of rates of extraction and depletion of natural resources and degradation of environmental quality. These processes along with the rapid increase in human population have begun to threaten the dynamic processes of the biosphere's human life-support system and contributed to the current massive extinction (e.g., Sarokin and Schulkin, 1992), substantially reducing the quality of human life.

Innovations in agricultural technologies allowed the use of the biosphere's resources to increase production of food based on both per capita and per unit land area. The Industrial Revolution brought similar increases in the productivity of labor, rapidly intensifying human use of biosphere resources. In both agriculture and industry, use of biosphere resources was expanded through direct utilization and transformations of natural resources, utilization of biosphere services (e.g., heat, moisture, wind, and light), and the utilization of biosphere reserves (e.g., streams, rivers, bodies of water, or atmosphere as waste sinks). Importantly, this last type of utilization owed its origins to and represents a manifestation of the extent of human abstraction from nature that had evolved and of the human acceptance of dominance over ecosystem processes, even in the absence of human knowledge and control of the dynamics of the biosphere. In fact, the Industrial Revolution directly expanded both the extent of human abstraction from nature and the need to dispose of enormous volumes of waste in central locations. The process of human abstraction from nature can be viewed as a natural product of the centralization of labor in factory settings. This process of industrial centralization mimicked the impacts of expanded agricultural land use on occupation and transformation of habitats as urbanization exploded.

As large parts of the global biodiversity are destroyed, many species, both described and unknown, have become extinct (Wolfe, 1987). Extinction is escalating at a rate unprecedented in the history of life. If this trend continues, a significant proportion of the global biodiversity (as much as 25–30% or as many as 500,000 species) will perish by the year 2000 (Myers, 1979b, 1983, 1989b; Lovejoy, 1980; Sayer and Whitmore, 1991). For example, the rate of species loss from deforestation alone is about 10,000 times greater than the rate of natural extinction prior to the appearance of humans on this planet (Silver and DeFries, 1990). Reid (1992) estimated the loss at 2–8% of the global biodiversity in the next 25 years if the current rate of deforestation continues.

By habitat destruction, environmental degradation, and ecosystem fragmentation technological humans are directly responsible for the current mass extinction (e.g., Saunders et al., 1991; Myers 1989a; Peters, 1994). Furthermore, these impacts of technological society are likely to get worse as more people are added to the planet (Brown, 1992). In 1992 the world population of 5.4 billion grew by a record number of 92 million (Brown et al., 1992). If this trend continues, the 1960 population of 3 billion will double by the year 2000. Global economic output, which increased almost five-fold during the last 30 years, will increase at an even higher rate. With the projected growth of the human population and related economic development (Western, 1989), accelerating damage to the global environment is likely, and the destruction of biodiversity will continue worldwide (Lovejoy, 1980; Myers, 1989a,b). This trend may ultimately threaten the planet's capacity to support life and sustain human civilization, unless drastic measures are made now to reverse the process.

Extinction and ecological dysfunction

The loss of species is not the simple loss of a collection of species. With the impoverishment of biodiversity we are losing the most basic resources and related dynamic processes that support our own life-support system, much of our natural heritage, as well as unknown benefits of past evolutionary innovations (Myers, 1979b; Wilson, 1988). This also means that we are losing food resources, such as marine fish stocks and wild germplasms, and the material basis for all human activities and needs, such as agriculture, recreation (fishing and hunting), ecotourism, and forestry.

Because of the intricate ecological roles they play in ecosystem processes, the loss of species can change the selection environment of the ecosystem, thus affecting the interactions of the remaining component species (Vermeij, 1986; Robinson et al., 1992) and altering the dynamic paths of the biosphere. Extinction of a species may threaten its associated species; when host animals or plants become extinct, their parasites, particularly those that are obligate, permanent, or host-specific, will also become extinct. Similarly, if a species becomes extinct in an ecosystem, other ecological associates may also perish. For example, many foraging bees are specific to particular types of flowering plants. If these bees disappear, the loss of these pollinators and foragers will directly affect the survival of their host plants. The loss of species may also disrupt or destroy many ecological services. The loss of many decomposers, such as blow flies, will dramatically reduce the rate of decay and recycling, which will result in mountains of slowly decomposing animal carcasses and vegetable matter (Swift and Anderson, 1989). Furthermore, along with the loss of species in a guild, the dynamic ecosystem process will be disrupted (Pimm, 1991), and if disruption is severe enough, the ecosystem may collapse (Samways, 1989).

Habitat destruction, environmental degradation, and the resulting extinction of species make it difficult to understand and predict the dynamics of ecosystem processes (e.g., Ravera, 1979). Without adequate knowledge of biodiversity and ecological assessment of component species in ecosystems, it is neither possible to understand precisely how the ecosystem functions, nor to realistically predict how the ecosystem with decreased biodiversity or major species substitution will behave in changing environments mostly caused by human activities (e.g., Myers, 1994). Ecosystems require time to adjust and reorganize in response to the massive loss of species (Raup, 1988, 1991;

Jablonski, 1991). However, contemporary mass extinctions taking place in a short time at a high rate do not permit the ecosystem to make necessary ecological adjustments, ecosystem reorganizations, and long-term evolutionary responses for sustaining its dynamic processes (Briggs, 1991a,b).

The challenges to humanity

Development and human paradox

Economic development and growth are the outcomes of the technological zeal and material aspirations of our society, resulting in part from the necessity of meeting the needs of a rapidly expanding human population. Such development has been possible because of human capacity to utilize and manipulate natural processes and the environment. Yet, technological humans have been unable to understand, control, predict, or prepare for the variety of grave biological, environmental, and social problems that now face humanity.

The technological advances of the past 100 years have improved the biological and cultural capacity of humankind. Many diseases such as smallpox, polio, typhus, and plague have been nearly eradicated, and infant mortality has decreased from 155 per 1000 in 1950 to 63 per 1000 in 1991 (Brown et al., 1992). Medical advancement has increased life expectancy in industrial nations, reaching more than 75 years of age by 1983. With the control of devastating diseases, improved health care, increased longevity, and better diet, the world human population has doubled from 2.565 billion in 1950 to 5.409 billion in 1991, with recent animal increases of 92 million (Brown et al., 1992). People of the world are increasingly more affluent. The Gross World Product (GWP) increased from $3.8 trillion in 1950 to $18.7 trillion in 1991 (1987 dollars). In other words, the Gross World Product, as a measure of economic well-being, doubled in 40 years from $1,544 per capita in 1950 to $3,477 per capita in 1991 (Brown et al., 1992).

To support economic development and the rapidly increasing human population, the production of energy, food, and materials had to grow rapidly, requiring accompanying expansion of human cultural capacity. For example, world food grain production has risen from 631 million metric tons in 1950 to 1,194 million metric tons in 1971 and 1,696 million metric tons in 1991 – an increase of more than two and one-half times in 40 years (Brown et al., 1992). Likewise, commercial energy production increased by 14% for the decade from 1979 to 1989 and similarly, the production of metals and other necessary commodities has also grown (World Resources Institute, 1992). These expansions in output of consumable goods and services have been facilitated by rapid evolution of human institutions and cultural practices to accommodate what has now become a global scope of utilization and destruction of ecosystems and the biosphere by humanity. Human institutions have evolved from local to global levels allowing global access to and exchange of ecosystem and biosphere resources.

Along the way we have modified the evolutionary forces that controlled our cultural capacity and limited our life expectancy. As a result, human populations have aged greatly worldwide. People age 65 or older constituted less than 1% of the human population of 1.6 billion in 1900. Since then, the age structure has changed drastically, and by 1992 people over 65 or older accounted for 6.2% of the population of 5.4 billion. It is predicted that by 2050 this age group will make up about 20% of the population, totaling at least 2.5 billion (Olshansky et al., 1993). This change in age structure will strain many social programs, such as retirement, and overburden the financial capacity of society to pay for health-care costs. For example, each U.S. citizen will spend an average of $3600 on health care this year, or about $940 billion for the United States as a whole. The skewed age structure impacts the job market, housing, transportation, energy costs, retirement pattern, and many other aspects of life (Olshansky et al., 1993). The clear implication of this process for the impact of technological society on biodiversity and the biosphere is to further accelerate the intensity of human demands on the biosphere. As the human population ages, the active, productive population is increasingly required to produce sufficient output to sustain both their own growing demands as well as those of less productive aged individuals. From this perspective, aging of the population results in increased pressure on the biosphere at an extent and rate that exceeds increased demands associated with population growth through births.

Although total global food production continues to increase, there are ominous signs that food security is declining. In 1991 the world per capita grain production stayed the same. Similarly, the world grain stocks declined from 102 days in 1987 to 66 days for 1991, and the total area in grain production increased only slightly (World Resources Institute, 1992). In the face of erosion and degradation processes that threaten world agricultural systems, to meet the future world need for food grains may be a difficult and challenging task.

Likewise, the modest increase in world energy production has not met rapidly increasing energy requirements. While commercial energy production increased by 14% between 1979 and 1989, commercial energy consumption increased by 18% and energy requirements by 22% (World Resources Institute, 1992). Continued consumption of energy and metals is continually depleting the remaining reserves. At the same time, world global energy consumption has contributed greatly to atmosphere CO_2 concentrations, which increased from

315 parts per million in 1959 to 355 parts per million in 1991. Burning fossil fuels has increased its contribution to global carbon emissions 3.6 times since 1950, reaching 5,854 million metric tons by 1991. All these factors have contributed greatly to global climate changes (Budyko and Izrael, 1991; Brown et al., 1992; World Resources Institute, 1992).

Modern technology has continued to produce new synthetic chemicals for diverse commercial and domestic uses, while many products once considered safe are now banned because of their environmental and public health hazards. Chlorofluorocarbons (CFCs), for example, were hailed as safe and inexpensive refrigerants in 1930 but have been recently identified as major culprits in the depletion of the ozone layer. Use of CFCs is now being rapidly reduced and will be completely phased out by 1999 under the Montreal Protocol (Brown et al., 1992). Although in the early 1940s DDT (dichlorodienyltrichloroethane) was hailed as an outstanding insecticide and used widely, it was completely phased out by 1972 in the United States because of its serious biological and ecological hazards. Subsequently, DDT and many other organophosphate pesticides are mostly banned worldwide. Nevertheless, DDT and other subsequent groups of pesticides along with a combination of control strategies have reduced pest densities and helped to eradicate certain vector-borne diseases such as malaria in much of the temperate world. Paradoxically, despite all human effort to control pests with new chemical pesticides during the past 100 years, no single pest species has been eradicated and the percentage of crop damage due to pests remains fairly constant.

Population growth and economic development necessitate the construction of more roads and highways, resulted in paving over, draining, or permanently destroying natural filtering systems in forests, grasslands, wetlands, and prime farmlands. Extensive paved surfaces and buildings contribute to excessive runoff of industrial and domestic wastes. Rainwater picks up excesses of chemical fertilizers, pesticides, and animal wastes from agricultural production and lawn management, as well as other domestic and industrial pollutants, such as heavy metals, petroleum-based products, and air-borne acids. Rivers eventually discharge these pollutants into estuaries and oceans, causing serious pollution problems: oyster beds are destroyed, and fishes and other marine animals are contaminated and killed with toxicants (e.g., Horton, 1993, on Chesapeake Bay).

While the GWP is rapidly increasing, economic disparities among rich and poor nations are also widening. Per capita GNPs now range from a high of $30,070 in United Arab Emirates to a low of $80 for Bhutan (Kurian, 1984). In 1980, when the average worldwide GNP per capita was $2,430, only 20 nations out of 171 had a GNP per capita over $10,000, and 54 had less than $600. The pattern also shows great disparities in living conditions, education, and health (Southwick, 1985). This economic pattern will add to global socio-political challenges to the current use of ecosystems by our technological society.

Evolution and sustainability

A Crow Indian spiritual leader, Burton Pretty On Top, Sr., recently stated in his teaching of North American Indians that humankind has a sacred responsibility to take care of this world as God is within all things (Letters, *Time,* May 31, 1993, p. 9). Safeguarding the world environment is our ultimate responsibility for two reasons. The human species, as a part of the biosphere, must be a responsible participant in sustaining the viability of the planet Earth for life, and such a human role is required for the sustenance of human civilization.

Zeal for enhanced standards of living through economic development and technological advancement has begun to overtake the ecological capacity of the biosphere and challenge the biological and intellectual limits of humans. We have changed all the things surrounding us that provided the basis for our materialistic prosperity and technological advancement. Our lifestyles and landscapes have permanently changed. Our world is crowded with 5.5 billion people whose needs and economic aspirations are rapidly increasing. The natural resources upon which past economic growth was based are rapidly becoming depleted or polluted. The global environment is rapidly deteriorating. Economic development and urbanization has made us increasingly detached from nature, resulting in the loss of biophilia. The costs we incur for economic affluence go well beyond those reflected by market prices and must be viewed as including the costs of ecosystem services and utilization of natural resources and ecological processes. Although these costs are largely unknown, they are likely to be enormous. At a scale that cannot be repaid by values created by the current processes of our technological society, new approaches are needed to finance these ecological costs.

The challenge we face is not only to reduce environmental pollution and preserve biodiversity but also to make fundamental changes in the aspirations, approaches, and trends of our technological society. The world no longer has the luxury of treating superficial symptoms and trends without attacking the basic causes of all these environmental and human problems, with natural resources being depleted, environmental quality deteriorating, and the human population increasing, a salient question is how to provide basic needs and satisfy economic aspirations of all the people throughout the world while controlling the current trends and sustaining the basic dynamics of ecosystem processes. All human enterprises including economic development, food production, and health care improvement must be sustainable, because all the resources upon which these activities depend are finite and no longer naturally renewable. To sustain humanity, we must immediately find ways to build sustainable societies.

Humanity and biodiversity: opportunity

This chapter has aimed to provide a statement of the challenges faced by humanity and its current technological society. Both the role of humanity in the biosphere and the role of biodiversity in human life point to an inextricable interdependency that has yet to be clearly recognized within the context of current technological society. The imperative to recognize this interdependency has evolved as human dominance in utilization of material resources and ecosystem processes in the biosphere appeared to challenge both the survival of humanity and the biosphere itself. As evidenced by the enormous scale, scope, and speed of the current mass extinction, the imperative demands immediate redress by current technological society. Although the magnitude of this challenge would seem clear, our human view of the challenge is obscured by the inadequacy of our knowledge, the myopia of our perspectives, and the imperatives of our existence. The chapters in this volume attest to the conclusion that it is paramount that humanity accepts the challenge to resolve the paradoxical relation between humanity and biodiversity.

The imperative for resolution of the paradox of biodiversity and humanity through conservation of biodiversity and the biosphere will require wisdom drawn from multicultural and interdisciplinary bases (e.g., Master, 1991; Soulé, 1991). This effort will require a thorough reconsideration of humanity's relationship with biodiversity and the biosphere, which goes so far as to question the logic and appropriateness of human dominance where human control and prediction of the dynamics of impacted biological systems are infeasible. Every sector of the world community must participate in the global effort, requiring a generous infusion of energy and financial contributions.

New science and technologies are needed to conserve biodiversity and the biosphere (e.g., Mitsch, 1993; Odum, 1994). New technologies will protect endangered and threatened species, enhance habitat restoration, and encourage landuse planning and other conservation related enterprises, while finding the way to sustainably feed, clothe and provide shelter for all of the human population. Conserving biodiversity and the biosphere will require a new synthesis of knowledge, technology, and cultural practices and values as it aims to save both endangered species and the dynamics of ecosystem processes (e.g., McNaughton, 1989). Biodiversity research must become an anticipatory and preventive science, in contrast to conservation biology, which has been the science of crises and preservation. To conserve biodiversity for sustainable ecosystems, we must know: a) what remains of the global biodiversity, b) what ecological roles different component species play in the ecosystem, and c) how ecosystem processes function, particularly under conditions of reduced biodiversity or considerable species substitution.

Biodiversity conservation will require both basic and applied research at different levels of biological hierarchy from genes to landscapes (e.g., Soulé, 1989; WRI, IUCN and UNEP, 1992). Research geared for short-term as well as long-term strategies involves activities from salvage inventory and habitat restoration to landscape planning. As our current knowledge of biodiversity does not permit an accurate prediction of the process of biodiversity rehabilitation, research must be focused on the structure and function of biodiversity and ecosystems. Technology needs to be developed for restoring or rehabilitating habitats and biodiversity.

As current mass extinction is a manifestation of the development of our technological society, strategies for conserving and restoring biodiversity and the biosphere must include both short-term actions and long-term programs in research and education involving a broad spectrum of disciplines and expertise. Diverse economic and environmental conditions dictate different research priorities, although common priorities can be defined as those identified by the Society for Conservation Biology (Soulé and Kohn, 1989). Effective implementation of biodiversity conservation strategies cannot be accomplished without involving all humankind; this is the only way for us to persist as a viable species on this planet.

The current state of our planet reflects an underlying paradox between humanity and biodiversity. It represents the most serious challenge faced by humanity, and provides a historic opportunity for technological humans, as demonstrated by the 1992 Earth Summit at Rio de Janeiro (Haas et al., 1992; Parson et al., 1992). We must feed, clothe, and shelter the entire human population while safeguarding the dynamic process of Earth's life-support system. The challenge is to find and adapt cultural means through which the present paradox is resolved. Clearly, any strategy to meet these apparently conflicting imperatives must include biodiversity conservation based on ecological justice (Katz, 1994). The anthropogenic causes of extinction must be reduced and destructive processes must be reversed. Achieving these goals will preserve the remainder of global biodiversity, allowing its study and its sustainable and equitable use by future generations of humankind.

Acknowledgments

We have greatly benefited in the final preparation of this chapter through discussion and critical reviews of many drafts by many colleagues too numerous to name here to whom we wish to express our gratitude. Special thanks are due to Les Lanyon, Carl Mitcham, Steve Thorne, G. L. Storm, Shelby Fleischer, and Joe Slusark for their comprehensive review of the manuscript with constructive suggestions.

References

Aldrich, W. (1993). Flower power. A lifetime devoted to learning how people react to plants. *Centre Daily Times,* Sunday, June 14, 1992, 1E, 6E.

Balandrin, M. F., Klocke, J. A., Wurtele, E. S., & Bolinger, W. H. (1985). Natural plant chemicals: Sources of industrial and medicinal materials. *Science* **228,** 1154–1160.

Birch, C. & Cobb, J. B., Jr. (1990). *The Liberation of Life.* Denton, TX: Environmental Ethics Books.

Bormann, F. H. & Kellert, S. R. (eds.). (1991). *Ecology, Economics, Ethics: The Broken Circle.* New Haven: Yale University Press.

Boussienquet, J. (1991). Problems of assessment of biodiversity. In *The Biodiversity of Microorganisms and Invertebrates: Its Role in Sustainable Agriculture,* ed. D. L. Hawksworth, pp. 31–36. Wallingford, UK: C.A.B. International.

Briggs, J. C. (1991a). A Cretaceous–Tertiary Mass Extinction? Were most of Earth's species killed off? *Bioscience,* **41**(9), 619–624.

Briggs, J. C. (1991b). Global species diversity. *J. Nat. Hist.,* **25,** 1403–1406.

Brown, L. R. (1992). *State of the World. A Worldwatch Institute Report on Progress Toward Sustainable Society.* New York: W. W. Norton & Company.

Brown, L. R., Flavin, C., & Kane, H. (1992). *Vital Signs 1992. The Trends that are Shaping our Future.* New York: W. W. Norton & Company.

Budyko, M. I. & Izrael, Yu A. (eds.). (1991). *Anthropogenic Climate Change.* Tucson: Univ. Arizona Press.

Bull, A. T. (1991). Biotechnology and biodiversity. In *The Biodiversity of Microorganisms and Invertebrates: Its Role in Sustainable Agriculture,* ed. D. L. Hawksworth, p.p. 205–218. Wallingford, UK: C.A.B. International.

Cloud, D. (1983). The biosphere. *Scientific American,* September 1983, 176–189.

Dobzhansky, Th. (1962). *Mankind Evolving. The Evolution of Human Species.* New Haven: Yale Univ. Press.

Erwin, T. L. (1982). Tropical forests: their richness in Coleoptera and other arthropod species. *Coleopterists' Bulletin,* **36,** 74–75.

Erwin, T. L. (1983a). Beetles and other arthropods of the tropical forest canopies at Mancus, Brazil, sampled with insecticidal fogging techniques. In *Tropical Rain Forests: Ecology and Management,* ed. S. L. Sutton, T. C. Whitmore, & A. C. Chadwick, pp. 59–75. Oxford, UK: Blackwell.

Erwin, T. L. (1983b). Tropical forest canopies: the last biotic frontier. *Bull. Entomol. Soc. Am.,* **29**(1), 14–19.

Erwin, T. L. (1991). How many species are there? Revisited. *Cons. Biol.,* **5**(3), 330–333.

Franklin, J. F. (1988). Structural and functional diversity in temperate forests. In *Biodiversity,* ed. E. O. Wilson, pp. 166–175. Washington, D.C.: National Acad. Press.

Gaston, K. J. (1991a). The magnitude of global insect species richness. *Cons. Biol.,* **5**(3), 283–296.

Gaston, K. J. (1991b). Estimates of the near-imponerable: A reply. *Cons. Biol.,* **5**(4), 564–566.

Gray, J. & Shear, W. (1992). Early life on land. *Amer. Scientist,* **80,** 444–456.
Gregg, N. T. (1992). Sustainability and politics: The cultural connection. *J. Forestry,* July 1992, 17–21.
Groombridge, B. (ed.). (1992). *Global Biodiversity, Status of the Earth's Living Resources* (compiled by World Conservation Monitoring Center). London, UK: Chapman and Hall.
Haas, P. M., Levy, M. A., & Parson, E. A. (1992). Appraising earth summit: how should we judge UNCED's success? *Environment,* **34**(8), 6–11, 26–33.
Hargrove, E. (1994). The paradox of humanity: two views on biodiversity and landscapes. In *Biodiversity and Landscapes,* ed. K. C. Kim & R. D. Weaver, pp. 173–186. New York: Cambridge.
Holt, R. D. & Gaines, M. S. (1992). Analysis of adaptation in heterogeneous landscapes: implications for the evolution of fundamental niches. *Evol. Ecol.,* **6,** 433–447.
Horton, T. (1993). Chesapeake Bay: Hanging in the balance. *National Geographic,* **183**(6), 2–35.
Jablonski, D. (1991). Extinctions: A Paleontological Perspective. *Science,* **253:** 754–757.
Jacobsen, J. E. & Firor, J. (eds.). (1992). *Human impact on the Environment: Ancient Roots, Current Challenges.* Boulder: Westview Press.
Katz, E. (1994) Biodiversity and ecological justice. In *Biodiversity and Landscapes,* ed. K. C. Kim & R. D. Weaver, pp. 61–74. New York: Cambridge.
Kaufman, L. & Malloy, K. (eds.). (1986). *The Last Extinction.* Cambridge, Mass.: MIT Press.
Kim, K. C. & Knutson, L. (1986). Scientific bases for a national biological survey. In *Foundations for A National Biological Survey,* ed. K. C. Kim & L. Knutson, pp. 3–22. Lawrence, KS: Assoc. Syst. Coll.
Kurian, G. T. (1984). *New Book of World Rankings.* New York: Facts on File Publications.
Lovejoy, T. E. (1980). A projection of species extinctions. In *Global 2000 Report to the President.* V. 2. The Technical Report, Council on Environmental Quality and U.S. Department of State, pp. 328–332. Washington, D.C.: U.S. Government Printing Office.
Malde, H. E. (1966). Environment and man in arid America. In *Human Ecology. Collected Readings,* ed. J. B. Bresler, pp. 104–119. Reading, MA: Addison-Wesley.
Margulis, L. & Olendzenski, L. (ed.). (1992). *Environmental Evolution: Effects of the Origin and Evolution of Life on Planet Earth.* Cambridge, Mass.: The MIT Press.
Markle, G. E. & Robin, S. S. (1985). Biotechnology and the social reconstruction of molecular biology. *Bioscience,* **35**(4), 220–225.
Master, L. L. (1991). Assessing threats and setting priorities for conservation. *Cons. Biol.,* **5,** 559–563.
May, R. M. (1986). How many species are there? *Nature,* **324,** 514–515.
May, R. M. (1988). How many species are there on earth? *Science,* **241,** 1441–1449.
May, R. M. (1990a). How many species? *Philos. Trans. Roy. Soc.* b, **330,** 293–304.
May, R. M. (1990b). Taxonomy as destiny. *Nature,* **347,** 129–130.
Mayr, E. (1978). Evolution. *Scientific Amer.,* **239**(3), 47–55.
Mayr, E. (1982). *The Growth of Biological Thoughts: Diversity, Evolution, and Inheritance.* Cambridge, MA: Harvard University Press.
McNaughton, S. J. (1989). Ecosystem and conservation in the Twenty-first Century.

In *Conservation in the Twenty-first Century,* ed. D. Western & M. C. Pearl, pp. 109–130. New York: Oxford Univ. Press.

Meadows, D. H., Meadows, D. L., & Randers, J. (1992). *Beyond The Limits, Confronting Global Collapse, Envisioning a Sustainable Future.* Post Mills, Vermont: Chelsea Green Publ. Co.

Meggers, B. J. (1966). Environmental limitation on the development of culture. In *Human Ecology. Collected Readings,* ed. J. B. Bresler, pp. 120–145. Reading, MA: Addison-Wesley.

Mitsch, W. J. (1993). Ecological engineering. A cooperative role with the planetary life-support system. *Environ. Sci. Technol.,* **27**(3), 430–445.

Myers, N. (1979a). Conserving our global stock. *Environment,* **21,** 25–33.

Myers, N. (1979b). *The Sinking Ark: A New Look at the Problem of Disappearing Species.* Oxford: Pergamon Press.

Myers, N. (1983). *A Wealth of Wild Species.* Boulder, CO: Westview.

Myers, N. (1989a). *Deforestation Rates in Tropical Countries and Their Climatic Implications.* London: Friends of the Earth.

Myers, N. (1989b). Major extinction spasm: Predictable & inevitable? In *Conservation in the Twenty-first Century,* ed. D. Western & M. C. Pearl, pp. 42–49. New York: Oxford Univ. Press.

Myers, N. (1994). We Do Not Want to become Extinct: The Question of Human Survival. In *Biodiversity and Landscapes,* ed. K. C. Kim & R. D. Weaver, pp. 133–150. New York: Cambridge.

Norse, E. A., Rosenbaum, K. L., Wilcove, D. S., Wilcox, B. A., Romme, W. H., Johnston, D. W., & Stout, M. L. (1986). *Conserving Biological Diversity in Our National Forest.* Washington, D.C.: The Wilderness Society.

Norton, B. G. (1994). Thoreau and Leopold on Science and Values. In *Biodiversity and Landscapes,* ed. K. C. Kim & R. D. Weaver, pp. 31–46. New York: Cambridge.

Noss, R. F. (1990). Indicators for monitoring biodiversity: a hierarchical approach. *Cons. Biol.,* **4,** 355–364.

Odum, E. P. (1989). *Ecology and Our Endangered Life-Support Systems.* Sunderland, Mass.: Sinauer Associates.

Odum, E. P. (1994). "Emergy" Evaluation of Biodiversity for Ecological Engineering. In *Biodiversity and Landscapes,* ed. K. C. Kim & R. D. Weaver, pp. 339–359. New York: Cambridge.

Oldfield, M. L. (1989). *The Value of Conserving Genetic Resources.* Sunderland, Mass.: Sinauer Assoc.

Olembo, R. (1991). Importance of microorganisms and invertebrates as components of biodiversity. In *The Biodiversity of Microorganisms and Invertebrates: Its role in sustainable agriculture,* ed. D. L. Hawksworth, pp. 7–16. Wallingford, UK: C.A.B. International.

Olshansky, S. J., Carnes, B. A., & Cassel, C. K. (1993). The aging of the human species. *Scientific American* (April 1993), **268**(4), 46–52.

Orr, D. W. (1992). Some thoughts on intelligence. *Cons. Biol.,* **6**(1), 9–11.

Parson, E. A., Haas, P. M., & Levy, M. A. (1992). A summary of the major documents signed at the earth summit and global forum. *Environment,* **34**(8), 12–15, 34–36.

Patrusky, B. (1986). Mass extinctions: the biological side. *Mosaic,* **17**(4), 2–13.

Peters, R. L. (1994). Conserving Biological Diversity in the Face of Climate Change. In *Biodiversity and Landscapes,* ed. K. C. Kim & R. D. Weaver, pp. 105–132. New York: Cambridge.

Peters, R. L. & Lovejoy, T. E. (eds.). (1992). *Global Warming and Biological Diversity.* New Haven: Yale Univ. Press.
Pilbeam, D. (1984). The descent of hominoids and hominids. *Scientific American,* March 1984, 94–101.
Pimm, S. L. (1991). *The Balance of Nature.* Chicago: Univ. Chicago Press.
Ponting, C. (1991). *A Green History of the World: The Environment and Collapse of Great Civilizations.* New York: Penguin Books U.S.A.
Randall, A. (1994). Thinking about the Value of Biodiversity. In *Biodiversity and Landscapes,* ed. K. C. Kim & R. D. Weaver, pp. 271–385. New York: Cambridge.
Raup, D. M. (1988). Extinction in the geologic past. In *Origins and Extinctions,* ed. D. E. Osterbrock & P. H. Raven, pp. 109–119. New Haven, Conn.: Yale Univ. Press.
Raup, D. M. (1991). *Extinction: Bad Genes or Bad Luck.* New York: W. W. Norton.
Raup, D. M. & D. Jablonski. (1993). Geography of End-Cretaceous Marine Bivalve Extinctions. *Science,* **260,** 971–973.
Ravera, O. (ed.). (1979). *Biological Aspects of Freshwater Pollution.* New York: Pergamon Press.
Reid, W. V. (1992). How many species will there be? In Tropical Deforestation and Species Extinction, ed. T. C. Whitmore & J. A. Raven, pp. 55–73. London: Chapman and Hall.
Robinson, G. R., Holt, R. D., Gaines, M. S., Hamburg, S. P., Johnson, M. L., Fitch, H. S., & Martinko, E. A. (1992). Diverse and contracting effects of habitat fragmentation. *Science,* **257,** 524–526.
Rolston II, H. (1994). Creation: God and Endangered Species. In *Biodiversity and Landscapes,* ed. K. C. Kim & R. D. Weaver, pp. 47–60. New York: Cambridge.
Samways, M. J. (1989). Insect conservation and the disturbance landscape. *Agriculture, Ecosystem and Environment,* **27,** 183–194.
Sanders, W. T. & Webster, D. (1994). Preindustrial Man and Environmental Degradation. In *Biodiversity and Landscapes,* ed. K. C. Kim & R. D. Weaver, pp. 77–104. New York: Cambridge.
Sarokin, D. & Schulkin, J. (1992). The role of pollution in large-scale population disturbances. Part 2: Terrestrial Populations. *Environ. Sci. Technol.,* **26**(9), 1694–1701.
Saunders, D. A., Hobbs, R. J., & Marguler, C. R. (1991). Biological consequences of ecosystem fragmentation: a review. *Conserv. Biol.,* **5,** 18–32.
Sayer, J. A. & Whitmore, T. C. (1991). Tropical moist forests: Destruction and species extinction. *Biol. Conserv.,* **55,** 199–213.
Schopf, J. W. (1993). Microfossils of the early Archean Apex Chest: New evidence of the antiquity of life. *Science* **260,** 640–646.
Siever, R. (1983). The dynamic earth. *Scientific American,* September 1983, 46–55.
Silver, C. S. & DeFries, R. I. S. (1990). *One Earth One Future: Our Changing Global Environment.* Washington, D.C.: National Acad. Press.
Soulé, M. E. (1989). Conservation biology in the Twenty-first Century: Summary and Outlook. In *Conservation for the Twenty-first Century,* ed. D. Western & M. C. Pearl, pp. 297–303. New York: Oxford Univ. Press.
Soulé, M. E. (1991). Conservation: Tactics for a constant crisis. *Science,* **253,** 744–750.
Soulé, M. E. & Kohn, K. A. (eds.). (1989). *Research Priorities for Conservation Biology.* Washington, D.C.: Island Press.

Southwick, C. H. (ed.). (1985). *Global Ecology*. Sunderland, MA: Sinauer Associates.

Stern, D. C., Young, O. R., & Druckman, D. (eds.). (1992). *Global Environmental Change: Understanding the Human Dimensions*. Washington, D.C.: National Academy Press.

Swift, M. J. & Anderson, J. M. (1989). Decomposition. In *Tropical Rain Forest Ecosystems, Biogeography and Ecological Studies*. [Ecosystem of the World No. 14B.] Ed. H. Lieth & M. J. A. Werger, pp. 547–569. Amsterdam: Elsevier.

Tudge, C. (1989). The rise and fall of *Homo sapiens sapiens*. *Phil. Trans. R. Soc. Lond.* B, **325**, 479–488.

Tukey, H. (1994). Urban Horticulture: A Part of the Biodiversity Picture. In *Biodiversity and Landscapes*, ed. K. C. Kim & R. D. Weaver, pp. 361–370. New York: Cambridge.

Turner II, B. L., Clark, W. C., Kates, R. W., Richards, J. F., Mathews, J. T., & Meyer, W. B. (1990). The Earth As Transformed By Human Action. *Global and regional changes in the biosphere over the past 300 years*. Cambridge: Cambridge Univ. Press.

Vermeij, G. J. (1986). The biology of human-caused extinction. In *The Preservation of Species: The Value of Biological Diversity*, ed. B. G. Norton, pp. 28–49. Princeton, NJ.: Princeton Univ. Press.

Walker, B. H. (1992). Biodiversity and ecological redundancy. *Cons. Biol.*, **6**, 18–23.

Weaver, R. D. (1990). *The Economics of Plant Germplasm*. Background Paper. The Board on Agriculture, National Research Council, Washington, D.C. (unpublished).

Weaver, R. D. (1994). Market-Based Economic Development and an Assessment of Conflict. In *Biodiversity and Landscapes*, ed. K. C. Kim & R. D. Weaver, pp. 307–324. New York: Cambridge.

Weiskel, T. C. (1990). *Cultural values and their environmental implications: an essay on knowledge, belief and global survival.* Am. Assoc. Advanc. Sci. Annual Meeting 1990, New Orleans.

Western, D. (1989). Population, resources, and environment in the Twenty-first Century. In *Conservation for the Twenty-first Century*, ed. D. Western & M. C. Pearl, pp. 11–25. New York: Oxford Univ. Press.

White, L. Jr. (1967). The historical roots of our ecological crisis. *Science*, **155**, 1203–1207.

Williams, P. H., Humphries, C. J., & Vane-Wright, R. I. (1991). Measuring biodiversity: Taxonomic relatedness for conservation priorities. *Augt. Sept. Bot.*, **4**, 665–679.

Wilkes, G. (1989). Germplasm Preservation: Objectives and needs. In *Biotic Diversity and Germplasm Preservation, Global Imperatives*, ed. L. Knutson & A. K. Stoner, pp. 13–41. (Beltsville Symposium Agricultural Research). Dordrecht, Netherlands: Kluwer Academic Publishers.

Wilkes, G. (1994). Germplasm Conservation and Agriculture. In *Biodiversity and Landscapes*, ed. K. C. Kim & R. D. Weaver, pp. 151–170. New York: Cambridge.

Wilson, E. O. (1984). *Biophilia*. Cambridge, Mass.: Harvard Univ. Press.

Wilson, E. O. (1985a). The biological diversity crisis. *Bioscience*, **35**(11), 700–706.

Wilson, E. O. (1985b). The biological diversity crisis: A challenge to science. *Issues in Science and Technology*, **11**(1), 22–29.

Wilson, E. O. (eds.). (1988). *Biodiversity.* Washington, D.C.: National Academy Press.
Wilson, E. O. (1992). *The Diversity of Life.* Cambridge, Mass.: The Belknap Press of Harvard Univ. Press.
Wolfe, S. C. (1987). *On the Brink of Extinction: Conserving the Diversity of Life.* Worldwatch Paper 78. Washington, D.C.: Worldwatch Inst.
The World Resources Institute. (1992). *World Resources 1992–1993.* New York: Oxford University Press.
WRI, IUCN & UNEP. (1992). *Global Biodiversity Strategy. Guidelines for Action to Save, Study, and Use Earth's Biotic Wealth, Sustainably and Equitably.* World Resources Institute (WRI), The World Conservation Union (IUCN), and United Nations Environment Programme (UNEP).

Part two

Human values and biodiversity

2

Thoreau and Leopold on science and values

BRYAN G. NORTON

When we ask, "What is the value of biodiversity?" we can expect that respondents, assuming that they answer the question at all, will answer in one of two quite different ways. Let us sketch these alternatives.

Some answers are mainly economic, emphasizing the actual and potential *uses* of living species. To this group, the value of biodiversity will be stated in quantifiable terms (Randall, 1988). This approach is utilitarian and anthropocentric. It measures value as contributions to human welfare. And it is "reductionistic" in the sense that it reduces to dollars all of the apparently disparate values and uses associated with wild species. Reductionists discuss the value of biodiversity by trying to put fair prices on its uses; they are most comfortable with the language of mainstream, neoclassical microeconomics. Natural objects, on this approach, are simply "resources" for human use and enjoyment. One characteristic of this approach, which makes it attractive in decision processes, is that it promises an aggregation of values: the contribution of nature to human welfare is made commensurable and interchangeable with other human benefits. This approach, therefore, holds open the possibility of a bottom-line figure that tells us what we should do in complex policy decisions; we should have exactly as much preservation of biodiversity as society is willing to pay for, given competing social needs.

Other approaches employ moral terminology and insist that we have an obligation to protect all species, an obligation that transcends economic reasoning and trumps our mere interests in using nature for our own welfare (Ehrenfeld, 1978; Rolston, 1988). These moralists limit human activities using nature by appeal to obligations that are independent of human welfare. Moralists do not believe our obligations to protect nature can be traded off against other obligations. Their language is a moral and sometimes a theological one. Moralists, who believe that wild species have "rights" or "intrinsic value" – value independent of human interests and consciousness – recognize our obligations to protect other species as prima facie commands; they posit at least a strong presumption against trading them off against values based in human welfare. John Muir, first president of the Sierra Club and a passionate advocate of preserving nature in its multiple forms, once said: "The battle we have fought and are still fighting for the forests is a part of the eternal conflict between right and wrong, and we cannot expect to see the end of it" (Fox, 1981, p. 107).

I call the question of the value of biodiversity, when posed as a choice between these approaches, "The Environmentalists' Dilemma" (Norton, 1991; Ehrenfeld, 1978) because most commentators have assumed that we should give one answer or the other – either our obligation is to save natural resources *for* future consumption, or we should save nature *from* consumption and for its own sake (see, for example, Passmore, 1974). I argue below that this is a *false* dilemma – the works of Henry David Thoreau show a way between the horns of the Environmentalists' Dilemma, though specific application of Thoreau's moral insight did not emerge until it was given expression, by Aldo Leopold, in the vernacular of community ecology. I will describe and advocate a system of values that follows in the well-chosen footsteps of Thoreau and Leopold.

Thoreau's transformative values

Thoreau's *Walden,* as well as his other writings, is sprinkled with analogies and metaphors drawn from wild species and applied to human life. I will begin by citing two passages in which Thoreau uses insect analogies to make points about people. First, in the most explicitly philosophical chapter of *Walden,* "Higher Laws," Thoreau notes that entomologists of his day had recognized that some insects in their "perfect state" (after transformation from the larval into the winged state) are "furnished with organs of feeding, [but] make no use of them." More generally, he noted, all insects eat much less in their perfect state. Thoreau applies this to human society: "The abdomen under the wing of the butterfly still represents the larva. . . . The gross feeder is a man in the larval state; and there are whole nations in that condition, nations without fancy or imagination, whose vast abdomens betray them" (Thoreau, 1960 [1854], p. 146).

This analogy illustrates Thoreau's peculiarly *dualistic* conception of human nature. Ostensively, Thoreau is explaining that hunting, and carnivorous habits more generally, are appropriate for the young, that these are a necessary *stage* in the individual's evolution (and in a culture's evolution), and that the urges to indulge in these practices will give way to higher sentiments and the abandonment of killing and meat-eating.[1] But it is obvious that carnivorous habits symbolize "gross feeding" more generally, and that Thoreau's intent is to portray materialistic consumerism as an immature developmental stage of the person.

Thoreau was a dualist not in the Cartesian sense, whereby two substances, mind and body, coexist in tandem through time, but in a dynamic or emergent sense. There exists a prior, primitive instinct that tempted Thoreau, upon seeing a woodchuck cross his path, "to seize and devour him raw." But Thoreau sensed in himself and in others also an instinct "toward a higher, or as it is named, spiritual life" (p. 143). Dynamic dualism, as Thoreau describes it, sees human nature as tensionally stretched between an older, primitive, "rank and savage" self and an emergent, higher, spiritual self in which the person's relationship with his surroundings changes from a consumptive one to a contemplative one. His economics – the first chapter of *Walden* – explains how one can retain and enhance one's creativity by living a simple, nonmaterialistic life.

Thoreau embraced both aspects of his humanity ("I love the wild not less than the good"), but he leaves no doubt that the emergent instinct toward a spiritual and perceptual relationship with nature is a "higher" instinct than the consumptive one, which is based in our animal nature: "The voracious caterpillar when transformed into a butterfly and the gluttonous maggot when become a fly content themselves with a drop or two of honey or some other sweet liquid" (p. 143). Note that, in this and other passages, Thoreau casually mixes moralism with description.

Thoreau confidently espouses a distinction between "higher" and "lower" satisfactions, much as John Stuart Mill advocated in his argument that it is "better to be Socrates dissatisfied than a fool satisfied" (Mill, 1957, Chapter II). Thoreau, however, is more specific than Mill; he advocates a life of simplicity and contemplation, of freedom from dependence on material needs. Thoreau therefore goes beyond Mill in proposing a substantive characterization of "higher" satisfactions. The important point for our present purposes is that Thoreau describes the benefits of the transformation to higher values in terms of human maturation and fulfillment of potential, as improvements *within* human consciousness, not in terms of obligations *to* nature and *extrinsic* to human consciousness.

Thoreau's dualism regarding human nature has three rather unusual aspects: (1) It is a *dynamic* dualism. The emergence of the spiritual aspect of the person represents a transformation from a "lower," primitive state to a "higher," more spiritual one. (2) It involves a comprehensive shift in perception and consciousness. In the immature, "larval" state, the person relates to nature physically, as a consumer; nature is thereby perceived in this immature state of the person as mere physical stuff, raw material, resources. In the second, perfect state, the person has undergone a perceptual transformation and now relates to nature nonconsumptively. Physical needs are minimized; the result is a life of "fancy and imagination." The world of nature is no longer seen as mere raw material for consumption; it is now seen as alive, soulful, and inspirational. (3) Thoreau explains this dynamic emergence of the higher self with an insect analogy, which illustrates the power of natural objects to teach us about human nature and simultaneously introduces the Aristotelian idea that the higher self is implicit within the lower state just as the butterfly is implicit in the caterpillar. In using organic analogies, Thoreau is emphasizing the power of nature study – observation – to hasten an inherent, systematic change in perceptual relations between the person and the natural world. Observation of dynamically unfolding natural processes, according to this view, hastens the evolution of a higher consciousness which is a potential of the human spirit. This perceptual and conceptual shift coincides with a deeper and more important shift in metaphysical assumptions, as well as a changing pattern of specific needs and values. Thoreau's dualism is therefore perceptual and psychological, but it is also behavioral. Posttransformational individuals will consume less.

In the penultimate paragraph of *Walden,* Thoreau again uses an insect analogy. He tells an anecdote of a strong and beautiful bug which gnawed its way out of an old table, "hatched perchance by the heat of an urn." It grew, he noted, from an egg that had been deposited many years before in the living apple tree. Thoreau explains the analogy explicitly, saying that the egg represents the human potential to live a free life of beauty, a potential which can only be released after the spirit of independence gnaws through "many concentric layers of woodenness in the dead dry life of society." So Thoreau returns to the dualistic idea at the end of *Walden.* The urn, I think, represents *Walden* itself – it is intended to provide the catalyst for truly human individuals to "hatch," to achieve the potential for individual freedom implicit in each person, and to escape the life of "quiet desperation" Thoreau saw as the plight of his neighbors. The book encourages the reader to begin gnawing through layers of social custom and expectations of material success toward individual freedom (p. 221).

Thoreau stood at a crossroads in American thought in the sense that he can be aptly described as both a transcendentalist and as a naturalist (Miller, 1968). Looking backward, he owed clear debts both to Puritanism and to Emerson's idealistic pantheism. But his boyhood days were filled with "nature studies" long before he knew of Emerson's philosophical glorification of nature. Thoreau's transcendentalism therefore represented from the start an uneasy compromise between the earthly love of natural events and a heady preoccupation with intuitions of transcendence.

Whether or not Thoreau rejected idealistic transcendence explicitly, I am arguing that he developed a theory of perceptual and psychological (world-view) change that could stand independently of idealism – the only transcendence involved is the transition from one state of consciousness to another. This shift away from commitments to "pure" insight unsullied by preconceptions and presuppositions puts him more in the tradition of contemporary naturalism than in the tradition of pure idealism. The validity of the change in world-view will be demonstrated not by pure reason but by improved experience of those who have undergone the transformation.

Thoreau's program was to describe living nature in such a way as to evoke analogies that will set in motion a shift to a higher, less materialistic and consumptive set of needs and style of life. And these analogies were essential to his ambitious project of reforming his contemporaries, of freeing them from slavery to commercialism and removing the "quiet desperation" from their lives.

The dynamics of nature and the dynamics of consciousness

Thoreau ultimately rejected Emerson's Platonism because he could not understand nature in terms of fixed essences. Thoreau thus chose Heraclitus over Plato: "All is in flux." All things in nature, including human beings and the values they live by, are constantly changing in a great interrelated whole. Thoreau's dynamic dualism required the rejection of Emerson's world of fixed forms and real essences – human nature itself shifts in response to its changing environment.

Heraclitean dynamism was abundantly confirmed by Thoreau's experience: *Walden* is a spring → summer → fall → winter guide to the dynamic transformations of nature, and it ends with the second spring. The return of spring is symbolic of the individual's resurrection from the "death" of social conformity and of rebirth into spirituality. In response to the beautiful bug that gnawed its way out of the table, Thoreau says: "Who does not feel his faith in a resurrection and immortality strengthened by hearing this?" (p. 221).

Thoreau thought he saw a way out of the trap of materialistic consumerism and its concomitant understanding of nature as raw resources, and hence the symbolic "urn," *Walden,* is Thoreau's spiritual legacy, a catalytic agent to unlock human potential. Human beings have an individual genius, a potential to be good, to live perceptually rich and consumptively simple lives; consciousness and imagination represent for Thoreau the chance humanity has to achieve transcendence, to become the butterfly free of addiction to consumption. But these instincts to a "higher" life of freedom are crushed within "the dead dry layers of society," (p. 221) the indoctrination in consumerism that Thoreau foresaw so clearly would become the fate of moderns.

And here Thoreau turns moralist and social philosopher by uniting his theory of world-view change with his social criticism. We experience *wonder* in observing nature, by being still to let other living things provide lessons, in the form of analogies. Thoreau believed we can learn how to live by observing wild species. Careful attention to natural analogies can make us wise.

Thoreau's dynamism, and the multiple layers of symbols it offered, was at its heart organicist. By this I mean that the world-view he advocated was governed on all levels by a dynamic, organic metaphor. In describing the break-up of the ice on Walden Pond in spring, he said: "Who would have suspected so large and cold and thick-skinned a thing could be so sensitive? . . . [T]he earth is all alive and covered with papillae. The largest pond is as sensitive to atmospheric changes as the globule of mercury in its tube" (p. 201).

The change in perception that accompanies the understanding of the natural world as alive and soulful, a shift in governing metaphors and world-view, also encourages the perceiver to experience a change in values – natural objects are transformed, within the new world-view, into objects of contemplation and inspiration, not mere objects of exploitation. Posttransformational consciousness experiences nature within an organicist world-view, a world of dynamic change and development, a world in which natural objects have a spiritual as well as a material value.

Thoreau's analogical method is therefore blatantly moralistic; one might even say "naive."[2] Thoreau chose observation as knowledge in the service of wonder – it is wonder at nature's intricacy, complexity, and economy that drives us to a changed understanding of ourselves and our place in nature. The plausibility of Thoreau's transformational theory therefore depends on his frank acceptance of the value-ladenness of facts – "Our whole life is startlingly moral" (p. 148).[3] Accordingly, he was not surprised to find significance in facts on the eating habits of insects. The sense of wonder inspired by these value-charged facts leads to deeper insights than those of botany or zoology. Because nature's facts have this significance, they can build moral character.

But Thoreau was not simply naive; he went far beyond quaint proverbs – he recognized that the adoption of a new, organic world-view would represent a shift toward more holistic thinking. Thoreau thought that his anecdotes and analogies could act as a catalyst for world-view transformation in his readers. But he also recognized that the change results more directly from the rediscovery of a sense of wonder. Thoreau thought that if he could induce people to develop their perception through patient and sensitive observation of nature, he could also induce them to reject mechanism and with it materialistic consumerism. Once the new world-view was embraced, he believed a new perceptual and evaluative relationship with nature would also emerge. Thoreau saw the conversion to the new world-view as a transformation in attitudes, values, desires, and demands as well as beliefs. The vehicle of this transformation is a change in the form of perception that emerges within the new world-view. Careful and patient observation is the trigger that initiates the life-long process of seeking beauty and truth in the dynamic natural world.

It is tempting to ridicule Thoreau's experiments in moral reasoning as naive-sounding analogies, comparing them to Aesop's fables. The temptation is increased by the writings of naturalists, such as Annie Dillard (1974) and David Quammen (1985), who have undermined the naive use of individual analogies by exhibiting the horrors and perversities (when viewed from a human perspective) that coexist with beauty in nature. Thoreau's simple analogies therefore sound quaint today. A sympathetic reader could nevertheless credit him with recognizing the holistic nature of world-view change, coupling this with a clear understanding that economics is as much about "managing" our preferences as it is about fulfilling them. The process that Thoreau undertook at Walden Pond was action, catalyzed by an intuitive wonder at the observation of nature, in service of liberation. Thoreau saw clearly that ethical development will be a dynamic process that will reach to the deepest assumptions of the modern world-view.

Thoreau did not cast his learn-from-nature analogies into the story of his experiences at Walden Pond without offering a supporting theory. He argued cogently that nature's analogies are illuminating on all levels, that the analogies from insects are just a part of a larger transformation in world-view, and that an organic, holistic metaphor is demanded if we are to understand both nature and human consciousness. Holism, the view that the whole is more than the sum of its parts, applies to understanding no less than to physical systems. A change in world-view will require an intuitive leap, an integrative act of creation catalyzed by the sense of wonder; it follows that the new consciousness cannot be expressed, must less justified, in the immature consciousness. The transformation is in this sense nondeterministic and requires an intuitive spark as much as observation and logic.

Thus, while he emasculated Emerson's theory of intuition regarding timeless essences, Thoreau remained true to his transcendentalist past in an important sense: the intuitive leap to a new, holistic world-view cannot be understood in a mechanistic system of psychology or logic. What Thoreau learned from nature was dynamism, the view that becoming is more fundamental than being, which entails that the system of nature is not deterministic; its most basic law is the law of creative activity. The striving to create goes beyond what can be known by a deductive process. Thoreau was an indeterminist who believed that the whole, both in nature and consciousness, is more than the sum of the parts.

Because the shift to the new consciousness cannot be described in the mechanistic models of nature popular in the modern age, Thoreau could not promise rational proof, but only catalyze a process. He never achieved, and despaired of achieving, an "objective" proof of the truth as it is experienced in evolving systems; and therefore he recognized the crucial role of an aesthetic, extralogical leap into a new world-view with more expressive concepts.

Thoreau also foresaw with remarkable clarity the conclusions that have emerged in contemporary physical theory and in contemporary philosophy of science. The breakdown of the logical positivists' program of a unified, reductionistic science, philosophical analyses that explain and justify this breakdown (Quine, 1969; Sellars, 1956), and conclusions in the new physics (Prigogene and Stengers, 1984), have all converged on the conclusion that there exists no complete and consistent theory that can represent the world on all levels simultaneously. The truth, it follows, cannot all be of the form of deductions from pure facts. Thoreau therefore anticipated the epistemological problems of postmodernism – he struggled to explain how one can explain and justify a change in world-view if there is no single, correct "description" of the physical world.

The problem of justification is, of course, tied to the problem of objectivity. Thoreau recognized that a demonstration of the inadequacy of mechanism would require a richer vocabulary than the one of deterministic science. Hence his experiments with analogies as ways to enlist scientific description in the larger, intuitive task of recognizing our proper "place" in the larger systems of nature. Thoreau's understanding of world-view change could not be explained in the objectivist tradition of Descartes and Newton; he needed the language of aesthetic creativity as much as the language of descriptive science. Thoreau recognized that dynamism and change is nature's most profound lesson and, by applying that insight to human consciousness and cultural development, Thoreau provided a plausible model of world-view change. He avoided serious mystical commitments by treating intuitions of holism as intuitive leaps that provide insight about our place in natural systems. He therefore reduced intuition to wonder, to a catalytic process that sets in motion a systematic rethinking of facts, and avoided commitment to intuitively justified knowledge – scientific or moral – of the physical world.

The environmentalists' dilemma revisited

By combining the religious idea of a moral transformation with the more mundane idea of consumption trimmed to fit context, Thoreau avoided the

Environmentalists' Dilemma. Thoreau's commitment to simplicity was *both* a call to live in harmony with nature economically *and* a response to the alienation evident in the quiet desperation characteristic of industrial society. The analogy of hunting as appropriate for young individuals and for "young" societies, but inappropriate for mature individuals and mature societies, represents a linkage of psychological maturation with social maturation. Nature thus holds more than economic value to humans – it is a sacred talisman, the honored symbol and guide leading humanity toward spiritual and material freedom. Nature exhibits a "higher" value than satisfying consumptive wants and needs. This does not mean that natural objects will no longer be used; but they will be "used" appropriately, with respect, and with a sense of awe at their sacred powers to transform consciousness and values.

Thoreau, like Darwin, recognized that the key to following nature, is to strive to fit our needs to nature's demands, rather than by altering nature to serve our unexamined demands. An ethic that takes into account our "place" in nature, a larger whole of which we are parts, will also be more satisfying. In this sense, "cultural survival" is determined, in a dynamic world, by appropriateness to the larger context that sets the conditions for survival. Morality is not determined by fixed moral principles, knowable by pure intuition; morality is determined, rather, by situation, and requires analogical insight based on careful observation of constantly changing situations.

Thoreau escaped the Environmentalists' Dilemma by insisting that nature has didactic value, value that stands outside the aggregated demands expressed by individuals in our "immature" society, but which is represented in worldview changes that will reshape those very demands. If individuals achieve freedom through transformation, they will also adopt a new, nonconsumptive lifestyle appropriate to the postmodern world. Sensitive observation of nature and an appropriate sense of wonder at nature's complex organization sensitizes us to the "whole" of which we are a part. Thoreau therefore anticipated the insight of Carl Jung, who once said that he never succeeded in helping any patient who was not convinced he was a part of some larger whole.

Nature's value is manifest within human consciousness and experience, and implies no commitment to values that are defined outside human consciousness. Thoreau said: "I am not interested in mere phenomena, though it were an explosion of a planet, only as it may have lain in the experience of a human being" (Richardson, 1986, p. 309). Observation, understanding, and appreciation of nature are inseparable parts of a process of change by which our lives are illuminated and seen in a new way. The process represents a constant transformation to a new form of perception and action, perception and action that seeks harmony with a larger, dynamic whole, of which we are a part.

Thoreau left Walden Pond when he discovered that, in two-and-a-half years, his "spontaneous" walks were cutting pathways in the woods, imposing patterns on nature, rather than finding them there. And so he left Walden Pond "for as good a reason as [he] went there." He had gone there to observe and react to stimuli he sensed within nature, rather than to manipulate nature and its resources to serve his own preferences. Thoreau's paths, radiating outward from his cabin, showed him that this project of living within nature is a worthy ideal, but one that cannot be fully achieved. Humans inevitably reconstruct their habitat from their own perspective. Even the "pure" experiencer, Thoreau in his cabin, altered the natural context.

Thoreau, the optimist, did not despair: "I learned this, at least, by my experiment: that if one advances confidently in the direction of his dreams, and endeavors to live the life which he has imagined, he will meet with a success unexpected in common hours" (p. 214–215).

Thoreau's science

While Thoreau showed a way through the Environmentalists' Dilemma by recognizing the transformational role of observation in value shifts, he fell short, I think, of providing a clear connection between science and values. To say that scientific observations of living things suggest analogies which, in turn, "catalyze" a world-view change is to leave the role of science, especially theoretical science, as a mysterious black box in Thoreau's account.

Assessments of Thoreau's scientific acuity differ and are currently undergoing reassessment (Scholfield, 1992). Some critics have found little more than idiosyncratic description in Thoreau's journals (Richardson, 1986), but recent scholarship has revealed that Thoreau was actively working on a pioneering book in protoecology in the last years of his life. This book has been recently reassembled and published (Thoreau, 1993), and Gary Nabhan argues on its basis, that "More than any botanist of his time, Thoreau moved past the mere naming of trees – the nouns of the forest – to track its verbs: the birds, rodents, and insects that pollinate flowers or disperse seeds, and all the other agents that shape the forest's structure" (Nabhan, 1993).

Apparently, on the basis of this newly assembled and rethought evidence, Thoreau quite explicitly recognized that the forest, a dynamic system, had a "language" of its own, and that the transition from the immature state was both literary and scientific. Thoreau argued that the important place to look for insight from wild species is in their natural habitat, not on a dissecting table (Thoreau, 1960). He saw that one learns more important things by relating an organism to its environment than by dissecting an organism into parts. This indicates that Thoreau was on the right track, seeking the secret of life and its organization in the larger systems in which species live. Especially, he thought we learn more important things about human behavior, and the evaluation of it, by observing organisms in environments. He believed that, if he could unlock the code of nature's language, it would provide the key to a new, dynamic and scientific understanding of nature. The key prerequisite for this change to a more contemplative consciousness was development of a new "language" of human values based on analogies from the "language" of nature. Nabhan asserts that, on this basis, we can conclude that Thoreau never gave up his attempt to become a romantic poet: "Instead of turning his back on these literary traditions, Thoreau tried to incorporate them into his search for a language more difficult but more enduring: the language of the forest itself" (Nabhan, 1993).

Thoreau saw that humans are analogous to other animals in the levels of organic nature, and he recommended sympathy for the hare "which holds its life by the same tenure" as a human person (Thoreau, 1960, p. 144). Understanding the roles of animals, including ourselves, in larger systems can therefore teach us a new code of behavior. The core change, the heart of the new world-view, is the adoption of an organic metaphor for understanding nature.

Thoreau's early death, unfortunately, prevented him from completing his project, which would surely have developed in new directions as he reacted to Darwin's *On the Origin of Species,* which he first read only two years before his death. It is interesting to speculate whether he would have developed an epistemology not unlike the American pragmatists such as John Dewey or Charles Peirce. In a dynamic system, truth must be dynamic and adaptive or become irrelevant. Natural selection of world-views for adaptability and contribution to cultural survivability would be the missing piece in Thoreau's theory, the link to tie his observations of natural events to his moralistic analogies. It might have guided him toward a full-blown evolutionary epistemology and an evolutionary ethic. We learn from observation because our role in nature is functional in a larger system. Analogies – Thoreau's preferred moral method – would be relevant because all species, including humans, must adapt to a constantly changing environment, even while functioning as an element in the systems that compose that environment. The ultimate lesson learned from nature is therefore a re-organization of thought that expresses atomistic facts as parts of an integrated understanding of a dynamic whole.

Aldo Leopold and scientific contextualism

It was left to Aldo Leopold, born in 1882, twenty years after Thoreau's early death, to recognize and explain the importance of ecology and evolutionary theory to perceptual and ethical transformations. Leopold, who described ecology as the biological science that runs at right angles to evolutionary biology, chose cranes as his illustration: "our appreciation of the crane grows with the slow unraveling of earthly history. His tribe, we now know, stems out of the remote Eocene. . . . When we hear his call we hear no mere bird. We hear the trumpet in the orchestra of evolution. He is the symbol of our untamable past, of that incredible sweep of millennia which underlies and conditions the daily affairs of birds and men" (Leopold, 1949, p. 96). The essay, "Marshland Elegy," which is one of Leopold's finest, begins on the edge of a crane marsh, with the narrator hearing "[o]ut of some far recesses of the sky a tinkling of little bells," and carefully describes in dragging prose how a human experiences the spectacle of the arrival of the cranes. He describes how there are periods of silence and periods of growing clamor until at last, "[o]n motionless wing they emerge from the lifting mists, sweep a final arc of sky, and settle in clangorous descending spirals to their feeding grounds" (Leopold, 1949, p. 95). Note that Leopold avoids *direct* moralizing from cranes to humans (which would be similar to Thoreau's use of insect analogies), arguing that cranes teach us about evolution and our role in those processes. The emphasis is on broadening perception, not on providing moral maxims.

Over the next six pages, Leopold guides the reader back and forth among many scales of time. He moves abruptly from human, perceptual time to place the marsh in geological time: "A sense of time lies thick and heavy on such a place. Yearly since the ice age it has awakened each spring to the clangor of cranes" (Leopold, 1949, p. 96). After stepping out of time altogether to discuss the aesthetics of time, he stresses the ability of scientifically enlightened perception to transform the arrival of the cranes into a semireligious experience, and, simultaneously, to explain why the destruction of crane marshes is, if not a sin, at least a tragedy.

Leopold then plunges back into geological time, tracing the path of the last glacier, describing the formation of the pond, and then braking the time machine down to the pace of ecological time and describing the development of ecological conditions that allowed the cranes to find a niche in Wisconsin. He recognizes that they have survived many earlier, gradual transformations of their habitat, and then laments how, in so many marshes, they had succumbed to human alteration of their habitat in just a few generations. Initially, in an arcadian time farmers and cranes cohabitated harmoniously. But technology, avarice, and inappropriate land use destroyed crane habitat even as it impoverished the human inhabitants. The downward spiral, looked at in ecological time, represented a deterioration of both crane and human habitat. But humans, unaccustomed to think like a mountain, unable to perceive the value of wolves or cranes, harm themselves, both economically and spiritually, by failing to see the difference between nature's gradual changes and the accelerated pace of change in larger systems attendant upon the technology augmented economic activities of modern humans.

Leopold's aesthetic explanation: "Our ability to perceive quality in nature begins, as in art, with the pretty. It expands through successive stages of the beautiful to values as yet uncaptured by language. The quality of cranes lies, I think, in this higher gamut, as yet beyond the reach of words" (Leopold, 1949, p. 96). The cranes, linking as they do the various scales of our history, unlock our understanding of time and our origins, and exhibit to us our "place" in nature. Leopold concludes that "The ultimate value in these marshes is wildness, and the crane is wildness incarnate" (Leopold, 1949, p. 101). The value of wildness to us is that it illustrates the multilevelled complexity, the dynamically stable system that has enough constancy to allow organization through evolution, and yet enough change to foster nature's creativity. The cranes act as living metaphors that locate our own species and its cultures in its larger context; they contribute to the larger-scale change in our world-view, a change that can be described metaphorically as embracing an organicist world-view, but experienced and made truly meaningful only in a crane marsh or some other such wild place. Here, Leopold follows Thoreau in his emphasis on the ways in which observation can catalyze changes in perception and in world-view. The moralizing, here, is indirect: a change in perception, aided by ecology, helps us to place ourselves in a larger dynamic, as evolved animals, and we consequently see that our destruction of the crane habitat is wrong – it cuts us loose from our evolutionary and cultural history.

When Leopold introduces his land ethic near the end of *A Sand County Almanac,* he returns to these potent evolutionary and ecological themes. "The extension of ethics to [land and to the animals and plants that live there] is, if I read the evidence correctly, an evolutionary possibility and an ecological necessity" (Leopold, 1949, p. 203). These passages hark back to passages written in 1923 in which Leopold expressly argued that a culture will be judged according to its treatment of the land it lives upon and that a society unable to sustain itself on its land will "be judged in 'the derisive silence of eternity'." Expressing this idea in the terms of an evolutionary epistemology, Leopold explicitly linked cultural survival with truth: "Truth is that which prevails in the long run" (Leopold, 1979, p. 141; Norton, 1988).

Leopold's land ethic represents the fruition of Thoreau's breakthrough in moral theory. Our society, even as it has tamed the wilderness and damaged natural systems, has created a society so productive that it can consciously forbear from further destruction. Change to a new world-view and a new form of life, one that is harmonious with its context, one that uses nature even as it recognizes the higher values embodied in the complex systems that form our ecological context, is now an "evolutionary possibility." Understanding of our ecological role clarifies our "place" in life's larger enterprise, and establishes our moral bearings in the changing world we face.

In this way, ecology is the key to a new morality. To understand our role in the biological world is to reject hubris: "a land ethic changes the role of *Homo sapiens* from conqueror of the land-community to plain member and citizen of it" (Leopold, 1949, p. 204). The goal, after the transformation to a new perspective, is to understand the role of our cultures in the ecological communities they have evolved within.

The central idea of Leopold's land ethic is that the land is a complex *system composed of many levels and subsystems that change according to many rates of speed* (Norton, 1991). Humans armed with conscious goals and powerful technologies can disrupt that system. But the same dynamic system of causes that created consciousness can, in the face of changing conditions, create a *new* consciousness. In the tradition of Thoreau, Leopold thought that the best antidote to disruptive behavior is a transformation in human consciousness and a new style of perception, perception that is informed by ecology and by evolution, and perception that recognizes, according to these sciences, the true role of the human species in the natural order. These changes, in turn, result in a very different conception of environmental "management." This new conception, informed by ecological science and tempered by the humility appropriate not to conquerors but to "plain citizens" of ecological communities, can be aptly called "scientific contextualism."

Scientific contextualism is a sort of holism. It is holism because it understands human activities, world-views, and ethics as a part of the evolving systems of nature. Science thus informs ethics by describing the appropriate role of humans in the system. Observation, as Thoreau so clearly recognized, unlocks treasures of understanding and self-understanding. But Leopold, who saw more clearly the role of ecological science, was able to conceptualize the metaphor of organicism as a systems approach to management and while he often anthropomorphised animals he also "animalized" humans by insisting that we are, like every other species, extruded into a niche, and therefore cannot destroy that niche – our natural context – with impunity. Human activities, including economic ones, represent subsystems within a larger ecological and physical whole, or context.

But contextualism is a *limited* holism; it reifies no single model as reality and need assume no superorganism that intentionally organizes the complex systems of nature (Wallace and Norton, 1992). According to scientific contextualism, many different models, which are no more than technical analogies, are useful in differing contexts. The difficult part is to choose analogies/models; in particular, it is difficult to conceive environmental problems on the correct scale and from the right perspective. Leopold's brilliant simile, "thinking like a mountain," is an exercise in choosing the correct *scale* for analysis and management. He first thought, while supporting predator eradication programs, was that deer/wolves/hunters formed an equilibrium system and that if he removed wolves, hunters would increase their take and create a new system with greater human utility. But he learned from experience that the larger ecological community, the mountain ecosystem, which changes slowly over millennia, was thrown into an accelerated pace of change by his actions. The vegetative cover – the skeleton and skin of the mountain "organism" – was destroyed and erosion began to set in. The key, Leopold concluded, is to see our activities as changing subsystems that function within a larger whole (Leopold, 1949).

Contextualism is a sort of "poor-man's systems theory." It proclaims no single model that can capture and relate all phenomena on all levels to all others. Models are seen more modestly as tools of the understanding and there is an implicit recognition that systems of different scales will be chosen to deal with different problems. Choice of the proper model will depend *both* upon social purposes (values) *and* scientific understanding, and will involve experiments, both social and ecological. But the experiments must be conducted with great care. Their purpose is to learn what nature is telling us, not merely to manipulate nature for human uses.

The American naturalist tradition, if not burdened by unreasonable epistemological requirements – such as strict adherence to a separation of science and ethics, or a belief in intuited, timeless moral principles – provides an interesting and plausible moral theory. This moral theory most centrally involves a commitment to world-view transformations that are only partially objective, and which evolve organically in reaction to changing values, concepts, and theoretical beliefs. But the criterion by which to judge such transformations must be improved human experience, not some abstract and timeless principle. The need for a transformation in consciousness is evident both in the illness and alienation of modern society and in the illness and deterioration of the ecological context. As Thoreau saw, the two problems have a common solution in the phenomenon of world-view change triggered by a sense of wonder at nature's complexity.[4]

Notes

1 Thoreau stressed a point that has been a recurring theme of naturalists since – sport hunting is appropriate for a young man, but a more mature individual will become a less consumptive naturalist. See, for example, Leopold, 1949, pp. 168–176.

2 As did Holmes Rolston, III, upon reading an earlier version of this paper.
3 I have discussed transformational theories and arguments in more detail in Norton 1987, especially chapters 10 and 11.
4. Portions of this paper appeared in "Thoreau's Insect Analogies" in *Environmental Ethics,* 13, 253–251.

References

Dillard, A. (1974). *Pilgrim at Tinker Creek.* New York: Harper & Row.

Ehrenfeld, D. (1978). *The Arrogance of Humanism.* New York: Oxford University Press.

Fox, S. (1981). *John Muir and His Legacy.* Boston: Little, Brown and Company.

Krutch, J. W. (1948). *Henry David Thoreau.* New York: William Sloane Associates.

Leopold, A. (1949). *A Sand County Almanac.* Oxford: Oxford University Press.

Leopold, A. (1979). Some fundamentals of conservation in the Southwest. *Environmental Ethics,* **1,** 131–148.

Mill, J. S. (1957) (originally published 1861). *Utilitarianism.* Indianapolis, IN: The Bobbs-Merrill Company, Inc.

Miller, P. (1968). Thoreau in the context of international romanticism. In *Twentieth Century Interpretations of Walden,* ed. R. Ruland. New York: Prentice-Hall.

Norton, B. G. (1987). *Why Preserve Natural Variety?* Princeton: Princeton University Press.

Norton, B. G. (1988). The constancy of Leopold's land ethic. *Conservation Biology,* **2,** 93–102.

Norton, B. G. (1991). *The Unity of Environmentalists.* New York: Oxford University Press.

Passmore, John (1974). *Man's Responsibility for Nature.* New York: Charles Scribner's Sons.

Prigogine, I. & Stengers, I. (1984). *Order Out of Chaos: Man's New Dialogue with Nature.* New York: Bantam.

Quammen, D. (1985). *Natural Acts.* New York: Schocken.

Quine, W. V. (1969). Epistemology Naturalized. In *Ontological Relativity and Other Essays.* New York: Columbia University Press.

Randall, A. (1988). What Mainstream Economists Have to Say about the Value of Biodiversity. In *Biodiversity,* ed. E. O. Wilson. National Academy Press, Washington, D.C.

Richardson, R. (1986). *Henry Thoreau: A Life of the Mind.* Berkeley: University of California Press.

Rolston, H. (1988). *Environmental Ethics: Duties to and Values in the Natural World.* Temple University Press,Philadelphia.

Scholfield, E. (1992). *A Natural Legacy: Thoreau's World and Ours.* Golden, CO: Fulcrum Publishing.

Sellars, W. (1956). Empiricism and the Philosophy of Mind. *Minnesota Studies in the Philosophy of Science, Vol. 1.* Minneapolis, MN: University of Minnesota Press.

Thoreau, H. D. (1960). *Walden and "Civil Disobedience."* New York: NAL Penguin Inc. (Originally published 1854 and 1848.)

Wallace, R. & Norton, B. G. (1992). "The Policy Implications of Gaian Theory," *Ecological Economics,* 6, 103–118.

3

Creation: God and endangered species

HOLMES ROLSTON, III

Religious value and the God Committee

When the United States Congress lamented the loss of species, they declared that species have "esthetic, ecological, educational, historical, recreational, and scientific value to the Nation and its people" (*Endangered Species Act of 1973,* sec. 2a). Religious value is missing from this list. Perhaps Congress would have overstepped its authority to declare that species carry religious value. But for many Americans this is the most important value. Christians or Jews will add that these species are also of religious value, and not only to Americans but to God. Defending the freedom of religion, guaranteed in the Constitution, Congress might well have insisted that the species of plants and animals on our landscape ought to be conserved because such life is of religious value to the Nation and its people.

Though God's name does not appear in the Endangered Species Act itself, it does occur in connection with the Act. The protection Congress authorized for species is quite strong in principle. Interpreting the Act, the U.S. Supreme Court insisted "that Congress intended endangered species to be afforded the highest of priorities" (*TVA vs. Hill,* 174). Since "economic" values are not among the listed criteria either but must sometimes be considered, Congress, in 1978 amendments, authorized a high-level, interagency committee to evaluate difficult cases. This committee may permit human development at the cost of extinction of species. In the legislation, this committee is given the rather nondescript name "The Endangered Species Committee," but almost at once it was nicknamed "the God Committee." The name mixes jest with theological insight and reveals that religious value is implicitly lurking in the Act. Any who decide to destroy species take, fearfully, the prerogative of God.

In the practical conservation of biodiversity in landscapes, concerned with habitat, breeding populations, DDT in food chains, or water flows to maintain fish species, it might first seem that God is the ultimate irrelevancy. In fact, when one is conserving life, ultimacy is always nearby. The practical urgency of on-the-ground conservation is based in a deeper respect for life. Extinction is forever; and, when danger is ultimate, absolutes become relevant. The motivation to save endangered species can and ought to be pragmatic, economic, political, and scientific; deeper down it is moral, philosophical, and religious.

Adam, Noah, and the prolific Earth

Genesis! Take that word seriously. In the Hebrew stories, the "days" (events) of creation are a series of divine imperatives that empower Earth with vitality. "The earth was without form and void, and darkness was upon the face of the deep; and the Spirit of God was moving over the face of the waters. And God said, 'Let there be . . . ' " (Genesis 1.2–3). "Let the earth put forth vegetation." "Let the earth bring forth living things according to their kinds" (Genesis 1.11, 24). "Let the waters bring forth swarms of living creatures" (Genesis 1.20). "Swarms" is, if you wish, the Biblical word for biodiversity.

A prolific Earth generates teeming life, urged by God. The Spirit of God is brooding, animating the Earth, and Earth gives birth. As we would now say, Earth speciates. When Jesus looks out over the fields of Galilee, he recalls how "the earth produces of itself" (Mark 4.28) spontaneously (in Greek: "automatically"). God reviews this display of life, finds it "very good," and bids it continue. "Be fruitful and multiply and fill the waters in the seas, and let birds multiply on the earth" (Genesis 1.22). In current scientific vocabulary, there is a dispersal, conservation by survival over generations, and niche saturation up to carrying capacity. The fauna is included within the covenant. "Behold I establish my covenant with you and your descendants after you, and with every living creature that is with you, the birds, the cattle, and every beast of the earth with you" (Genesis 9.5). In modern terms, the covenant was both ecumenical and ecological. Earth is a promised planet, chosen for abundant life. Adam's first job was to name this swarm of creatures, a project in taxonomy.

The Bible also records the first Endangered Species Project – Noah and his ark! That story is quaint and archaic, as much parable as history, teaching how God wills for each species on Earth to continue, despite the disruptions introduced by humans. Although individual animals perish catastrophically, God has an "adequate concern and conservation" for species – the species come through. After the Flood, God reestablishes "the covenant which I make between me and you and every living creature that is with you, for all future generations" (Genesis 9.12–13). Humans are to repopulate the earth, but not at threat to the other species; rather, the bloodlines must be protected at threat of divine reckoning (Genesis 9.1–7). The Biblical authors had no concept of genetic species but used instead the vocabulary of bloodlines. The prohibition against eating the blood is a sign of respect for these bloodlines.

The Endangered Species Act and the God Committee are contemporary events, but it can be jarring to set beside them these archaic stories. The stories are not only archaic in being couched in outmoded thought forms; they are archaic in that they are about aboriginal truths. The Noah story is antiquated genre, but the Noah threat is imminent today and still at the foundations. The story is a kind of myth teaching a perennial reverence for life. The ancient myth has, for the first time ever, become tragic fact. Humans have more understanding than ever of the speciating processes, more predictive power to foresee the intended and unintended results of their actions, and more power to reverse the undesirable consequences. If there is a word of God here, emerging out of the primordial past, it is "Keep them alive with you" (Genesis 6.19).

Indeed, these primitive stories sometimes exceed the recent legislation in the depths of their insights. Noah is not told to save just those species that are of "esthetic, ecological, educational, historical, recreational and scientific value" to people. He is commanded to save them all. These swarms of species are often useful to humans, and on the Ark clean species were given more protection than others. Noah was not simply conserving global stock and here, man is not the measure of things. The Noah story teaches sensitivity to forms of life and the biological and theological forces producing them. What is required is not human prudence but principled responsibility to the biospheric Earth, to God.

Today, preservation of species is routinely defended in terms of medical, agricultural, and industrial benefits. Other species may be indirectly useful for the resilience and stability they provide in ecosystems. High-quality human life requires a high diversity of species. However, such humanistic justifications for the preservation of species, although correct and required as part of endangered species policy, fall short of Noah's environmental ethics. These are good reasons but not the best, because they do not value these species for what they are in themselves, under God. These reasons are inadequate for either Hebrew or Christian faith, neither of which is simply humanistic about species. Facing the next century, turning the millennium, there is growing conviction among theologians that theology has been too anthropocentric. The nonhuman world is a vital part of Earth's story.

Biology and theology are not always easy disciplines to join, and we shall have more to say about that. One conviction they do share is that the ecosystemic Earth is prolific. Seen from the side of biology, this is called speciation, biodiversity, selective pressures for adapted fit, maximizing offspring in the next generation, niche diversification, species packing, and carrying capacity. Seen from the side of theology this trend toward diversity is a good thing, a godly thing. This fertility is sacred. Endangered species raise the "God" question because they are one place we come near the ultimacy in biological life. Earth is valuable, able for value, a system that generates valuable life. This genesis is, in biological perspective, "of itself," spontaneous, autonomous; and biologists find nature to be prolific, even before the God question is raised. Afterward, theologians wish to add that in such a prolific world, explanations may not be over until one detects God in, with, and under it all.

Resources and sources

The Genesis stories quickly mix human resources with divine sources. "Behold, I have given you every plant yielding seed which is upon the face of all the earth, and every tree with seed in its fruit; you shall have them for food" (Genesis 1.29). Placed in a garden, the couple are commanded "to till it and keep it" (Genesis 2.10). "The Lord God made for Adam and his wife garments of skins, and clothed them" (Genesis 3.21). After the Flood, animals are given as food. So there is no contesting that the biodiversity on the Genesis landscape includes an ecology that supports an economy. The story is about sources as much as resources. In terms of the two kinds of values missing from the Endangered Species Act, the economic values are recognized but entwined with religious values. If some of these species are good for food (or medicine or industry), Genesis warrants saving them on such account, but Genesis teaches this inseparably from a more central teaching that the values carried by species are vitally sacred.

Christians have often and admirably focused on economic values, insisting on political provision for jobs, food, shelter, and health care. In endangered species policy, the values that Christians wish to defend are often the more foundational and vital. Perhaps God wills a good life in a promised land; but without its fauna and flora, the land cannot fulfill all its promise.

One cannot look to the market to produce or protect the multiple values that the Nation and its people enjoy from the myriad species inhabiting the landscape, since many of these values carried by species are not, or not simply, economic ones. A pristine natural system, with its full complement of species, is a religious resource, as well as a scientific, recreational, aesthetic, or economic one. So we can call these species resources if we like, but there is more. If they are nothing but our human resources, it seems to profane them, to forget the pleasure that their Creator takes in this creation.

That explains why, confronting wildness, humans know the sense of the sublime. We get transported by forces awe-full and overpowering, by the signature of time and eternity. Being among the archetypes, a landscape, a forest, or a sea swarming with its kinds is about as near to ultimacy as we can come in the natural world – a vast scene of birth and death, sprouting, budding, flowering, fruiting, passing away, passing life on. We feel life's transient beauty sustained over chaos. Nature, swarming with its kinds, is a wonderland. "Praise the Lord from the earth you sea monsters and all deeps, fire and hail, snow and frost, stormy wind fulfilling his command! Mountains and all hills, fruit trees and all cedars! Beasts and all cattle, creeping things and flying birds!" (Psalm 148.8–9) "Thou crownest the year with thy bounty; the tracks of thy chariot drip with fatness. The pastures of the wilderness drip, the hills gird themselves with joy, the meadows clothe themselves with flocks, the valleys deck themselves with grain, they shout and sing for joy" (Psalm 65.11–13).

In contrast with the surrounding religions from which Biblical faith emerged, the natural world is disenchanted; it is neither God, nor is it full of gods; but it remains sacred, a sacrament of God. Though nature is an incomplete revelation of God's presence, it remains a mysterious sign of divine power. The birds of the air neither sow nor reap yet are fed by the heavenly Father, who notices the sparrows that fall. Not even Solomon is arrayed with the glory of the lilies, though the grass of the field, today alive, perishes tomorrow (Matthew 6). There is in every seed and root a promise. Sowers sow, the seed grows secretly, and sowers return to reap their harvests. God sends rain on the just and unjust. "A generation goes, and a generation comes, but the earth remains forever" (Ecclesiastes 1.4).

Randomness and creativity

But it is not always easy to join biology and theology. To put the problem in a contrasting pair of keywords: Is Earth by "design" or "accident"? Before Darwin, the world seemed well designed, species were adapted for their niches, fixed in kind, going back to an original special creation. Just as watches indicated a watchmaker, rabbits indicated a Rabbitmaker. After Darwin, there are random, blind mutations, the survival of the accidentally better adapted, and the evolution of species. There was no original creation at all, rather a billion years of accident and groping. Rather than God's first creating and subsequently preserving all of Earth's teeming species, species have come and gone in a constant and sometimes catastrophic turnover. All species, *Homo sapiens* included, are here by luck. Earth is not a watch, but a jungle; not a well-designed Eden, but a contingent chaos. Jacques Monod, a Nobel prizewinner, has claimed that natural history is "an enormous lottery presided over by natural selection, blindly picking the rare winners from among numbers drawn at utter random" (Monod, 1972, p. 138). Recently, David Raup has put catastrophism back into paleontology (Raup, 1988), and Stephen Gould has learned from the Burgess shale that the species on Earth, however wonderful, are chance riches and accidental life (Gould, 1989). If so, there can be no connection between God and species of whatever kind, much less endangered species.

Since we are touching creation and ultimacy, to keep the full picture in focus, we should notice that in physics, cosmologists have been finding this universe spectacularly fine-tuned for life. Hundreds of microphysical and astronomical phenomena, both contingencies and necessities, have to be almost exactly what they are if life is to be possible. Examples include the charges on electrons and protons, the strengths of the four binding forces, the scales, distributions, and ages of the stars, the expansion rate of the universe, the proportions of hydrogen and helium, and the structures of many heavier elements. Even before there is life, we already get a pro-life universe (Leslie, 1989).

If the contingencies and necessities of physics make life possible, so also do its indeterminacies. Just these microphysical indeterminacies provide the openness upon which a biological organism can superimpose its program. The organism is fine-tuned at the molecular level to nurse its way through the quantum states by electron transport, proton pumping, selective ion permeability, and so on. The organism interacts with the microphenomena (somewhat analogously to the way physicists participate in their observations), catching the random fluctuations to its advantage, setting up from above the conditions of probability. Through its biochemistry it shapes the course of the microevents that constitute its passage through the world. Physics frees the world for the adventure of biology.

The difference between physics and biology is that biology is a historical science, where cumulative discoveries are coded into the organism over time. The laws of physics and chemistry are the same on Jupiter, on Mars, or in the galaxy Andromeda. But genetic coding, the cytochrome-*c* molecule, the citric acid cycle, photosynthesis, trilobites, dinosaurs, and grizzly bears are peculiar to Earth. They incorporate elements of randomness, but even more they represent creative achievements on Earth, now coded into the DNA and expressed in these species. Perhaps we are beginning to see that "accident" is not the full story; there is valuable creativity at work on our planet.

George Wald, also a Nobel prizewinner, differs with Monod: "This universe breeds life inevitably" (Wald, 1974, p. 9). Manfred Eigen, still another Nobel laureate, concludes, "that the evolution of life . . . must be considered an *inevitable* process, despite its indeterminate course" (Eigen, 1971, p. 519). Melvin Calvin, still another Nobel laureate, concludes that life evolves "not by accident but because of the peculiar chemistries of the various bases and amino acids. . . . There is a kind of selectivity intrinsic in the structures." Far from being random, life is "a logical consequence" of natural principles (Calvin, 1975, p. 176).

Despite the prolife world in physics, there is not much in the atoms themselves that enables us to predict that they will organize themselves in this remarkable way. Given chemistry as a premise, there is no deductive or inductive logic by which biology follows as a conclusion. Still there is this remarkable story to tell; and, when it happens, though it is no inference, neither does it seem nothing but accident. There seems to be some creativity intrinsic in the Earth by which these elements order themselves up to life. The story goes from zero to five million species in five billion years, passing through perhaps one billion species en route. By some mixture of inevitability and openness, given the conditions and constants of physics and chemistry, together with the biased Earth environment, life will somehow both surely and surprisingly appear. Once upon a time there was a primitive planetary environment in which the formation of living things had a high probability. In other words, the archaic Earth was a pregnant Earth. We may need not so much interference by a supernatural agency as the recognition of a marvelous endowment of matter–energy with a propensity toward life. Yet this endowment can be congenially seen, at a deeper level, as the divine creativity.

Where once there was but matter and energy, there appeared information, symbolically encoded, and life. There emerged a new state of matter, neither liquid, nor gaseous, nor solid, but vital. Randomness does not rule out creativity; randomness plus something to catch the upstrokes, something to code them and pass them on to the future, yields creativity, at the same time that it puts adventure, freedom, drama, and surprise into the storied evolutionary course.

The word "design" nowhere occurs in Genesis. There is divine fiat, divine doing, but the mode is an empowering permission that places productive autonomy in the creation. "Let Earth bring forth. . . . " Biologists cannot deny this creativity; indeed, better than anyone else, biologists know that Earth has brought forth the natural kinds exuberantly over the millennia. The better question is not so much whether these creatures have *design* in the Craftsman/Architect–artifact/machine sense as whether they have *value.* Do they have inherent goodness? A thing does not have to be directly intended to have value. It can be the systemic outcome of a problem-solving process. If it results from such creativity, it is a valuable achievement.

Struggle and perceptual perishing

Perhaps the contrasting words that separate biology and religion are not "design" and "accident" but "good" versus "evil." Darwin once exclaimed that the evolutionary process was "clumsy, wasteful, blundering, low, and horribly cruel" (quoted in de Beer, 1962, p. 43). That is utter antithesis to the Genesis verdict of "very good." The governing principle is survival in a "nature red in tooth and claw" (Tennyson, *In Memoriam A. H. H.*, Part LVI, Stanza 4). The wilderness contains only the thousandth part of creatures that sought to be, but rather became seeds eaten, young fallen to prey or disease. The wilderness swarms with kinds, as Genesis recognizes, but is a vast graveyard with a hundred species laid waste for one or two that survive. Blind and ever urgent exploitation is nature's driving theme, the survival of the fittest. George Williams, a foremost student of natural selection, concludes, "The cosmos stands condemned. The conscience of man must revolt against the gross immorality of nature" (Williams, 1988). Biologists are not altogether comfortable with the word "struggle," often preferring the notion of "adapted fit." Still, plenty of "struggle" remains in biology, and can it be godly?

The truth is that biological creativity is logically entwined with struggle and perishing. Life is the first miracle that comes out of nature, and death comes inevitably in its train. For an organism things can go wrong just because they can go right; a rock or a river never fails, but then again neither can ever succeed. In biology, we are not just dealing with causes and effects but with vitality and survival. A rock exists on its own, having no need of its environment, but an organism has welfare and interests; it must seek resources. Generation means regeneration. Life decomposes and recomposes. Religion, monotheism included, seldom teaches that creativity is without struggle. Life is green pastures found in the valley of the shadows, a table prepared in the midst of enemies (Psalm 23). By the third chapter Genesis is teaching that we eat our bread in sweat and tears.

In physics and chemistry there is no history refolding itself into compounding chapters; that comes with the evolutionary epic. There is also no suffering; that too comes in biology. With all life there is duress; and, with the evolution of sentience, there is suffering. Conservation in physics and chemistry is a foregone conclusion, for example, conservation of energy, mass, baryon number, or spin. Conservation in biology is vital and contingent. Life can be lost; indeed in higher forms individual life invariably is lost, although by reproducing and speciating life is conserved over the millennia.

The death of earlier creatures makes room for later ones, room to live, and, in time, to evolve. If nothing much had ever died, nothing much could have ever lived. The evolutionary adventure uses and sacrifices particular individuals, who are employed in, but readily abandoned to, the larger currents of life. Evolution both overleaps death and seems impossible without it. The element of struggle is muted and transmuted in the systemic whole. Something is always dying, something is always living on.

In this perspective, biology and religion draw closer together. Israel is the rose of Sharon, blooming in the desert, the shoot budding forth out of the stump of Jesse. The root meaning of "Israel" is to struggle. Life is gathered up in the midst of its throes, a blessed tragedy, lived in grace through a besetting storm. Israel's founding historical memory is the Passover observance, a festivaı of the renewal of spring and of exodus releasing life from the powers that suppress it. Christianity intensifies this renewal in adversity with its central symbol: the cross. There is dying and rising to newness of life. Life is cruciform.

The grass, the flower of the field, is clothed with beauty today and gone tomorrow, cast into the fire. The sparrow is busy about her nest, and sings, and falls, noticed by God. There is trouble enough with each new day, and, beneath that, some providential power by which life persists over the vortex of chaos. We find life handed on, through ills and all, by wisdom genetically programmed, as well as in the cultural heritage of our forebears. The secret of life is only penultimately in the DNA, the secret of life ultimately is this struggling on to something higher. We dimly comprehend that we stand the beneficiaries of a vast providence of struggle that has resulted in the panorama of life. Just that sense of ongoing life, transcending individuality, makes life at the species level a religious value. Speciation lies at the core of life's brilliance, and to confront an endangered species, struggling to survive, is to face a moment of eternal truth.

Nature, law, and grace

Paradoxically, past the suffering, life is a kind of gift. Every animal, every plant has to seek resources, but life persists because it is provided for in the system. The swarms of creatures are not so much an ungodly jungle as a divinely inspired Earth. "Design" is not the right word; it is a word borrowed from mechanics and their machines. Genesis is the better word, with "genes" in it, the gift of autonomy and self-creation. Designed machines to not have any interesting history; clocks have no story lines. But organisms must live biographies, and such a story continues for several billion years. Such an Earthen providing ground is, in theological perspective, providential. Providential adventures do not so much have design as pathways. In grace accompanying a passage through history, there must be a genetic pathway available along which there can be a lineage of descent, ascent, exploration, and adventure. Monotheists who take genesis seriously do not suppose a *Deux ex machina* that lifts organisms out of their environment, redesigns them, and reinserts them with an upgraded design. Rather they find a divine creativity that leads and lures along available routes of Earth history.

Laws are important in natural systems, but natural law is not the complete explanatory category for nature, any more than are randomness and chance. In nature, beyond the law is grace. There is creativity by which more comes out of less. Science prefers lawlike explanations without surprises. One predicts, and the prediction comes true. But, nevertheless, biology is full of unpredictable surprises. Our account of natural history will not be by way of implication, whether deductive or inductive. There is no covering law (such as natural selection), plus initial conditions (such as trilobites), from which one can deduce primates, any more than one can assume microbes as a premise and deduce trilobites in conclusion. Nor is there any induction (expecting the future to be like the past) by which one can expect trilobites later from procaryotes earlier, or dinosaurs still later by extrapolating along a regression line (a progression line!) drawn from procaryotes to trilobites. There are no humans invisibly present (as an acorn secretly contains an oak) in the primitive eucaryotes, to unfold in a lawlike way. All we can do is tell the epic story – eucaryotes, trilobites, dinosaurs, primates, persons who are scientists, ethicists, conservation biologists – and the drama may prove enough to justify it.

In only seeming contrast to Adam and Noah, who are trustees of the creation, Job rejoices in how the nonhuman creation is wild, free from the hand of man. "Who has let the wild ass go free? Who has loosed the bonds of the swift ass, to whom I have given the steppe for his home, and the salt land for his dwelling place? . . . He ranges the mountain as his pasture, and he searches after every green thing" (Job 39.5–8). [Even in Biblical times, the wild ass was an endangered species; nevertheless it persisted in Palestine until 1928, when it became extinct.] "Is it by your wisdom that the hawk soars, and spreads his wings toward the south? Is it at your command that the eagle mounts up and makes his nest on high? On the rock he dwells and makes his home in the fastness of the rocky crag. Thence he spies out the prey; his eyes behold it afar. His young ones suck up blood; and where the slain are, there is he" (Job 39.26–40.2). "The high mountains are for the wild goats; the rocks are a refuge for the badgers. . . . The young lions roar for their prey, seeking their food from God. . . . O Lord, how manifold are thy works! In wisdom hast thou made them all" (Psalm 104.18–24).

Though outside the hand of man, the wild animals are not outside either divine or biological order. The Creator's love for the creation is sublime precisely because it does not conform to human purposes. That God is personal as revealed in human cultural relations does not mean that the natural relationship of God to hawks and badgers is personal, nor should humans treat such creatures as persons. They are to be treated with appropriate respect for their wildness. The meaning of the word "good" and "divine" is not the same in nature and in culture. Just as Job was pointed out of his human troubles toward the wild Palestinian landscape, it is a useful, saving corrective to a simplistic Jesus-loves-me;-this-I-know, God-is-on-my-side theology to discover vast ranges of creation that have nothing to do with satisfying our personal desires.

What the wildlands with their swarms of species do "for us," if we must phrase it that way, is teach that God is not "for us" humans alone. God is "for" these wild creatures too. In Earth's wildness there is a complex mixture of authority and autonomy, a divine imperative that there be communities (ecosystems) of spontaneous and autonomous ("wild") creatures, each creature defending its form of life. A principal insight that Biblical faith can contribute to conservation is to take the concept of wildlife "sanctuaries" in national policy to its logical and religious conclusion. A wildlife sanctuary is a place where nonhuman life is sacrosanct, that is, valued in ways that transcend human uses. In that sense Christian conviction wants sanctuaries not only for humans, but also for what wild lives are in themselves and under God. Since there is hardly a stretch of landscape in our nation not impoverished of its native fauna and flora, we want sanctuaries especially for endangered species.

Religious persons can bring a perspective of depth to biological conservation. Species are a characteristic expression of the creative process. The swarms of species are both presence and symbol of forces in natural systems that transcend human powers and human utility. Generated from earth, air, fire, and water, these fauna and flora are an archetype of the foundations of the world. We want a genetic account in the deeper sense. The history of Earth, we are claiming, is a story of the achievement, conservation, and sharing of values. Earth is a fertile planet, and in that sense, fertility is the deepest category of all, one classically reached by the category of creation. This creative systemic process is profoundly but partially described by evolutionary theory, a historical saga during which spectacular values are achieved and at the core of which the critical category is value, commonly termed "survival value," better interpreted as valuable information, coded genetically, that is adapted for, apt for "living on and on" *(survival),* for coping, for life's persisting in the midst of its perpetual perishing. Such fecundity is not finally understood until seen as divine creativity.

This history has been a struggling through to achieve something higher, to better adapted fit and more complex and diverse forms, and there is no particular cause to assume that the grim accounts of it are the adult, biologically correct ones, and the gracious, creative, charismatic ones childish, naive, or romantic. Or, shifting the meaning of "romantic" to its original sense, life is a romance, an epic of vital conflict and resolution producing rich historical novelty.

Religious conservation biologists

Whatever you may make of God, biological creativity is indisputable. There is creation, whether or not there is a Creator, just as there is law, whether or not there is a Lawgiver. Some biologists decline to speak of creation, because they fear a Creator lurking beneath. Well, at least there is genesis, whether or not there is a Genitor. Ultimately, there is a kind of creativity in nature demanding either that we spell nature with a capital N, or pass beyond nature to nature's God. Biologists today are not inclined, nor should they be as biologists, to look for explanations in supernature, but biologists meanwhile find a nature that is super! Superb! Science teaches us to eliminate from nature any suggestions of teleology, but it is not so easy for science to talk us out of genesis. What has managed to happen on Earth is startling by any criteria. Biologists may doubt whether there is a Creator, but no biologist can doubt genesis.

Ernst Mayr, one of the most eminent living biologists, concludes, "Virtually all biologists are religious, in the deeper sense of the word, even though it may be a religion without revelation. . . . The unknown and maybe unknowable instills in us a sense of humility and awe" (Mayr, 1982, p. 81). "And if one is a truly thinking biologist, one has a feeling of responsibility for nature, as reflected by much of the conservation movement" (Mayr, 1985, p. 60). "I would say," concludes Loren Eiseley, at the end of *The Immense Journey*, "that if 'dead' matter has reared up this curious landscape of fiddling crickets, song sparrows, and wondering men, it must be plain even to the most devoted materialist that the matter of which he speaks contains amazing, if not dreadful powers, and may not impossibly be . . . 'but one mask of many worn by the Great Face behind'" (Eiseley, 1957, p. 210).

Annie Dillard, a poet, once found herself terrified at the evolutionary ordeal, although she too can, in other moments be amply religious about it. Overlooking the long, odious scene of suffering and violence, she cries out: "I came from the world, I crawled out of a sea of amino acids, and now I must whirl around and shake my fist at that sea and cry shame" (Dillard, 1974, p. 177). Must she? There is nothing shameful about amino acids rising out of the sea, speciating, swarming over Earth, assembling into myriads of species, not the least of which is *Homo sapiens,* with mind to think and hand to act. If I were Aphrodite, rising from the sea, I think I would turn back, reflect on that event, and rather raise both hands and cheer. And if I came to realize that my rising out of the misty seas involved a long struggle of life renewed in the midst of its perpetual perishing, I might well fall to my knees in praise.

J. B. S. Haldane was asked by some theologians what he had concluded from biology about the character of God. He replied that God had an inordinate fondness for beetles, since he made so many of them. Haldane went on to say that the marks of biological nature were its "beauty," "tragedy," and "inexhaustible queerness" (Haldane, 1932, pp. 167–169). My experience is that this beauty approaches the sublime; the tragedy is perpetually redeemed with the renewal of life, and that the inexhaustible queerness recomposes as the numinous. "Nature is one vast miracle transcending the reality of night and nothingness" (Eiseley 1960, p. 171).

Biology produces many doubts; here are two more. I doubt whether you can be a conservation biologist without a respect for life, and the line between respect for life and reverence for life is one that I doubt that you can always recognize. If anything at all on Earth is sacred, it must be this enthralling creativity that characterizes our home planet. If anywhere, here is the brooding Spirit of God. Whatever biologists may make of the mystery of life's origins, they almost unanimously conclude that the catastrophic loss of species that is at hand and by our hand is tragic, irreversible, and unforgivable. Difficult to join though biology and theology sometimes are, they are difficult to separate in their respect for life. Earlier we worried that the processes of creation might be ungodly. But faced with extinction of these processes, biology and theology quickly couple to reach one sure conclusion. For humans to shut down Earth's prolific creativity is ungodly.

References

Calvin, Melvin (1975). Chemical evolution. *American Scientist,* **63,** 169–177.

de Beer, Gavin (1962). *Reflections of a Darwinian.* London: Thomas Nelson and Sons.

Dillard, Annie (1974). *Pilgrim at Tinker Creek.* New York: Harper & Row.

Eigen, Manfred (1971). Selforganization of matter and the evolution of biological macromolecules. *Die Naturwissenschaften,* **58,** 465–523.

Eiseley, Loren (1975). *The Immense Journey.* New York: Vintage.

Eiseley, Loren (1960). *The Firmament of Time.* New York: Atheneum.

Gould, Stephen Jay (1989). *Wonderful Life: The Burgess Shale and the Nature of History.* New York: W. W. Norton.

Haldane, J. B. S. (1932, 1966). *The Causes of Evolution.* Ithaca: Cornell University Press.

Leslie, John (1989). *Universes.* New York: Routledge.

Monod, Jacques (1972). *Chance and Necessity.* New York: Random House.

Mayr, Ernst (1982). *The Growth of Biological Thought.* Cambridge, MA: Harvard University Press.

Mayr, Ernst (1985). How biology differs from the physical sciences. In *Evolution at a Crossroads,* ed. David J. Depew and Bruce H. Weber. Cambridge, MA: MIT Press.

Raup, David M. (1988). Changing views of natural catastrophe. In *The Great Ideas Today.* Chicago: Encyclopedia Britannica.

United States Congress, *Endangered Species Act of 1973.* 87 Stat. 884.

United States Supreme Court, *Tennessee Valley Authority vs. Hill* (1978). 437 US 153.

Wald, George (1974). Fitness in the universe: choices and necessities. In *Cosmochemical Evolution and the Origins of Life,* ed. J. Oró et al. Dordrecht, Netherlands: D. Reidel.
Williams, George (1988). Huxley's evolution and ethics in sociobiological perspective. *Zygon,* **23,** 383–407.

4

Biodiversity and ecological justice

ERIC KATZ

Introduction

The title of this essay requires explanation. The idea of "biodiversity" is rarely conjoined with the idea of "justice." This is because "biodiversity" is a scientific concept, whereas "justice" is, in part, normative. So I will have to explain why I believe the conjunction makes sense. The use of the adjective "ecological" to modify the concept of justice also requires explanation. Once again I have combined a scientific idea – that of ecology – with the normative idea of justice. Does *this* combination make sense? Is it possible to think of a system of justice that is ecological as well as being normative, political, social? Or is this a hopeless jumble of incompatible and contradictory ideas?

I will show that this seemingly contradictory combination of scientific and normative concepts is necessary for a full understanding of the moral dimensions of biodiversity. I will argue that only by developing a system of justice that can properly be called "ecological justice" can we morally justify policies that preserve the biodiversity of the planet. If we remain trapped in the traditional categories of normative thought we will be unable to justify acceptable and necessary environmental policies.

Two preliminary warnings: First, it must be emphasized that this essay is an exercise and argument in applied moral philosophy. It is an examination and criticism of a set of normative beliefs underlying various kinds of environmental policy. Philosophical arguments have their own standards of proof, which differ in important ways from the standards of scientific proof. It is probably impossible to "prove" in an "objective" way that a position in moral philosophy is "true" – instead, philosophical rigor involves an examination of the coherence, consistency, and implications of a given position. It is from that starting point that this exercise in applied moral philosophy proceeds.

Second, it also must be emphasized that this argument involves a sharp criticism of conventional value judgments – concerning both their meaning in moral philosophy in general and their specific application to environmental issues. I argue for views that lie outside the major traditions of Western thought, precisely because the major traditions have failed to deal with the contemporary environmental crisis. But in what is a pleasant paradox, the views I represent actually constitute the mainstream of that branch of applied moral philosophy called "environmental ethics." Little more than twenty years old, the field of environmental philosophy offers a continuous reexamination of conventional opinions about the meaning and basis of ethical judgments regarding both humans and the natural environment.

A crisis in moral value

The threat to planetary biodiversity caused by the technological, economic, and environmental policies of the last half century is only partly a scientific problem. For a practitioner of moral philosophy, the threat to biodiversity is primarily a crisis in moral values. Traditional assumptions about value and the normative principles that shape moral life need to be rethought and modified. In light of increasingly complex environmental and social problems, these traditional views are, at best, inadequate; at worst, they are contributing causes of the environmental crisis.

The traditional "enlightened" interpretation of the environmental crisis is that humanity must now acknowledge the mutual interdependence of human society and the biological and ecological processes of the natural environment. The survival of individual human beings, and human civilization itself, requires the preservation of diverse biological systems and environments. Failure to recognize this interdependence could lead to a state of neobarbarism, the collapse of organized modes of local and international cooperation and stability, as a violent competition for the scarce resources necessary for human survival dominates social policy. This message – the warning that human civilization will collapse as a consequence of environmental destruction – originated, in recent years, in the work of social critics such as Paul and Anne Ehrlich, Barry Commoner, Garrett Hardin, Murray Bookchin, Norman Myers, and Rachel Carson, among others. The message is no longer the work of "extremist" critics; it is accepted, at least on the surface, as a truism. The consciousness of the environmental crisis and the need for preservation, conservation, and recycling as a requirement of human survival have spread throughout the population at large.

Why is this a problem? Why should we not rejoice that this once "extreme" position regarding the preservation of the natural environment has become the "mainstream" traditional view? What could be wrong with saving humanity?

The problem involves the determination of moral value. Any response to the environmental crisis requires the implementation of new social policies and new obligations on the part of human agents, institutions, and governments. As such, this response (or set of responses) makes explicit normative and ethical assumptions regarding human action and value. Whether they realize it or not, environmental policymakers, governmental officials, and ecological scientists are making moral decisions; these decisions reflect a (generally) unarticulated and uncritical vision of life and value. It is one task of the moral philosopher to articulate and to examine the values that are expressed through the implementation of these new policies. What values, in short, serve as the basis of the policies of conservation of the natural environment and preservation of planetary biodiversity?

The current response to the environmental crisis expresses values that are blatantly anthropocentric: the main source of value lies in the continued existence of human life and civilization. The natural world – with its ecosystems, species, individual entities – is valued for its service to humanity, its instrumental use for the preservation of human goods. The possibility that the natural world could be valued for its own sake, that it would have a good of its own worth preserving, is hardly considered at all, and rarely plays any important part in the determination of policy.

An anthropocentric value system that only regards natural processes as important for human survival cannot serve as the basis of a comprehensive environmental policy. Anthropocentrism, in its narrow formulations of egoism, economic expediency, and utilitarianism, has been the primary force in the creation of the environmental crisis. Broadening the concept, to include the instrumental importance of the natural world for the prevention of human extinction, is hardly adequate as a solution. It merely restates the problem of the environmental crisis: why is all value based on human goods and interests? (e.g., Ehrenfeld, 1978).[1]

The real solution to problems in environmental policy lies in a specific transformation of values – the transcendence of human-based systems of ethics and the development of an "ecological ethic." Humanity must acknowledge that moral value extends beyond the human community to the communities within natural systems (Leopold, 1970; Taylor, 1986; Rolston, 1987; Callicott, 1989; Wenz, 1988).[2] It is for this reason that the concept of "biodiversity" must be linked with "ecological justice." And it is for this reason that the problem of a diminishing planetary biodiversity is a crisis in moral value. Policies that ensure the preservation of planetary biodiversity must express values derived from a nonanthropocentric moral system, a normative theory of justice that is "ecological," i.e., a theory not based merely on human goods and interests.

The problem of anthropocentrism

Why is an anthropocentric value system an inadequate basis for environmental policy? This essay considers two major arguments, one theoretical, one practical. The theoretical argument involves the *contingent* relationship between the promotion of human interests and the continued preservation of the natural environment (Krieger, 1973; Sagoff, 1974; Katz, 1979). Many human interests or goods are thought to be connected to the preservation of the natural world, connected in the sense that the satisfaction of the human interest or the production of the good derives from the continued existence of the natural processes. The environmentalist claims that the human interest or good requires the existence of the natural environment. But this connection is not necessary: human interests can probably be satisfied without nature, and certainly without a pristine nature.

A simple example is the human interest in *beauty*. It is claimed that natural environments and the biodiversity of the planet ought to be preserved because of the human need for beauty. A planet of diverse natural habitats and diverse individuals and species provides more opportunities for a necessary component of human life: appreciation of the beautiful. But the human interest in beauty can be satisfied in other ways. Urbanites, for example, New Yorkers such as myself, can satisfy their need for aesthetic stimulation and beauty by visits to the Metropolitan Museum or the Museum of Modern Art; they can view the impressive architectural designs of the city; with a heightened imagination, they can even find beauty in the dirt and debris of the urban landscape. The satisfaction of the need for beauty is thus only contingently connected to the preservation of the natural environment.

I am not claiming that humans have no interest in natural – as opposed to artifactual or cultural – beauty; I am arguing that the existence of natural beauty is not a necessary requirement for a complete human life. Thus, the environmentalist *argument* that natural diversity be maintained *because it satisfies a necessary human need for beauty* is a flawed argument: it does not provide an adequate justification for the preservation of the natural world.

This example concerning the human interest in beauty can be generalized to include all arguments for environmental preservation that are based on the satisfaction of human interests. Any such "human interest" argument is an *instrumental* argument: natural processes, environments, species, etc., are preserved for their instrumental use to humanity. The problem with instrumental arguments for preservation is that they deemphasize the intrinsic value of the object being preserved. Since the only value that matters is the use value, an adequate substitute that provides the same use will be valued just as highly as the original object. The original need not be preserved, for the instrumental use and the satisfactions derived thereby are provided by other means (Katz, 1985).

It is clear that I am using a very broad notion of instrumental value. On my view, any entity or process that provides a benefit for human beings has an instrumental anthropocentric use value. Money, power, and pleasure are clearly instrumental goods – but so are beauty, friendship, and spiritual wonder. The point is that if humans preserve natural entities because of the benefits derived from the entities, their motivation and justification is instrumental, regardless of the kind of benefit being sought; this is different in fundamental ways from a justification based on moral obligation, which is not based on the maximization of benefits (Godfrey-Smith, 1979; Rolston, 1987).[3]

In sum, anthropocentric arguments for the preservation of natural systems fail to achieve their aim. Justifications of environmental policies that are based on the satisfaction of human interest overlook the possibility of adequate substitutes for the promotion of these goods. Since the intrinsic qualities and value of natural objects and systems are ignored, these cannot be used to justify preservation. The contingent use for humanity is all that matters.

It might be objected that certain interests and goods for humanity are not contingent, but necessary: for example, the interest in human survival. Although my need for beauty can be fulfilled by looking at a Vermeer at the museum rather than a sunset over a wilderness lake in the Rockies, I cannot produce adequate substitutes for the food, water, and air I need to survive. Anthropocentric arguments for the preservation of natural environments gain force by focusing on the basic needs of human survival: the preservation of the biological cycles responsible for the production of clean food, air, and water.

There is some truth to this objection, but the argument does not take us very far. Given the increasing technological sophistication of the human race, it is unclear how many purely natural entities and processes we require for survival. We can create artificial foodstuffs, desalinate water, and purify air. How much of the natural world do we really require?[4] Because this is an open question, I am reluctant to base the justification of environmental policies of preservation on the necessity of natural soil, water, and air. The human race is surviving right now, despite the massive destruction we have imposed on natural systems and planetary biodiversity. So anthropocentric arguments that emphasize the connection between the preservation of nature and the survival of humanity are no less instrumental and contingent than those that emphasize other nonbasic human needs and interests (such as beauty). Survival arguments for environmental preservation are contingent on a given technological capability. If we have a technology that replaces nature we will no longer need to preserve it;

but we do not want to base the policy on the existence or nonexistence of a specific technology.

This theoretical argument involving contingency may not be convincing to ecological scientists and policymakers who daily investigate the connection between planetary diversity and the preservation of necessary biological cycles. Scientists are often skeptical of philosophical thought experiments. Consider, then, a second, more practical argument for the inadequacy of anthropocentric justifications of environmental policy. This argument involves the problem of ecological imperialism and the development of the Third World (Katz & Oechsli, 1993).

Consider the Amazon rainforest. The preservation of the rainforest is important, not only to preserve biodiversity but also to prevent environmental problems such as the "greenhouse" effect. The rainforest is home to millions of species with many possible instrumental uses for the betterment of human life. Destruction of the rainforest will eliminate the habitats for these species and cause their extinction. In addition, the burning of the wood from the forests increases the amount of carbon dioxide in the atmosphere, and removes carbon dioxide-consuming vegetation from the planet's surface. Both processes increase the likelihood of global warming.

Policymakers and scientists therefore urge the preservation of the rainforests, their removal from areas open to development for farming and industry. This is obviously the correct environmental position. But many environmentalists, including myself, are uncomfortable with this position: it is too similar to ecological imperialism. We in the industrialized North are urging the poorer nations of the nonindustrialized South to refrain from the economic development of their own resource base. After having destroyed our own areas of diverse natural resources in the pursuit of national and individual wealth, we suddenly realize the importance of these areas for the survival of humanity, and so we prevent the rest of the world from achieving our own levels of national and individual affluence. We reap the benefits from past ecological destruction and development. The poorer undeveloped nations now pay the price: being forced to preserve their natural environments for the sake of the world and the rest of humanity.

The correct environmental policy of rainforest preservation thus raises questions of political and moral justice. Is it fair to inflict the costs of preservation on the poorer nations of the world when the benefits of this preservation (biodiversity and the slowing down of global warming) are distributed throughout the world as a whole? Is it doubly unfair that the cause of the problem has been the unchecked industrial growth of the richer nations? These are rhetorical questions: the injustice of the present situation is too obvious to argue. Environmentalists who care about broader issues of justice are faced with a painful dilemma – development or preservation; neither option produces desirable results.

There is a way out of this dilemma, and the resolution of the problem provides a *practical* argument against anthropocentric value systems. The problem of justice arises here because the issue is framed exclusively in terms of human goods and benefits; it is a dilemma involving competing claims of human values. On one side are the benefits to be derived from the preservation of the rainforest: biodiversity, less carbon dioxide in the atmosphere, etc. On the other side are the benefits to be gained from economic development: increased wealth for individuals in the poorer nations of South America. We cannot have both benefits. The dilemma arises because we do not know how to balance these competing human claims.

In thus framing the problem, we ignore the intrinsic value of the *rainforest itself.* The preservation or development of the natural environment is here conceived, as usual, as instrumentally useful for humanity, nothing more. The problem involves determining which instrumental-use value is greater and/or fairer. But the problem disappears once we focus our attention past the narrow anthropocentric interests of use and consider the rainforest, the natural environmental system, itself. Consider an analogy with two businessmen, Smith and Jones, who are arguing over the proper distribution of the benefits and costs resulting from a prior business agreement between them. If we just focus on Smith and Jones and the issues concerning them, we would want to look at the contract, the relevant legal precedents, and the actual results of the deal, before rendering a decision. But suppose we learn that the agreement involved the planned murder of a third party, Green, and the resulting distribution of his property. At that point the issues between Smith and Jones cease to be relevant; we no longer consider it important who has claims to Green's wallet, overcoat, or Mercedes. The competing claims become insignificant in light of the intrinsic value and respect due to Green.

This kind of case is analogous to the conflict over the development of the rainforest, and indeed most other environmental problems. The difference is that instead of an exclusively human case, the third party here is the rainforest. As soon as we realize that the intrinsic interests of the rainforests are relevant to the conflict of competing goods, then the claims of both the developers and the preservationists lose force. What matters is the rainforest, not the economy of Brazil or the survival of humanity.

The dilemma over the third-world development of natural environments is thus a *practical* example of the crisis in value. If we remain within the framework of anthropocentrism, we view this problem as an impossible balancing of competing human claims; but if we transcend anthropocentrism, and view the natural environment as valuable in itself, then the problem dissolves. This provides a powerful argument for abandoning our traditionally exclusive reliance on human values, goods, and interests in the determination of environmental policy.

It might be objected that I have misstated the problem: the dilemma between development and preservation of the Amazon rainforest may be more apparent than real. Development, it can be argued, will only produce short-term economic benefits that will not help the nonindustrialized nations and indigenous peoples of the region. The real goal of policy is neither preservation nor development per se, but sustainable development, the creation of an economy that uses and replenishes the natural environment of the region. There is merit in this objection, and so I remain open to possible empirical solutions to the dilemma of justice in third-world environmental policy. Nevertheless, a shift in values will be required for the successful implementation of any environmental policy. Even a policy of sustainable development will have to be based on the intrinsic respect for that which is being sustained, the natural environment.

The need for an ecological ethic

My criticisms of anthropocentric value theory are based in part on a vision of moral value that extends beyond the human community to embrace the entire natural world. The failure of anthropocentric justifications of environmental policy shows that it is necessary to develop such a transhuman or nonanthropocentric ethic. But how do we develop a nonhuman ethic? On what concepts or models can it be based? Is there any possibility of demonstrating the validity of such a radical value system?

For me, the focus of moral concern and the determination of moral value must lie in the idea and the concrete existence of community. It is within communities that we perceive and acknowledge moral obligations and relationships. It is within and for communities that we act beyond the narrow confines of self-interest. Altruism and self-sacrifice only make sense within the context of communal relationships.

The origin of this view of community within the history of human-based ethical systems can be traced back to Plato and Aristotle. For Aristotle, human beings could only live an excellent life, a life of virtue, if they lived in the *polis*, the political and social community. The various social relationships that existed in the polis were the source of moral obligation: all moral value had its foundation in the functions of the social community.[5] For Plato, at least in *The Republic*, the role of community was even more important. In establishing the ideal state the good for the community was the supreme good: the well-being of the whole society had precedence over individual interests and needs.[6] The community as such was thus the primary focus of moral value and obligation.

Is it now possible to use this traditional notion of community in the establishment of a nonanthropocentric ethic? All that is required is to acknowledge that biological systems, ecosystems, natural environments, bioregions, etc., are communities in some relevant sense. Do natural systems establish mutually interdependent relationships among the members of the systems? Do various entities in the systems work towards common goals in a kind of natural cooperation? Is there greater value in the whole system than in the individual members?

These are crucial questions for the development of an environmental ethic, and many ecological theorists and environmental philosophers have debated them (e.g., Brennan, 1984, 1988; Norton, 1987; Cahen, 1988). Although there are many differences between biological and cultural communities, the idea of an ecosystem as analogous to a moral and social community of human beings is a powerful analytical tool for the development of ethical ideals. The idea of community as a metaphor for the illumination of ethical concepts, values, and obligations was useful to Plato 2500 years ago. Now the notion of community can be extended to include natural ecosystems, as the naturalist Aldo Leopold did 45 years ago.

Leopold used the notion of community as the heart of his seminal essay "The Land Ethic." Because community was the source of all ethical obligations, Leopold argued for the existence of a broader sense of community. "The land ethic simply enlarges the boundaries of the community to include soils, waters, plants, and animals, or collectively: the land." (Leopold, 1970, p. 239). This natural community includes human beings in their interactions with the nonhuman natural world. "A land ethic changes the role of *Homo sapiens* from conqueror of the land-community to plain member and citizen of it. It implies respect for his fellow-members, and also respect for the community as such." (Leopold, 1970, p. 240). This respect for the natural community and its members is the source and justification of moral obligations. The land ethic provides a nonanthropocentric foundation for policies of environmental preservation and conservation. Actions will be evaluated from the perspective of the natural community and its interests, not from the perspective of human interests and satisfactions. Leopold concludes: "A thing is right when it tends to preserve the integrity, stability, and beauty of the biotic community. It is wrong when it tends otherwise." (Leopold, 1970, p. 262). This moral imperative inspires what can be called an "ecological ethic," an ethic that derives its values from the nonhuman natural systems of the environment. The need for the establishment of an ecological ethic is apparent in the environmental crisis that engulfs us, and in the failure of traditional anthropocentric ethics to explain and to solve the crisis.

Biodiversity and the ecological ethic

An ecological ethic based on natural community can now be applied to the problem of biodiversity and the ethical justification of policies of preservation. The key analysis is the analogy between human and natural communities. Obligations and values inherent in strictly human communities should be found, on an analogical basis, in the natural ecosystemic community.

Consider the concept of *diversity* in human communities such as cities, universities, or classrooms. Although it is not an absolute good, we generally consider diversity in the human population to be a good worth preserving or developing. Different kinds of people, different ages, different cultural and racial heritages, all contribute to the well-being of the community and to the individuals contained therein. The community is stronger or more interesting since it has a wide variation of backgrounds to draw upon; and individuals benefit from the interaction of differing types. The kind of diversity that is beneficial is relevant to the kind of community: diverse age groups are important in a city or university, for example, but not in an elementary school classroom.

We can assume, therefore, that diversity in the natural community (biodiversity) is a similar good worth preserving and promoting. Diversity within natural ecological systems strengthens and makes more interesting the life of the member entities. It preserves the good of the community as a whole, since a diverse system provides more resources, more alternatives, for solving problems and responding to threats.

Biodiversity is thus an *instrumental* good for natural communities; it is useful for the preservation and promotion of the *intrinsic* values and goods found in natural systems. But this instrumental good should not be confused with the anthropocentric benefits previously offered as justifications for environmental policy, for here the primary goal is the continuation of the system itself, not the promotion of human goods.[7]

Pushing the analogy between human and natural communities, we can see the importance of global biodiversity, the diversity of systems spread throughout the planet. Human diversity is important and useful, not only within communities, but also *of* communities. Different kinds of communities strengthen an entire class. This is the justification for the varied mix of colleges and universities throughout the United States. The education one receives at a major university is different than the education received at a small college with limited enrollment; but both experiences are intrinsically valuable, and so a justifiable educational policy must preserve the diverse alternatives of college education. The same is true of a justifiable environmental policy: it must preserve the alternatives of biodiversity by preserving natural habitats and bioregions in their nonaltered states. This result will be beneficial for the natural systems themselves and for the planet (perhaps conceived as one large community).

This is not an argument for *absolute* diversity, either in the realm of human communities or in nature. We do not seek to multiply diverse kinds just for the sake of diversity; we do not seek to maximize diverse pathogens or other disease organisms. What we do seek to promote is a diversity of good or valuable entities. The value of these entities is determined by the relevant situational context.

In conclusion, it must be emphasized that this argument is not based on the scientific benefits to be derived from a biologically diverse habitat or community or planetary system. The point of this argument is that the benefits to be derived from biodiversity should not be conceived in exclusively human categories. An ecological ethic requires that planetary biodiversity be preserved, not as a pragmatic response to threats to human survival, nor as an instrumental betterment of human life, but as a basic moral obligation to the nonhuman members of our moral community. Our value system must be transformed, modifying the dominant concern of human interests. This transformation of value solves the moral crisis that has led to the environmental crisis. The imperative of preserving biodiversity derives from the moral structure of the natural communities of the planetary biosphere, communities in which humanity is both a "member and citizen."

Biodiversity and ecological justice

One problem remains: the just implementation of a nonanthropocentric ecological ethic. An ecological ethic creates disturbing results for policy. I am still bothered by the Amazon rainforest. A policy of preserving biodiversity based on an ecological ethic will require that the indigenous peoples of the poorer nonindustrialized countries refrain from the development of their national resources. The policy of preservation preserves not only the natural environment but also the economic and geopolitical status quo. Of course, the basis of the policy is now conceived differently. Preservation is required not as a means for maintaining human life and benefits, but as an ethical obligation to the natural community itself. But the policy of nondevelopment remains the same, with all the unjust implications for a fair world economic order. Must an ecological ethic be unjust? (Callicott, 1980, 1989).[8]

We thus return to the concept of "ecological justice" as an indispensable component of an ecological ethic. Any implementation of environmental policy must include not only the moral consideration of all members of the natural community, but also a fair distribution of the benefits and burdens resulting from the policy. Justice extends to all members of the moral community, however we have broadened the notion of community. Within this broadened human and natural community are the indigenous peoples of the third world who need to gain access to the realm of economic development. It is clearly unfair to deny them this access.

Thus, the implementation of a nonanthropocentric system of ecological justice will require a close examination and revision of environmental and developmental policies throughout the world. This examination will not proceed along the lines of a narrow comparison and trade-off of human benefits. An anthropocentrically based policy with an enlightened view of the threat to planetary biodiversity still leaves the poorer nations of the world in the position of shouldering the major burdens of environmental preservation. A truly global ecological ethic will view the problem in terms of the entire planetary system, both human and natural. From this all-encompassing perspective, it becomes incumbent upon the richer nations of the world, who have previously gained the benefits of environmental destruction and economic development, to pay their fair share in the preservation of a diverse planetary environment. In one sense, the richer developed nations owe "reparations" to both the nonindustrialized nations and to the natural community as such (Katz, 1986; Taylor, 1986; Wenz, 1988). Only by paying for the preservation of a diverse biosphere can a *just ecological order* be maintained on the earth.

Biodiversity and ecological justice are thus necessarily connected; my title is not, I think, a jumble of incompatible ideas. The preservation of planetary biodiversity will only be achieved by the transformation of human values. Our system of ethics has to include the notion of an ecological community; our system of justice has to include a global and nonhuman perspective. I believe that we are partway to that transformation; my hope is that the transformation will be completed before the diversity of the planetary system is destroyed.

Notes

1 The critique of anthropocentrism is a major theme of the field of environmental philosophy; it would be impossible to cite all the works that develop this theme. See as major examples, Callicott (1989), Rolstan (1987), Taylor (1986), and Wenz (1988).

2 The need to transform human-based ethical systems is the chief concern of what I consider "mainstream" environmental ethics. This discipline of applied moral philosophy follows the work of Aldo Leopold in the attempt to develop an ethic of ecological community. See Aldo Leopold, "The Land Ethic," in *A Sand County Almanac* (New York: Ballantine, 1970, rpt. of 1949 edition), pp. 237–264.
3 For a simple taxonomy of instrumental values in nature, which includes scientific knowledge and religious experiences as instrumental, see William Godfrey-Smith (1979). For a somewhat different view, see Rolston, *Environmental Ethics,* pp. 1–44.
4 As Martin Krieger writes, "Artificial prairies and wildernesses have been created, and there is no reason to believe that these artificial environments need be unsatisfactory for those who experience them." See Krieger, "What's Wrong with Plastic Trees," p. 453.
5 Surely this is one reason why the concept of *friendship* plays such a major part in Aristotle's ethics. All of Books VIII and IX concern the analysis of friendship. Different kinds of friendships determine different moral obligations: "what is just is not the same for a friend towards a friend as towards a stranger, or the same towards a companion as towards a classmate." (*Nicomachean Ethics,* 1162a 32).
6 In *The Republic* Plato has Socrates answer the objection that the guardians will not be happy without private property by re-emphasizing the idea that the welfare of the state as a whole is what matters: "in establishing our city, we are not aiming to make any one group outstandingly happy, but to make the whole city so . . ." (420 b); and again, "We should examine . . . whether our aim in establishing our guardians should be to give them the greatest happiness, or whether we should in this matter look to the whole city and see how its greatest happiness can be secured" (421 b). That this point is stressed twice in one page shows its crucial importance to Plato's ethic of community.
7 For a different argument concerning these instrumental goods, see Norton (1987), *Why Preserve Natural Variety?*
8 Even more than unjust, an ecological ethic seems at times to be misanthropic. As J. Baird Callicott (1980) writes, "The extent of misanthropy in modern environmentalism thus may be taken as a measure of the degree to which it is biocentric," i.e., focused on nonhuman natural values.

References

Brennan, A. (1984). The moral standing of natural objects. *Envir. Ethics,* **6,** 35–56.
Brennan, A. (1986). *Thinking about Nature: An Investigation of Nature, Value and Ecology.* Athens, GA: Univ. Georgia Press.
Cahen, H. (1988). Against the moral considerability of ecosystems. *Environmental Ethics,* **10,** 195–216.
Callicott, J. B. (1980). Animal liberation: a triangular affair. *Envir. Ethics,* **2,** 326.
Callicott, J. B. (1989). *In Defense of the Land Ethic: Essays in Environmental Philosophy.* Albany: SUNY Press.
Ehrenfeld, D. (1978). *The Arrogance of Humanism.* New York: Oxford Univ. Press.
Godfrey-Smith, W. (1979). The Value of Wilderness. *Environmental Ethics,* **1,** 309–319.
Katz, E. (1979). Utilitarianism and preservation. *Envir. Ethics,* **1,** 357–365.
Katz, E. (1985). Organism, community, and the "substitution problem." *Envir. Ethics,* **7,** 241–256.
Katz, E. (1986). Buffalo-killing and the valuation of species. In *Values and Moral*

Standing, ed. L. W. Sumner, pp. 114–123. Bowling Green, OH: Bowling Green State Univ. Press.
Katz, E. & Oechsli, L. (1993). Moving beyond anthropocentrism: environmental ethics, development, and the Amazon. *Envir. Ethics,* **15,** 49–59.
Krieger, M. (1973). What's wrong with plastic trees? *Science,* **179,** 446–455.
Leopold, A. (1970). The Land Ethic. In *A Sand County Almanac.* New York: Ballantine (reprint of 1949 edition).
Norton, B. G. (1987). *Why Preserve Natural Variety?* Princeton: Princeton Univ. Press.
Rolston, H. III. (1987). *Environmental Ethics: Duties to and Values in the Natural World.* Philadelphia: Temple Univ. Press.
Sagoff, M. (1974). On preserving the natural environment. *Yale Law Journal,* **84,** 205–267.
Taylor, P. W. (1986). *Respect for Nature: A Theory of Environmental Ethics.* Princeton: Princeton Univ. Press.
Wenz, P. S. (1988). *Environmental Justice.* Albany: SUNY Press.

Part III

Human processes and biodiversity

5

Preindustrial man and environmental degradation

WILLIAM SANDERS and DAVID WEBSTER

Introduction

One of the most intriguing scientific discoveries has been that of the great variety in the modes and ranges of behavior of human beings as members of organized societies, what anthropologists refer to as culture. Even as late as the 19th to the 20th centuries, when the science of anthropology evolved, many human populations still lived in a great variety of biotic and physical environments, with economies based on the exploitation of wild food resources, both plant and animal. The entire continent of Australia, for example, was inhabited by peoples with this type of exploitation when the continent was colonized by Europeans. Much of North and South America was occupied by hunters and gatherers in the 16th century, and in many areas they survived as late as the 19th. Scattered groups of hunters and gatherers were also found in a number of regions of the Old World. With a few exceptions, recent hunters and gatherers were organized into very small bands that shifted residence seasonally in response to factors affecting their food supply and had an essentially egalitarian social structure.

An even greater number of food producers, organized into relatively small-scale societies with an essentially egalitarian social structure, have also survived into the recent past. Many societies consisted of a single settlement, in this case, a permanent village rather than a nomadic band. In others, sets of villages were organized into larger, but still egalitarian societies, called tribes by anthropologists. Most tribes and villages were in a constant state of warfare, with only occasional periods of truce. In virtually all village and tribal societies that survived into the recent past, the utilization of the environment was relatively extensive. Hunting and gathering of wild plants and animals was still practiced and combined with an extensive system of cropping, called swidden. Swidden farming involved only a few years of cropping of fields followed by long periods of rest, in the tropical forest areas of the world, to some intermediate stage of forest succession. For both foragers and tribal farmers the impact of the humans on their environment was consequently low.

At the other end of the scale of cultural complexity were large states with social systems characterized by economic stratification, a professional ruling class, centralization of power, formal political institutions, specialized economic institutions, and varying degrees of division of labor, exchange, and levels of urbanization. Populations in such societies are dense and land use is very intensive, often reaching a situation where virtually all of the natural biota of huge areas has been removed and replaced by an artificial, human-controlled vegetation. In between the hunting and gathering bands and these complex societies was a continuum, in terms of size of socially organized groups, their internal complexity, degree of division of labor, population density, and level of intensification of land use – a set of factors that are functionally interrelated. Obviously, considering this range, there is a comparable range in the impact of human societies on their biotic and physical environments.

A notion common among those scholars who identify themselves as ethnoecologists is that the vast majority of these nonindustrial economies, often referred to as traditional economies, are relatively stable [see discussions of this issue in Hardesty 1977, and individual papers by Cowgill (1975), Blanton (1975), and Moran (1979)]. The idea is that cultural practices and beliefs, the product of centuries of experience in dealing with environmental problems, have evolved to maintain a kind of equilibrium with the landscape. In direct contrast to this position is the evidence revealed by archaeological data from studies of literally thousands of local areas. In many of these areas we find a full sequence of change from hunters and gatherers to tribal farmers to stratified societies; in others there is a shift from at least hunting and gathering bands to some intermediate level of society. All of this suggests that human adaptation is highly dynamic. Accompanying the evidence of change in the nature of society and economy is evidence of increasing intensification of use of the natural environment. Central to the position of ethnoecology is the idea that men are gifted learners who perceive problems and resolve them (Rappaport, 1967). This would have been particularly true in the past when the pace of population growth was very slow. An extension of this belief is the idea that this process of increasing knowledge and incorporation and use of it as a basis of making decisions involves certain concepts of conservation of resources.

We disagree with this position and argue that individual human beings, whatever their stage of society and economy, have never been concerned with long-range problems. Their response is always to the immediate problem. We furthermore argue that in many cases a series of decisions to resolve immediate problems triggered long-term difficulties that were not resolvable or were not resolved. With respect to the argument about population growth being slower in the past, we respond by turning the argument around. If populations in the past were generally increasing at extremely slow rates, as in fact archaeological and historical evidence indicates, then it is doubtful that the human actors would perceive a process that would engender problems in the distant future. They probably did not realize that their lives were changing in any measurable way. The only difference, from this perspective between preindustrial or nonindustrial and industrial humans is the enhanced ability of industrial humans to destroy their environment because of the greater range of resources used and the scale of such use. One can apply this principle to the sample of preindustrial societies and economies and say that hunters and gatherers have limited ability to alter and change their environment (because the population levels are so low), that tribal farmers have a somewhat greater capacity to impact on their biotic and physical environment, and intensive cultivators have the strongest impact of all. Intensive agriculture may not only have the affect of completely removing the wild vegetation over large areas, but may result in long-term degradation of soils and may even have an impact on the climate of the region.

Henry Lewis has been collecting information on the use of fire by hunters and gatherers living in a wide range of habitats and has dramatically demonstrated that even populations at this level of adaptation can have significant impact on, at least, the biotic environment, that they alter it to fit their own needs (Lewis, 1982). His data are most revealing with respect to the Australian Aborigines in the savannah regions of northeastern Australia. Here the Aborigines periodically and systematically set fires, in sequence, throughout their habitat to induce a vegetation growth much richer in plant foods, for both human beings and the wild fauna that they also exploit. They even have a special term for this procedure, "housecleaning," and refer to environments that are not yet treated in this manner as "dirty" environments. He finds many references by early explorers to the use of fire by hunters and gatherers all over the world. In most cases it is described as a technique to drive game. What his evidence suggests is that the purpose of fire is much more highly structured, and has a much more serious impact on the natural environment. In one sense, hunters and gatherers are incipient cultivators, since they alter the balance of various kinds of plants in their landscape, using techniques that favor plants most useful for human purposes. This new understanding of hunting and gathering is also very instructive, in that it suggests new leads as to how the economy finally shifted over to a fully agricultural one.

Evidence from archaeology indicates that this technique of land use by hunters and gatherers is ancient. There is clear evidence from pollen profiles, for example, that during the Mesolithic Period of England humans used fire to induce certain types of secondary vegetation that was much more productive of the plant foods used by human beings and by the animals that they hunted. One ecologist (Hopkins, 1965) has even argued that much of the savannah vegetation of Africa has been produced by human beings, using fire, when hunting and gathering was the primary economic system. Recent studies show that the density and variety of species of ungulates is the highest in the world in the African savannas, and that it is considerably higher than in the scrub forest that probably was present in the same area prior to human intervention. In fact, Hopkins goes further and argues that many of the grassland regions of the world are the product of human intervention, through the use of fire, and probably date from the period when hunters and gatherers were utilizing the regions.

Whether the manipulation of the vegetation by hunting and gathering peoples as described by Lewis has a deleterious affect on the natural environment is a matter of debate. Favoring some plants that are useful to human beings over others clearly alters the vegetation mix. The major question is, does this reduce biodiversity and, if so, does it also reduce the energy efficiency and productivity of the environment? Short-term studies like those of Lewis cannot resolve this debate.

Even swidden farmers practicing a long-fallow system of agriculture in which the natural vegetation is allowed almost to return to climax forest may induce long-term deleterious effects upon the soil resources of their environment. This is in sharp contrast with our usual impression of the impact of swidden farming on the environment. Ethnoecologists have argued that the relationship is a highly stable one and that productivity is not impaired (Conklin, 1957; Rappaport, 1967; Nigh, 1975). Recent data, however, from those areas in the Peten region in Guatemala where swidden farmers have been practicing this kind of agriculture for long periods of time, indicate that even a long-fallow system is not stable. The present-day Peteneros rest land the same length of time in fields located within a few kilometers of their villages as they do in more distant fields. The nearby lands, however, have been cropped for a much longer period of time and are now producing only half the yields of recently colonized fields.

A more important question, however, is the impact of intensive farming on the landscape. Do we have evidence that human populations, practicing traditional but intensive agriculture, have caused long-term destruction of their physical and biotic environments? Archaeology has revealed literally hundreds of cases of situations where populations of farmers increased at a fairly steady rate, reached a peak and then went through a rapid decline. In some cases, the area involved was only occupied once; in others there were several cycles of this kind of demographic history. The question is, is this cycling the product of the misuse of the environment or of some other factor? It is clear that in a very small area a political factor may be paramount, i.e., an area may become unsafe for settlement because of intensive warfare. But if the area is large and the period of decline is long, it is highly suggestive that something else is involved. We have a number of cases of large-scale abandonment of regions that seem to suggest ecological factors and processes at work.

One example is the great Anasazi collapse in Southwestern United States. Soon after 1000 A.D. large areas that were relatively densely settled were abandoned and were still not reoccupied when Europeans entered the area in the 16th century. In this case, while natural events, such as periods of severe droughts, operated as one of the factors to cause the abandonment of some local areas, it is also equally clear that human-induced factors affecting the water table, through deforestation and soil erosion, played a paramount role as well (Martin & Plog, 1973).

A more dramatic example, in the sense that it involved a much larger population, was the collapse of the ancient Sumerian civilization in lower Mesopotamia. The region, often referred to as the Cradle of Civilization, was the earliest area where human beings developed a complex society, in this case in a desert environment. The overall area was approximately 25,000 square kilometers with a population estimated at 200,000 around 2500 B.C. Following this period there was a catastrophic decline in population. Huge areas that were under intensive-irrigation agriculture were abandoned to the desert and became wastelands. Many have not been recultivated, even in the modern era. While a number of factors have been suggested to explain this abandonment of southern Sumer, one of the most convincing arguments has been a process of salinization of the soils through the very intensive practice of irrigation agriculture, a process similar to that occurring in central California today.

However, the most dramatic example of an ancient population going through a catastrophic decline, after centuries of successful adaptation to an environment, is that of the collapse of Classic Maya civilization in the Yucatan Peninsula.

Classic Maya civilization

The Yucatan Peninsula, located in what is today Mexico, Guatemala, and Belize, is a great shelf of limestone with a typical karst topography, an area of generally flat terrain, but with some zones of hills in the form of domes and ridges interspersed with low-lying swampy areas, little surface hydrography, and thin but fertile soils. The temperature regime is tropical, and rainfall, with the exception of the northwest quadrant, is high, ranging from 1000–3000 mm a year. It falls primarily in the summer and early fall (Vivo Escoto, 1964).

It was occupied by Maya speakers approximately 1000 B.C. or earlier. Population growth was sustained for the next 2000 years and accompanied by an evolution of society from self-sufficient, autonomous, egalitarian villages, to large, complex societies, either chiefdoms or states, with hereditary rulers and hierarchical social relationships within the society. During the final two centuries of the first millennium B.C., the Maya began to construct large centers with massive masonry buildings, in the form of palaces and temples, and continued to do so for the next 1000 years. Population reached a maximum around A.D. 800, at which time several million people lived in the area known as the Maya Lowlands. The period from A.D. 300–900 is referred to as the Classic period, and was one of the climactic developments in architecture, painting, monumental sculpture, ceramics, and certain intellectual developments, such as mathematics, writing, and astrology (Coe, 1966).

One of the most intriguing and as yet unresolved problems of Maya archaeology is the phenomenon of the collapse of the Classic period civilization. This collapse was signalled by an apparent sudden cessation in dated monuments after nearly 500 years (A.D. 300–800) of continuous erection of such monuments. It was also signalled by the end of the construction of temples and palaces in scores of major sites in the southern two-thirds of the Peninsula of Yucatan. Scholars have argued endlessly as to the nature, magnitude, and scope of the collapse, its geographic distribution, temporal aspects, social significance, and, most particularly, causes (Culbert, 1973). Essentially the explanations fall into two groups – one in which it is assumed that the collapse occurred only at the elite level, and the other in which is assumed that it affected the entire social system. The first case assumes that the peasant population remained in the area after the cessation of dated monuments and major construction; the second assumes a catastrophic decline in population. The various concepts and notions about the collapse can be summarized in the following outline:

I. Only the elite level collapses:
    A. Peasant revolts
    B. Invasion from outside the region
    C. Collapse of trade networks
II. Total system collapse:
    A. Ecological
        1. Catastrophic
            a. Earthquakes
            b. Sustained drought over a number of decades

c. Epidemic diseases – human
d. Epidemic diseases – crops

2. Processual:
   a. Succession of forest to grassland resulting in a type of vegetation that would not have been removable or controllable with the simple hand tools possessed by the ancient Maya, particularly considering the fact that they were made of stone.
   b. Soil erosion and nutrient depletion as part of a process of agricultural intensification
   c. Endemic diseases combined with nutritional stress

B. Religious prophecy
C. Breakdown of trade networks

Some of the above theories are interrelated, and the most convincing argument has always involved a number of variables in some systemic relationship. The assumption of an elite level collapse only is easily disposed of from two sources of data.

Archaeological research on settlement history at numerous major centers has demonstrated that a massive population decline accompanied or occurred soon after the cessation of monument erection and large-scale construction. The major question here is how long after the collapse of the elite level did the population decline?

Spanish accounts from the 16th century indicate that much of what is today the state of Yucatan was densely settled by Maya speakers. An exception was the area south of the Puuc range, where a dense Terminal Classic population resided (A.D. 800–1100), but virtually uninhabited in 1519 when Cortez arrived in Yucatan. A band of substantial population approximately 40–50 km wide was present along the west coast of the peninsula in what is today the state of Campeche. This zone merged with an area of much larger population in Tabasco to the southwest. The northern third of Quintana Roo was well settled and a narrow band of population extended south to the extensive swampy region around Espiritu Santo Bay. A detached area of dense population was concentrated around the shores of Chetumal Bay with small extensions into central Belize. To the south there was a band of relatively dense population along the foothills of the northern limestone ridges of highland Guatemala. The remainder of the peninsula, its inland heart, including the Department of Peten, inland Campeche, southern Yucatan and inland Quintana Roo was virtually a human desert, an area of some 100,000 square kilometers (Fig. 5.1). Exceptions were a few islands of population, the largest located around the lakes of the Peten. The total population of this interior area, however, could not have exceeded 100,000 people and was more probably no more than 50,000. In

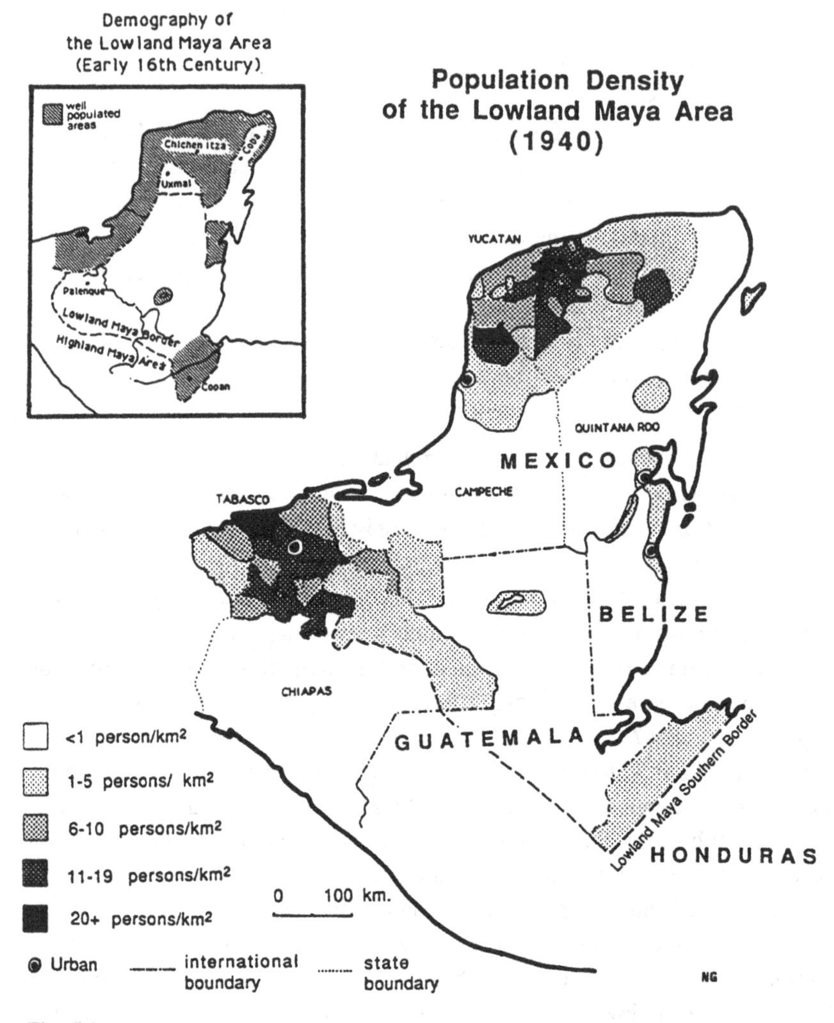

Fig. 5.1.

contrast, between 800,000 and 1,000,000 Maya lived along the peripheries of this region.

Evidence, therefore, is squarely in support of a massive decline of population in the central area and it is the magnitude of this decline that is striking. During Late Classic times (approximately A.D. 800) the area was densely settled by a population that we believe could not have been less than 3,000,000 people, possibly double this number, organized into at least 50 semiautonomous or autonomous polities. The most commonly held notion is that the population decline occurred over a period of less than a century after the termination of the erection of dated monuments and construction of major architecture in the various centers of these polities. The most puzzling aspect of the collapse has always been its apparent suddenness over such a wide area, an area characterized by political fragmentation. While other civilizations have risen and declined, none have involved a demographic disaster of the order of the Classic Maya. Accepting the fact of this demographic catastrophe, these questions need to be addressed. First, what happened to the population? Did it decline *in situ*, or did people emigrate, and at what rates? Second, what caused the decline? And third, why was there no recovery for 500–700 years following the collapse?

The most enduring explanation over the years has been an ecological one. The argument is that only a major problem of environmental adaptation could result in either mass migration or *in situ* decline of the magnitude that the demographic histories of sites suggest. Specific scenarios used to explain it, within an ecological framework, have varied. One presupposes a series of catastrophic events. Physical environmental factors that have been suggested are earthquakes, volcanic eruptions, and droughts. Biological factors include plant or human diseases.

Catastrophic physical environmental disasters are easily dismissed. Nowhere have earthquakes or volcanic eruptions destroyed such a large population, dispersed over such a large area; and in fact the Peninsula of Yucatan is one of the areas least prone to such events in all of Mesoamerica. A drought cycle would be a strange one indeed if it left all of the surrounding areas untouched, some with even lower annual rainfall today.

On the biological side, the possibility of human diseases is suggested by events after the Conquest, when native populations, particularly in the lowlands of Mesoamerica, declined within a century of the Conquest, at a rate supposedly comparable to the Classic Maya collapse. With respect to diseases, the one kind that could cause such a large population decline, over such a short period, would be epidemics. Tabasco and Veracruz suffered a population decline from 2.5 million people to 150,000 in less than one hundred years, due primarily to epidemics. However, these epidemics were exotic diseases introduced from the Old World, and we have no evidence of the presence of similar diseases in Classic times, nor would we be able to explain why the diseases would only affect the heartland Maya. Furthermore, there is no archaeological evidence of mass burials or grossly distorted age and sex distributions in cemeteries that would equate with an epidemic type disease. Crop diseases have been another explanation (Brewbaker, 1979).

The most convincing argument has always been a processual ecological disaster of anthropogenic origin, related to the unusually high population densities (in the hundreds of people per square kilometer) recorded in the surveys around the southern Maya sites. In this formulation, a number of variables may have operated, in conjunction, to produce, as the immediate affect, an *in situ* decline in population, emigration, or a combination of both.

The driving force behind the disaster, in this formulation, supposedly was runaway population growth, inducing a process of intensification of land use and increasing sedentism. Its long-term affect on the physical environment could have been a reduction of soil nutrients and, in hilly areas, soil erosion, resulting in a steady decline of productivity and actual loss of cultivable land. With respect to biological variables, the same process would lead to a decline in the nutritional status of the population, with consequent reductions in fertility or successful pregnancies in women, and increasing mortality, due to endemic diseases, as a result of a combination of increasing sedentism and reduction of nutritional quality.

The most attractive aspect of this system approach is that it helps explain why the decline occurred primarily in the core area. This is an area of high relief and wet climate, both conducive to the physical processes described above. It was also the area of the densest and earliest Classic Period population. The survival of population at the periphery could be the product of lower population density, drier climate, and flatter terrain, or a combination of these factors. All of the population that remained in the periphery resided either in zones with extensive flood plains that could sustain permanent cultivation, or in northern Yucatan, which is a flatter, more arid region.

The major argument against the processual theory is the apparent suddenness and simultaneity of events over such a large area. Recent research at the Classic Maya center of Copan may help to answer these questions.

The Copan Valley as a test case

Geographically, the Copan Valley in Honduras differs from most of the Maya lowlands, and in some features more closely resembles Highland Guatemala. First, it is a true valley with a permanent river, has a network of tributaries, some of which are also permanent, tracts of alluvium, surrounded by rugged mountainous terrain, a much more varied geology, and derivatively greater soil variability. On the other hand, it is only 600–900 meters above sea level, placing it in the tropical temperature zone, and has an annual precipitation of between 1500–2000 mm, falling within the humid tropical environment so characteristic of the southern Maya area. Finally, where the soil characteristics permitted, it was covered by a dense tropical forest. This combination of high temperature and rainfall engendered the same problems as in the southern Yucatan Peninsula – soil nutrient stability, a feature typical of tropical forest environments. In fact, the soil mosaic of the Copan Valley presented even more serious problems than most Maya lowlands.

Carrying capacity

In assessing the impact of ancient Maya farming on the landscape of the Copan Valley, a necessary step is the modeling of the carrying capacity of the region. This is a difficult and complex problem even when dealing with living populations, and it is immeasurably more so in the case of prehistoric ones. Essentially, the following evaluation is based on two conceptual schemes, one presented by William Allan (1967), the other by Ester Boserup (1965).

Allan developed a rigorous method of calculating carrying capacity for a number of native reserves in what was then British Africa. Basically, what he did was to establish man–land ratios that would ensure long term stable relationships between the two variables, that is, he assumed that any increase of the man–land ratio in the particular case would lead to long range deleterious affects on the landscape. His assumption was that most of his sample population had reached this critical point. The method consists of the following steps:

1. Establishment of a soil map of each region related to agricultural productivity of the staple crop used in that region.
2. Ascertainment of what percentage of each soil type defined was actually cultivable in terms of native crops and techniques; he called this the cultivable land factor.
3. The establishment of what he called the cultivation factor – the amount of land needed in a particular year to sustain a single person. This was an empirical measure based on each case. For example, in areas where the economy was relatively simple, it was the amount of land needed to feed the individual only; in more complex economic systems, this calculation would include surplus production for trade and taxation.
4. Establishment of the land use factor, i.e., how many plots of land the size of the cultivation factor would be needed to allow sufficient fallowing. A land use factor of 2, for example, would mean two plots, a land use factor of 3, three plots, etc.

Allan's scheme suggests that mature agricultural populations have a relatively closely tuned relationship with variety (in terms of agricultural potential) in their environment and that the range of the land-use factor found in each of his study areas was directly conditioned by variety in the fertility of the soils.

Boserup's (1965) model of agricultural adaptation is a more dynamic one. She argues that the more extensive approach to cultivation is virtually always more productive than the intensive one, in terms of the work input–crop yield ratio, and that colonists will always opt for an extensive pattern of land use. The subsequent history is one of gradual intensification of land use, i.e., reduction of land-use factor, as population growth creates shortages of land. The better soil areas of the environment will be colonized first, the more marginal areas last. Furthermore, areas of high productivity will be undergoing intensification of use relatively early in the process, at the same time that more marginal habitats will be cropped using a swidden system of farming. She would argue that what Allan found in his studies of African reserves was a stage in the intensification process, and that if population growth continued it would ultimately produce a pattern of intensive use of all land. The driving force behind this process, in her formulation, is population growth.

Boserup (1965) does not address the question of long term deleterious effects of the intensification process on soil fertility, Allan's (1967) primary concern.

The Proyecto Arqueologico, Phase I (PAC I), directed by Claude Baudez, included an ecological team, headed by B. L. Turner, that carried out extensive studies of the geomorphology, geology, pedology, biota, climatology, and present day agricultural use of the Copan Valley (Baudez, 1983). During the Proyecto Arqueologico, Phase II (PAC II), directed by William T. Sanders and David Webster, a number of follow-up studies of modern agriculture were conducted (Sanders, unpublished; Wingard, 1992).

On the basis of these studies, it is clear that the most important single factor determining agricultural productivity in the valley is soil variability, and this is related to the nature of the underlying geology. The most outstanding feature of the Copan habitat for agriculturists is its oasis-like quality. The drainage basin covers some 400 km^2. About 330 km^2, however, consists of steep, mountainous terrain, in which the bedrock consists of deep strata of very ancient, compacted volcanic ash. This material is severely leached and eluviated and these processes have produced a thin, sandy, acid soil of little or no use for agriculture, particularly one based on maize. The balance consists of a series of somewhat isolated pockets of alluvium, each pocket adjacent to foothills. The bedrock underlying the latter consist of various mixes of sedimentary rock. The fertility of the foothill areas varies in relationship to the ratio of limestone to other kinds of rock, and there are a few small areas of pure limestone, which produce the most fertile foothill soils. The alluvium is the most productive area but is not uniform because the river has downcut its bed and only half of the area receives periodic flooding. In order to reconstruct a stable model of degrees of intensiveness of land use, similar to that of Allan's, we made the following assumptions:

1. The ratios of kinds of soil found in the valley have been relatively stable over the past 2,000 years.
2. The staple crop was always maize.
3. Those fields in the active flood plain have the highest degree of stability in productivity. Sanders' (1972) studies indicate that the average yield in those fields without fertilization, and with land use factor of 1, is approximately 1400 kg per hectare.
4. This yield can be sustained in all the agricultural land in the valley with proper adjustments of the land-use factor. The prehistoric population would elect to achieve this maximal yield of productivity, to cut labor costs.
5. Two models of maize productivity are used here. One assumes that the genetically modern maize varieties were available by Coner times (A.D. 700–1200). The second assumes that genetic improvement has continued up to the 20th century and that maize grown during the Coner Phase produced only 5/7 of today's yield. Table 5.1 summarizes the results of our calculations. The Copan Valley as a whole has a carrying capacity of 12,000–17,000 people, assuming a stable relationship of people to the land.

Table 5.1 also shows what may have happened to the valley in terms of a process of intensification of the type suggested by Boserup. What we have done is progressively reduce the fallow period, starting with the most fertile soil areas, until virtually all land is cropped on an annual basis. The yields are based on Sanders' research. He collected historical information on land use in a series of 30 fields, scattered throughout the valley in different soil areas, that are intensively cropped today. There is an obvious process of deterioration of the soil as agricultural intensification occurs. More recently, a follow-up study involving a much larger sample accompanied by extensive soil mapping was completed by Wingard (1992). Theoretically, at the end of the process of intensification the valley could have supported a population of 16,000–22,000 people. It is also clear, however, from Wingard's study that most of the soils of the valley could not sustain this level of cropping for more than a few decades. Some present-day fields on hillsides are, in fact, abandoned because of unacceptably low crop returns and increases in labor costs of production. Furthermore, in some of these fields virtually all of the topsoil had been washed downslope during a period of only 10 years of intensive cropping. Much of the poorer land would, then, have passed out of cultivation, either as the product of nutrient depletion and/or soil erosion, and it is very probable on the basis of

Table 5.1. *Copan Valley: stages of agricultural intensification*

Land type	LUF	Yield	C.L.F.	Annually cropped	Population capacity
Active alluvial	1	1000	990	990	4950 (6930)
Ancient alluvial	2	1000	830	415	2075 (2905)
Intermontane basin	3	1000	500	170	850 (1190)
Piedmont	4	1000	3160	790	3950 (5530)
					11125 (16555)
Active alluvial	1	1000	990	990	4950
Old alluvial	1	800	830	830	3320
Intermontane basin	2	800	500	250	1000
Piedmont	3	800	3160	1090	4360
					13630 (19250)
Active alluvial	1	1000	990	990	4950
Old alluvial	1	800	830	830	3220
Intermontane basin	1	600	500	500	1500
Piedmont	2	600	3160	1580	4740
					14510 (20370)
Active alluvial	1	1000	990	990	4950
Old alluvial	1	800	830	830	3320
Intermontane basin	1	600	500	500	1500
Piedmont	1	400	3160	3160	6320
					16090 (22610)

Added components not included in calculations:
1. Soil erosion on piedmont, effects
   a. Faster rates of yield decline
   b. Reduction of cultivable-land factor
2. Effects of a century or more of use on high alluvium and intermontane basin productivity
3. Possibility of reduction of amount of prime land used for subsistence, conversion to cash crops
4. Possibility of double cropping, effects on soil
5. Expansion into pine forest areas with transport of these poor soils to lower elevations through erosion

Wingard's data that even the high alluvium would have continued to suffer a progressive decline in productivity after a century or two of cycles of permanent cropping, a process not represented in Table 5.1. By the end of the Coner Phase (A.D. 1200) the only viable agricultural land left, if the population had reached these potential levels, would have been the active flood plain, and this could have supported only 5,000–7,000 people.

The demography of ancient Copan

How does the archaeological picture – the history of Maya settlement – fit the intensification model? To answer the question one needs to first develop a series of methods to estimate population size at the peak of the Classic period and to track the growth and decline of the population.

The method used to reconstruct the ancient population for the Copan project included the following steps (Webster & Freter, 1990):

1. A surface study of the region to locate all prehistoric residential sites.
2. Dating of these sites by relative dating based on a program of test pitting, of a selected sample of sites, to obtain a sample of ceramics. (Relative dating means simply the dating of one site compared to another, i.e., which sites are older, younger, or contemporary. This is done by stylistic analysis of ceramics, which reflect changes in culture over time.)
3. Chronometric dating of sites. At Copan, this has included the use of Carbon-14 and archeomagnetic dating (i.e., of baked clay surfaces). More importantly, we have applied a method known as obsidian hydration dating. Over 2,200 obsidian blades have been dated from test pits and large-scale excavations.
4. Large-scale excavations of a selective number of sites were conducted to obtain information on the characteristics of residences, the uses of buildings and the size of roofed-over spaces. We then compared this data with a selected sample of living populations, roughly at the same stage of socio-cultural evolution, to ascertain the ratio between roofed-over space and the size of the resident population.

A major problem with the processual ecological explanation of the collapse has always been its assumed suddenness (Culbert, 1973). In the case of each Maya center we had a dynastic sequence, consisting of at least five or six rulers, often more, each of whom commissioned the erection of monuments with dates, and the construction of palaces, temples, ball courts, or other special structures. This behavior continued over a period of hundreds of years. Suddenly there was a cessation of monument erection and construction of major buildings in scores of major sites, all ostensibly within a few decades of each other. Shortly after the cessation of these elite activities, the population presumably disappeared over the entire region.

The first crack in this elegant edifice of historical reconstruction came when it was discovered that not all sites were abandoned simultaneously, even in the Peten. Some were abandoned, or at least the stelae were no longer erected, as early as A.D. 750, others lasted until A.D. 900. In the far northern periphery of the southern lowlands, in the Puuc range, sites did not reach their peak until after A.D 800 and did not collapse until A.D. 1000. Interestingly, however, in each local case population density reached hundreds of people per square kilometer just prior to the collapse, a strong suggestion that ecological processes were at work.

A major problem, however, remains the suddenness of the collapse at each site. It was admittedly difficult to date the final depopulation, since it occurred some time after the cessation of dated monuments. It was supposed that the ceramic phase associated with the final architectural phase ended shortly thereafter.

The major problem was that the final abandonment was related to ceramic styles and not to absolute dates. At Copan, for example, the Classic period was traditionally divided into three major phases: Bijac, A.D. –400, Acbi, A.D. 400–700; and Coner, A.D. 700–850 or 900. The last dated monument at Copan was A.D. 822. If these data were correct, the population doubled and disappeared at Copan during the Coner phase, a phase of only 150–200 years in length. The population, following this chronology, disappeared over a period of 30 to 100 years! There was some indication of continuing activities at the Main Group (the political heart of the kingdom) after A.D. 900, in the form of offerings of ceramics and gold ornaments placed at the center, and this phase was called Ejar. It was assumed, however, that only a few hundred people, at most, lived in the Copan Valley during this Terminal phase.

The obsidian hydration dating completely reversed our understanding of the nature of the collapse, and restored to prominence the ecological processual explanation of the collapse. What the data suggest, from a large sample of obsidian from many test pits, is the following:

1. A period of relatively stable population, never numbering more than 1,000 people from 1100 B.C. to A.D. 300, with a strong possibility that the valley may have been temporarily abandoned for several hundred years prior to the latter date.
2. This period was followed by one of rapid and sustained growth, the population doubling roughly every 85 years between A.D. 300–750, reaching a peak around A.D. 750 of 26,000–28,000.
3. A relatively stable period of 100 years, A.D. 750–850, during which time this peak was sustained.
4. A period of steady decline until approximately A.D. 1100, during which the population halved every century, finally disappearing shortly after A.D. 1200. Interestingly, the Coner ceramics did not change significantly over this long period of 500 years. This raised serious questions about the cessation date of the final ceramic phase at other Maya sites.

The overall population density of the Copan Valley between A.D. 750–850 was approximately 48–50 people per km^2. However, the actual portion of the landscape that was available to agriculture is only 70 km^2, resulting in a density of between 300–310 people per km^2 of agricultural land (see Fig. 5.2). Even this area, the zone of agricultural utilization, was not 100% cultivable, due to the presence of streams, unusually rugged slopes with very thin soils, areas too swampy to easily convert to agricultural land, and areas occupied by residential sites and other structures. When adjustments are made for these features, the total area of agricultural land is only 55 km^2, which means that the effective agricultural population density was 436–453 people per km^2 at the peak of the Classic Period.

On the basis of our model, the carrying capacity of the Copan Pocket (4,000–6,000 people), where virtually all of the population resided until the Coner Phase, was achieved between A.D. 550–600. In the decades immediately before A.D. 550 the first permanent colonization of parts of the Sesismil and Santa Rita areas, the zones nearest the Copan Pocket, occurred, clearly a reflection of emerging problems. Scattered dates from more distant areas indicate the beginning of cropping, presumably long-fallow swiddening, and seasonal occupation in those areas. Between A.D. 675–725 the population reached the carrying capacity (12,000–17,000) of the entire valley, and the permanent occupation of the peripheral areas was well under way. During the period from 750–850, the population reached a level that could only be sustained by annual cropping of all agricultural land.

The early colonization of Sesismil and Santa Rita, and initiation of long-fallow swidden cultivation in more peripheral areas, suggests that even as early as A.D. 600 the land resources of the pocket were inadequate. Further confirmation of the serious impact of growing population is the early decline of population in these same areas.

Social factors

Sociopolitical and economic processes exacerbated the problems of the Maya of Copan during the 8th century. One problem was the unusual distribution of population over the landscape. Of the 4509 structures found in surface surveys, approximately 3372 were found in the alluvial plain and adjacent foothills of the Copan Pocket, an area of only 24 km^2 (Figs. 5.3–5.5).

Of these, about 1000 were densely packed into two ward-like areas adjacent to the Main Group of Copan (the great palace–temple compound) and formed a kind of urban nucleus. In the remainder of the 46 km^2 area of agricultural land, the survey located only 1066 structures. While the Copan Pocket is the most

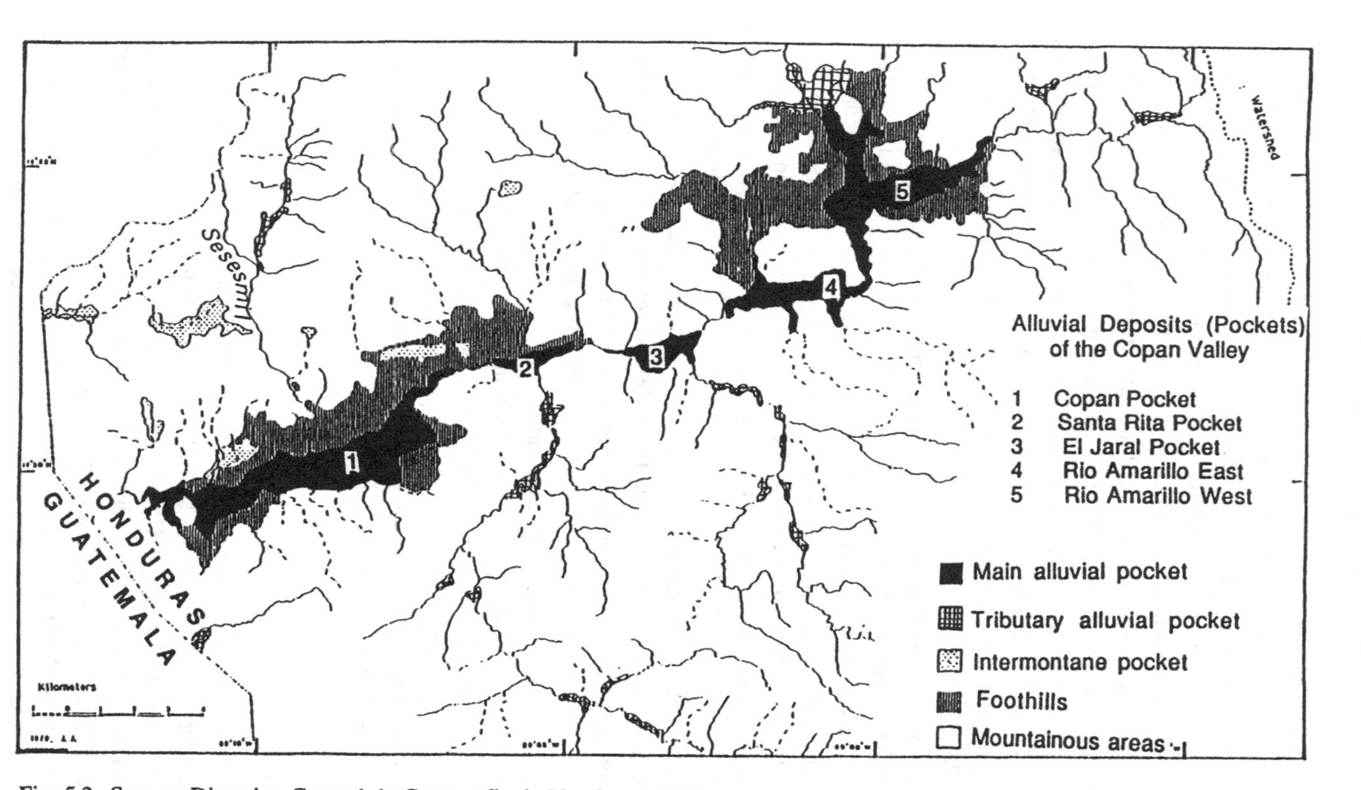

Fig. 5.2. Source: Direccion General de Cartografia de Honduras, 1979.

Fig. 5.3 The Urban Core at Copan (courtesy of William Fash). The El Bosque residential enclave can be seen west and southwest of the Main Group; the Las Sepulturas enclave is clearly strung out along the causeway on the northeast.

productive section of the Copan Valley, it is clear that the population residing there vastly exceeded the carrying capacity of that part of the valley by A.D. 750. The unusual concentration of population was primarily a product of political and social factors – the need for a relatively weak but centralized state to administrate and control the various rival segments within the political system, and the need for certain high-ranking individuals, the nobles, to remain as close to the center of power as possible in order to safeguard their interests. What the archaeological data reveal is a highly volatile political environment.

The problem of meeting subsistence demands would also have been exacerbated if more privileged land holders had converted some of the prime land of the valley to nonstaple crop production like cacao, or nonfood crops like

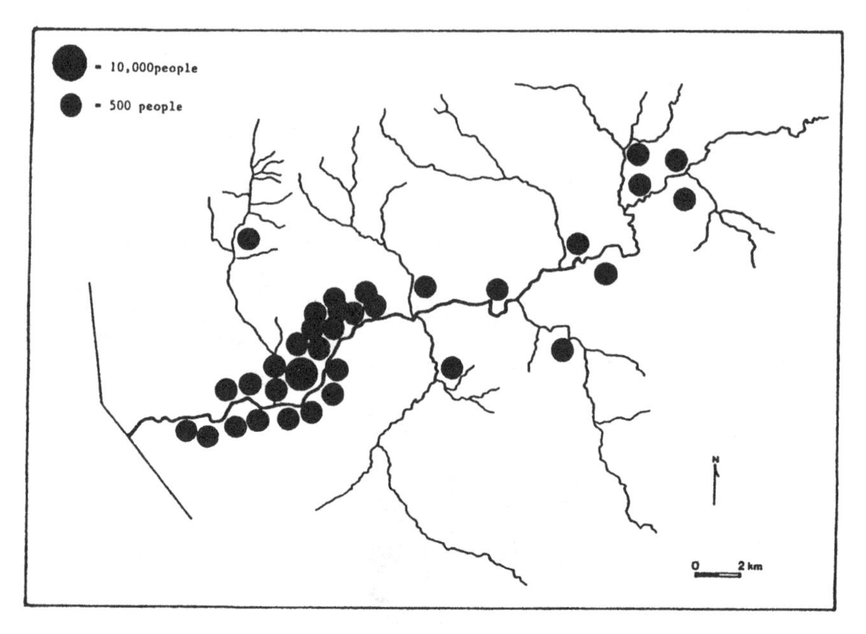

Fig. 5.4 Population distribution at its peak in the Copan Valley.

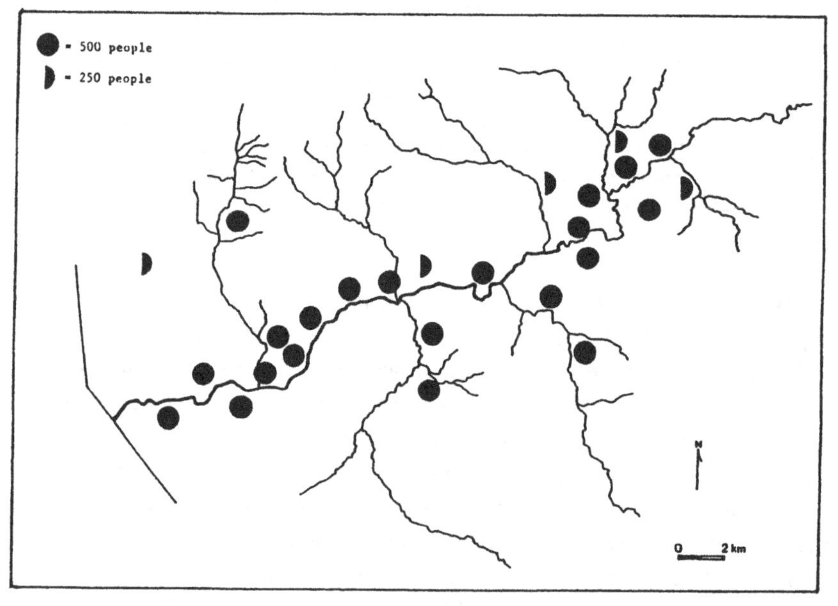

Fig. 5.5. Predicted population distribution if farmers were spaced on the landscape according to a carrying capacity model based on moderately intensive maize agriculture.

cotton, for export, a process analogous to the present-day conversion of much of the alluvial land to commercial tobacco production.

The Copan demographic history and studies of present-day agriculture strongly suggest processual ecological variables at work, exacerbated by economic stratification and a political system that mandated that the vast majority of the population reside in a relatively small section of the drainage basin, the Copan Pocket, and that these were the basic and primary causes of the collapse. Considering the potential error in estimating population from archaeological data, this conclusion needs to be tested by a variety of other kinds of data and research methods. In the following sections, this supporting data is summarized.

The pollen profile

One of the intermontane basins of the Copan Pocket, near the headwaters of a tributary called the Petapilla, is occupied by an extensive swamp. As part of the PAC II project, a pollen sample was extracted from the swamp to track the impact of Classic Maya occupation on the biotic landscape (Rue, 1986). The core has revealed corroborative data as to the nature of the Classic collapse at Copan and the level of intensity of use of the agricultural soils of the Copan Pocket. On the basis of two Carbon-14 samples, the level immediately above the base of our profile dates at approximately A.D. 1010 ± 60 years and the level 70 cm from the top of the core at A.D. 1350 ± 70 years.

If the Classic Maya collapse at Copan, in a political sense, had occurred sometime after A.D. 800 and the total demographic collapse and abandonment of the valley no later than A.D. 900, as the traditional chronology seemed to suggest, the pollen profile should have revealed evidence of succession back to tropical forest by A.D. 1000. What it actually shows is that the area was still being heavily utilized by the Maya, at least until the latter date. In fact, succession back to tropical forest did not begin until approximately A.D. 1200 and was not fully achieved until A.D. 1400. The zone from 80–130 cm below the surface is dominated by compositae and grasses, obviously weedy growth related to agricultural utilization, and ferns in the more swampy sectors of the pocket. The profile, therefore, confirmed the obsidian hydration dates as to a much later and slower rate of decline of the population of the valley. It also indicates that, as late as A.D. 1000–1100, the Petapilla pocket was very intensively cropped using a very short fallow system (i.e., with no more than 1–2 year rest period) (Fig. 5.6). The kind of vegetation revealed by the profile is precisely what one sees today in the very intensively utilized areas of the valley (a process revealed in the upper 30 cm of the soil profile).

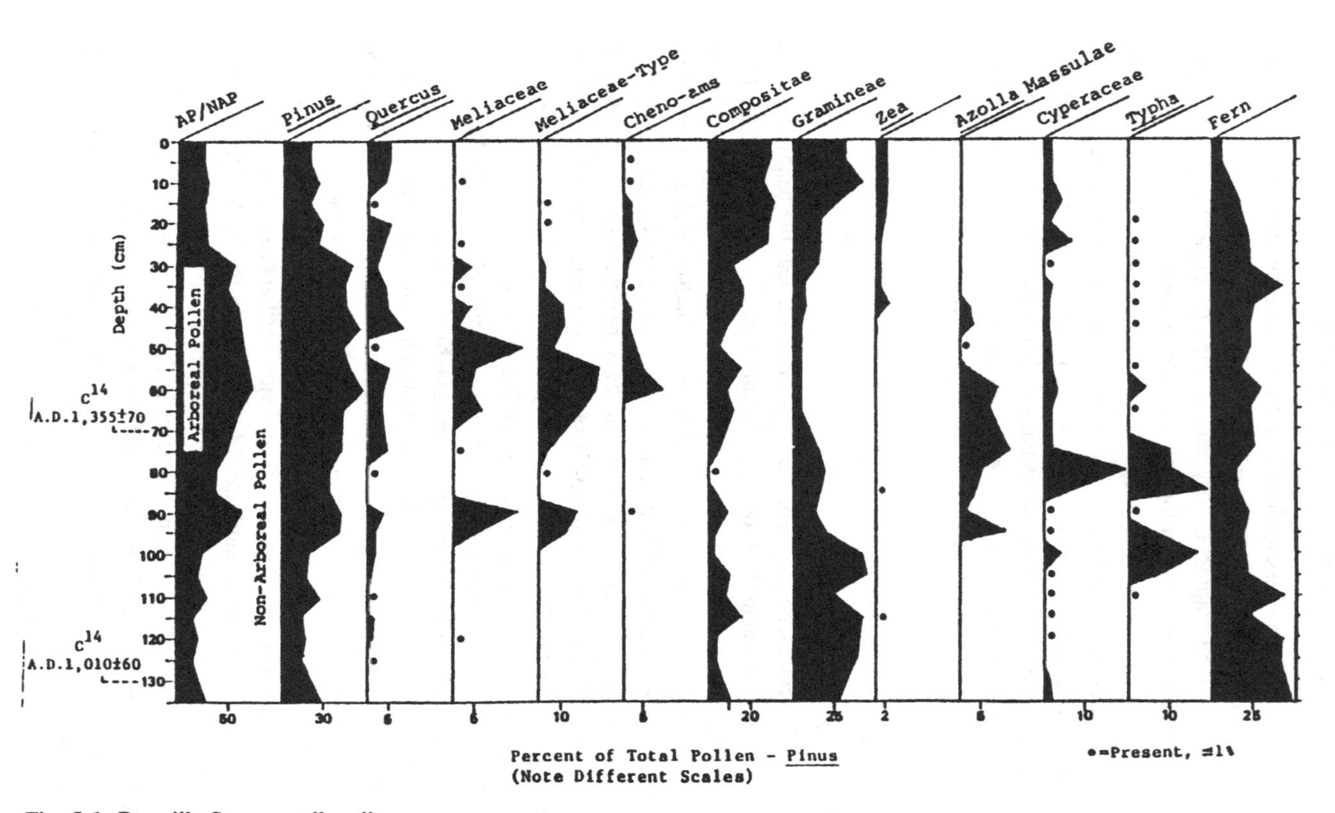

Fig. 5.6. Petapilla Swamp pollen diagram.

Soil erosion

Direct archaeological evidence of Classic period soil erosion is also now available. Between Sepulturas, the eastern ward of the urban core of Copan, and the Main Group is a zone several hundred meters wide consisting of deep soils and relatively flat terrain with only traces of prehistoric occupation; the latter occur primarily in the form of scattered groups of mounds that represent remains of residential compounds (Fig. 5.3). A raised causeway departs from the Main Group, disappears in this area and reappears again in the zone of dense habitation of Sepulturas. The causeway here has either been eroded away by flash floods or by erosion from the nearby slopes, or buried under sediments deposited by the river.

A test pit excavated along the edge of the highest mound in one of these occupational groups yielded clear evidence of prehispanic soil erosion (9M-101 in Fig. 5.7). The group was classified in the survey as relatively low status because of the modest height of this mound. The test pit, however, revealed startling new evidence as to the nature of the compound. The stone retaining wall of the substructure of the building turned out to be two meters high and built of well-cut masonry, suggesting a much higher status for the residents of the house compound. Apparently a residential compound of the Classic period was literally buried by soil up to the height of its substructure. As a result of this process, a Classic Maya building, almost certainly built between A.D. 750–800, was largely buried. Most of this soil had to have eroded off the nearby hills to the north, and erosion occurred even while the Maya still lived in or near the group, as well as after the building collapsed (a conclusion confirmed by obsidian hydration dates which cluster in the 8th–9th century A.D.). Several additional test pits excavated by PAC I archaeologists near other mound groups in the plain show the same pattern and history (Baudez, 1983). This massive erosional process undoubtedly reflects intensive land use of hillsides at Copan and is further evidence for environmental degradation.

The possibility then is very high that this apparently lightly occupied zone actually has numerous lower-status residences completely buried under the deep soils of the area, and that the missing section of the causeway was, in fact, buried by soil eroding from the nearby slopes.

At this point, we will reevaluate the processual model of the collapse of the Copan polity. The obsidian hydration dating demonstrates that the final ceramic phase at Copan, Coner, which spans the period of final growth and collapse, had a considerably longer duration than anyone anticipated, 400–500 years. This resolves the major problem of a processual ecological explanation, the supposed suddenness of the collapse. Accepting the ecological processual theory,

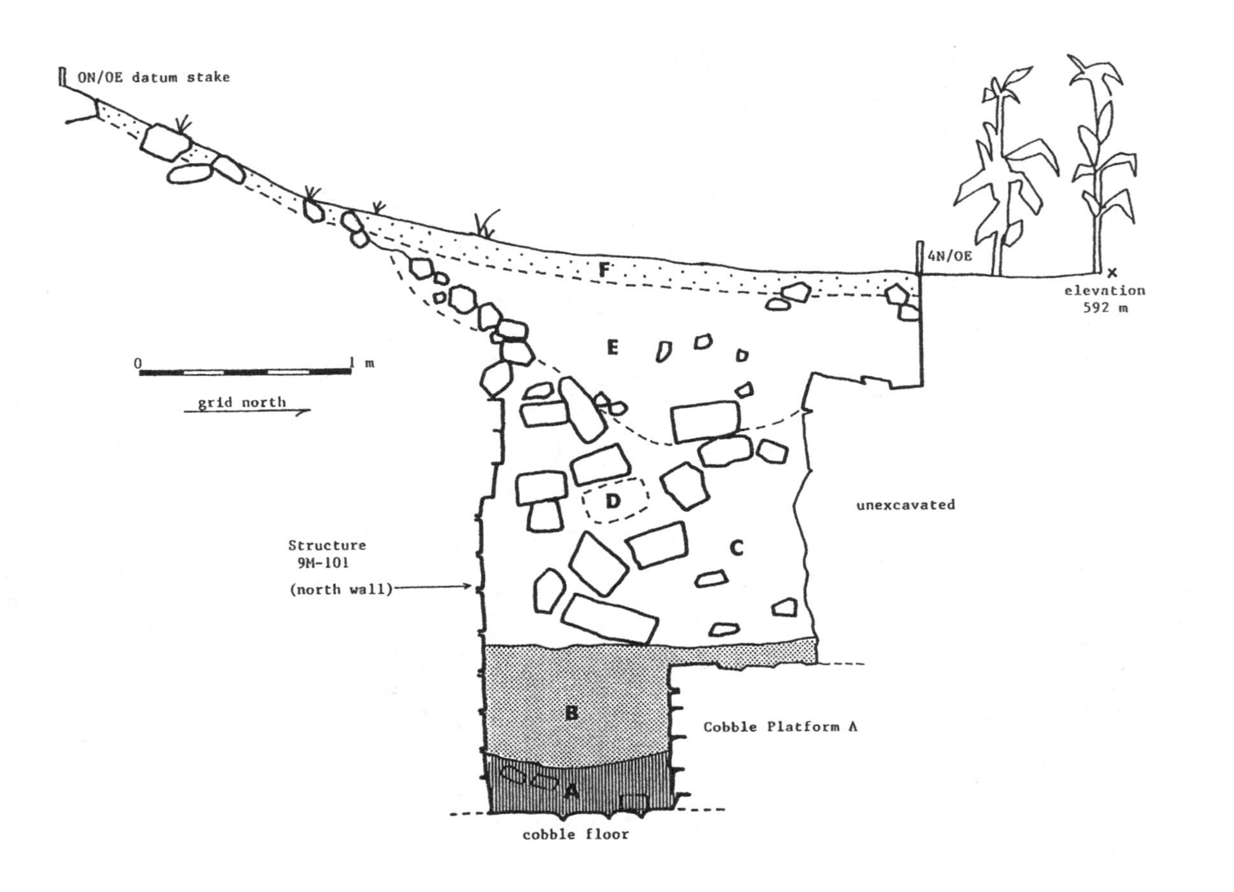
ON/OE datum stake
4N/OE
elevation
592 m
F
E
D
C
B
A
0
1 m
grid north
unexcavated
Structure
9M-101
(north wall)
Cobble Platform A
cobble floor

the remaining question is that of the specific causes and mechanisms that resulted in a population decline from the high of 26,000–28,000 achieved by A.D. 750, a level that was sustained for approximately 100 years, to the level found at A.D. 1050, approximately 5,000 people. The archaeological data on settlement patterns and demographic history, present-day agricultural productivity, the pollen profile, and direct evidence of soil erosion in the past all strongly support the evidence for a process of progressive decline in the capability of the valley to sustain a sufficiently large permanent population to underwrite a Maya site, the major argument in the processual theory.

The mechanisms of decline

The major remaining question is the precise mechanism that produced the population decline, i.e., whether it was a disjunction between fertility and mortality or due to emigration. Following the first possibility, the increasing pressure on the physical environment could have resulted in lower nutritional status of the population, causing a reduction in fertility in women, increase in the mortality of the population, or both. Most particularly, children would be more susceptible to endemic diseases. If migration was the mechanism, the population could have emigrated as pressure on the food supply became increasingly severe. Finally, both processes could have been at work.

One way of ascertaining which of these two explanations was paramount, or whether it was a mix of both, would be the study of human skeletons collected during excavations at the site. Fortunately our excavations and those of earlier

---

Fig. 5.7. Profile, West Wall of Trench 1, Op. 40, Str. 9M-101, Copan. Legend: A = dark brown soil with burned clay, charcoal, small inclusions of volcanic tuff, and many artifacts; B = fine, yellow, silty clay virtually devoid of artifacts except in upper levels; C = red-brown clayey silt filled with collapsed building stone and plaster fragments; D = fragment of facade sculpture from building; E = black silt with collapsed building stone; F = dark brown recent humas. The profile tells an important story. The original land surface was at the bottom of the trench. At some point the Classic Maya laid down a cobble floor, and then built two structures on top of it (Str. 9M-101 on the south and cobble platform A on the north). Next, the narrow corridor between these two structures accumulated about 20–30 cm of trashy, middenlike deposits, including sherds and obsidian tools. People obviously discarded rubbish here. Next, while the large building was still intact, about 70 cm of yellow silt was washed into the corridor. This apparently happened quite rapidly, since artifacts did not accumulate in it to any degree. The Maya were still using the area, because we found a huge flat stone (not shown) on top of the yellow silt layer, partly overlying the corridor and partly over platform A. Next the upperpart of Str. 9M-101 collapsed, creating deposit C. More soil continued to erode, forming deposit E, and, finally, the recent humus layer developed.

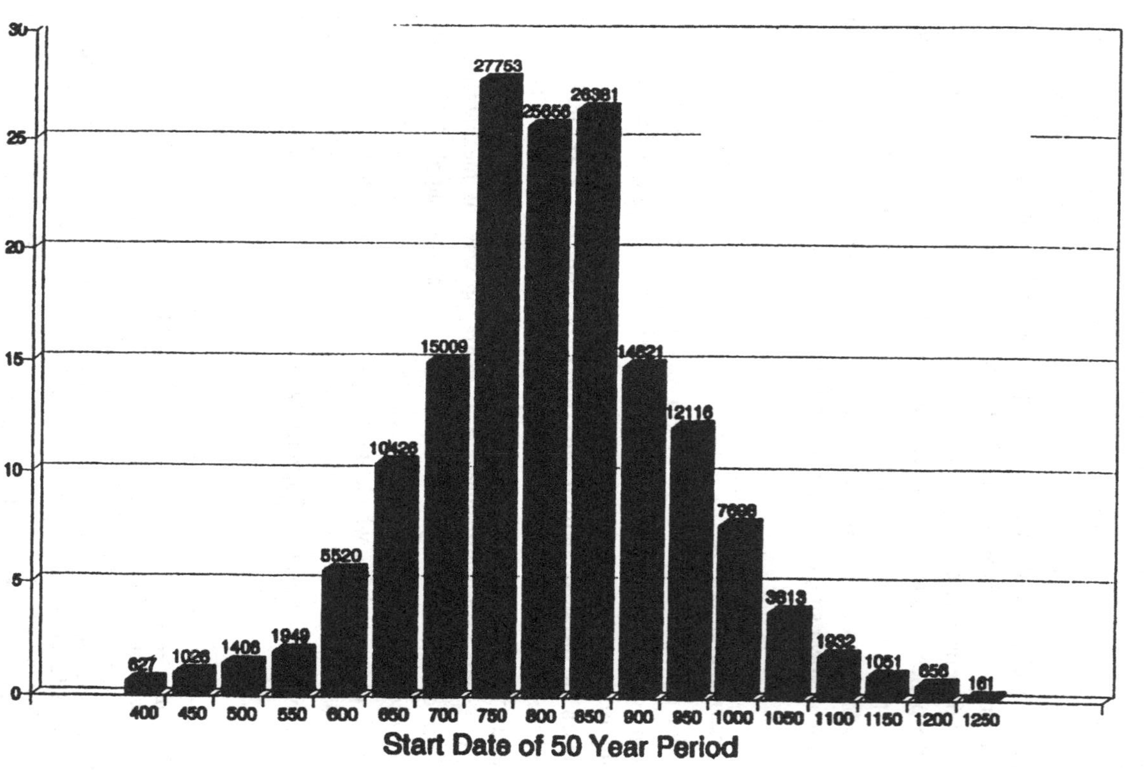
30
25
20
15
10
5
0
627
1026
1408
1949
5520
10426
15009
27753
25656
26381
14621
12116
7696
3613
1932
1051
656
161
400
450
500
550
600
650
700
750
800
850
900
950
1000
1050
1100
1150
1200
1250
Start Date of 50 Year Period

Fig. 5.8.

projects have yielded a very large sample – over 600 skeletons. The sample covers the entire time span of Copan, includes all social class levels, is from both urban and rural contexts, and both sexes and all ages are represented.

The analysis to date (Storey, personal communication; Whittington, 1989), has demonstrated certain trends. First, the population of Copan during the Coner phase was clearly under severe nutritional and pathological stress. Dental hypoplasia, a condition that indicates growth interruptions during childhood and adolescence, and osteoporosis, a condition that reflects iron-deficiency anemia, is found in over 80% of the sample, even including those from the elite residential compounds. This suggests that during the final phase of the history of Copan there was a problem with food, both in terms of quantity and balance of nutrients in the diet. Corollary studies, using different methods, by two other investigators (still in progress), David Reed (carbon isotope analysis of bone) and David Lentz (identification of plant remains found in the middens), suggest that the diet did in fact shift from one that was relatively wide ranging to one increasingly dependent on maize, perhaps 90% by Coner times.

The skeletal analysis to date tends to support the thesis that the population was under heavy stress and that *in situ* decline could have been a major factor.

Conclusion

In summary, archaeological data reveals a number of cases of substantial decline of ancient populations after centuries of sustained growth. Environmental degradation, the product of increasingly more intensive use of land, was almost certainly a major factor in these declines. In the case of the Copan Valley studies of present day land use, population history, pollen analysis, evidence of soil erosion in the past, and skeletal analysis are all mutually supportive of the hypothesis that runaway population growth and its disastrous impact on the soil resources of the valley was a major factor in the decline of the population and ultimate abandonment of the Copan Valley. The major question remaining is to what degree the evidence from Copan reflects conditions over much of the Classic Maya lowlands. The view of preindustrial man as the quintessential environmental conservationist is almost certainly a misguided and misleading view of human adaptation.

References

Alan, W. (1967). *The African Husbandman.* Edinburgh, Scotland: Oliver and Boyd.

Baudez, C. (ed.) (1983). *Introduction a la Arqueologia de Copan.* Tres tomos: Proyecto Arqueologica Copan, Secretaria de Estado, en el Despacho de Cultura y Turismo, Tegucigalpa d.c., Mexico.

Blanton, R. E. (1975). The Cybernetic analysis of human population growth. In *Population studies in Archaeology and Biological Anthropology,* a Symposium, ed. A. C. Swedlund. American Antiquity, Vol. 40, No. 2 Pt. 2. Mem. Soc. Am. Archeol. 30.
Boserup, E. (1965). *Conditions of Agricultural Growth.* Chicago: Aldine.
Brewbaker, J. L. (1979). Diseases of maize in the wet lowland tropics and the collapse of the Classic Maya civilization. *Economic Botany,* **33**(2), 101–118.
Coe, M. D. (1966). *The Maya.* Ancient Peoples and Places Series, Vol. 52. New York: Praeger.
Conklin, H. C. (1957). *Hanunoo Agriculture: A Report on an Integral System of Shifting Cultivation in the Philippines.* Rome: FAO United Nations, Forestry Development, Paper No. 12.
Cowgill, G. L. (1975). Population pressure as a non-explanation. In *Population Studies in Archaeology and Biological Anthropology,* a Symposium, ed. A. C. Swedlund. American Antiquity, Vol. 40, No. 2 Pt. 2. Mem. Soc. Am. Archeol. 30.
Culbert, T. P. (ed.) (1973). *The Classic Maya Collapse.* Albuquerque: University of New Mexico Press.
Hardesty, D. G. (1977). *Ecological Anthropology.* New York: Wiley.
Hopkins, B. (1965). *Forest and Savannah.* Ibadan and London: Heinemann.
Lewis, H. (1982). Fire, Technology, and Resource Management in Aboriginal North America and Australia. In *Resource Managers: North American and Australian Hunters and Gatherers,* ed. Williams and Huron. Boulder, Colorado: Westview Press.
Martin, P. S. & Plog, F. (1973). *The Archaeology of Arizona.* New York: Doubleday and American Museum of Natural History Press.
Moran, E. F. (1979). *Human Adaptability.* North Scituate, Mass.: Duxbury Press.
Nigh, R. B. (1975). *Evolutionary Ecology of Maya Agriculture in Highland Chiapas.* Ph.D. Dissertation. Stanford University.
Rappaport, R. A. (1967). *Pigs for the Ancestors,* Ritual in the Ecology of a New Guinea People. New Haven: Yale University Press.
Roys, R. (1957). The Political Geography of the Yucatan Maya. Publ. No. 613. Washington, DC: Carnegie Institution.
Rue, D. (1986). *The Palynological Analysis of the Prehispanic Human Impact in the Copan Valley.* Ph.D. Dissertation. Pennsylvania State University, University Park.
Sanders, W. T. (1972). Population, Agricultural history and societal evolution in Mesoamerica. In *Population Growth: Anthropological Implications,* ed. B. Spooner. Cambridge, Mass.: MIT Press.
Vivo Escoto, J. A. (1984). Weather and climate in Mexico and Central America. In *Handbook of Middle American Indians, Vol. 1,* ed. R. West. Austin: University of Texas Press.
Webster, D. L. & Freter, A. C. (1990). The demography of Late Classic Copan. In *Precolumbian Population History in the Mayan Lowlands,* ed. C. P. Culbert & D. S. Rice. Albuquerque: University of New Mexico Press.
Whittington, S. (1989). *Characteristics of Demography and Disease In Low Status Maya from the Classic Period Copan, Honduras.* Ph.D. Dissertation. Pennsylvania State University.
Wingard, J. (1992). *Soils and Classic Maya Settlement at Copan.* Ph.D. Dissertation, Pennsylvania State University.

6

Conserving biological diversity in the face of climate change

ROBERT L. PETERS

Introduction

Our understanding of how atmospheric composition affects global climate is still in its infancy, but an increasing body of knowledge suggests that rising concentrations of CO_2 and other anthropogenic polyatomic gases will raise global average temperatures substantially (National Research Council, 1983; Schneider & Londer, 1984; World Meteorological Organization (WMO), 1982). Associated with global warming will be regional and local changes in average temperature, in the distribution of hot and cold periods, and changes in a number of other chemical and physical variables, including precipitation, evaporation rates, sea level, and soil and water chemistry (Schneider et al., 1991).

We can infer how the biota might respond to climate change by observing present and past distributions of plants and animals, which are heavily determined by temperature and moisture patterns. For example, one race of the dwarf birch (*Betula nana*) can only grow where the temperature never exceeds 22°C (Ford, 1982), suggesting that it would disappear from those areas where global warming causes temperatures to exceed 22°C. Recent historical observations of changes in range or species dominance, as observed in the gradual replacement of spruce (*Picea rubens*) by deciduous species during the past 180 years in the eastern U.S. (Hamburg & Cogbill, 1988), can also suggest future responses. Insight into long-term responses to large climatic changes can be gleaned from studies of fossil distributions of, particularly, pollen (Davis, 1983; David & Zabinski, 1991; Webb, 1991) and small mammals (Graham, 1986, 1991).

Such observations tell us that plants and animals are very sensitive to climate. Their ranges move when the climate patterns change – species die out in areas where they were once found and colonize new areas where the climate becomes newly suitable.

We can expect similar responses to projected global warming during the next 50–100 years, including disruption of natural communities and extinction of populations and species. Even many species that are today widespread will experience large range reductions. Efficient dispersers may be able to shift their ranges to take advantage of newly suitable habitat, but most species will at best experience a time lag before extensive colonization is possible, and hence in the short-term will show range diminishment (see Webb, 1991, for a discussion of vegetation–climate "disequilibrium"). At worst, many species will never be able to recover without human intervention since migration routes will have been cut off by development or other human-caused habitat loss.

Although this paper will focus on the terrestrial biota, ocean systems may show similar shifts in species ranges and community compositions if warming of ocean water or alteration in the patterns of water circulation occur. For example, recent El Niño events demonstrate the vulnerability of primary productivity and species abundances to changes in ocean currents and local temperatures (e.g., Duffy, 1983; Glynn, 1984; Alexander, 1991; Ray et al., 1991).

The nature of the ecologically significant changes

Although the exact rate and magnitude of future climate change is uncertain, given imperfect knowledge about the behavior of clouds, oceans, and biotic feedbacks, there is widespread consensus among climatologists that ecologically significant warming will occur during the next century. For example, the National Academy of Sciences (NAS, 1987) concluded that both global mean surface warming and an associated increase in global mean precipitation are "very probable." Although perhaps more outspoken than most, Hansen et al. (1988a) have said "we can confidently state that major greenhouse climate changes are a certainty."

It is expected that within the next 40 years greenhouse trace gases in the atmosphere, including carbon dioxide, chlorofluorocarbons, and methane, will reach a concentration equivalent in warming capacity to double the preindustrial concentration of carbon dioxide. The National Academy of Sciences and others have estimated that this concentration of greenhouse gases will be sufficient to raise Earth's temperature by 3 ± 1.5°C (Hansen et al., 1988b, NAS, 1987; NRC, 1983; WMO, 1982; Schneider & Londer, 1984; Schneider et al., 1991). More recent estimates suggest the possibility that warmings as high as 4.2 ± 1.2°C (Schlesinger, 1989), or even 8–10°C (Lashof, 1989), are possible. Because of a time lag caused by thermal inertia of the oceans, some of this warming will be delayed by 30–40 years beyond the time that a doubling equivalent of carbon dioxide is reached (EPA, 1988), but substantial warming could occur soon – the Goddard Institute for Space Studies (GISS) model projects a 2°C rise by 2020 A.D. (Rind, 1989). The Intergovernmental Panel on Climate Change (IPCC; see Houghton et al., 1990), with a slightly more conservative estimate, predicts that the most likely rate of warming is "1°C above the present value (about 2°C above that in the pre-industrial period) by 2025 and 3°C above today's (about 4°C above pre-industrial) before the end of the next century." At this rate, given that even a 1°C rise can have a large effect on ecological systems, these systems are likely to experience substantial stress within the next 50 years. Warming could be faster or slower than these estimates – the exact rate will be heavily dependent upon whether or not people are successful in curtailing production of greenhouse gases and on the significance of what are now poorly understood feedback processes.

It should be stressed that although projections can be made about global averages, regional projections are much less certain (Schneider, 1988). However, it is known that warming will not be even, with the high latitudes, for example, likely to be warmer than the low latitudes (Hansen et al., 1988a). Regional and local peculiarities of typography and circulation will play a strong role in determining local climates.

For the purpose of discussion in this paper, I will take average global warming to be 3°C, since this is a commonly used benchmark, but it must be recognized that additional warming well beyond 3°C may be reached during the next century if the production of anthropogenic greenhouse gases continues to increase. I will also assume that 3°C warming will not be reached until 2070 A.D. Additional warming or faster warming would cause additional biological disruption beyond that laid out in this paper.

The threats to natural systems are serious for the following reasons. First, 3°C of warming would present natural systems with a warmer world than has been experienced in the past 100,000 years (Schneider & Londer, 1984). 4°C would make the earth its warmest since the Eocene, 40 million years ago (Barron, 1985; see Webb, 1991). This warming would not only be large compared to recent natural fluctuations, but it would be very fast, perhaps 15–40 times faster than past natural changes (Schneider et al., 1991). For reasons discussed below, such a rate of change may exceed the ability of many species to adapt. Even widespread species are likely to have drastically curtailed ranges, at least in the short-term. Moreover, human encroachment and habitat destruction will make wild populations of many species small and vulnerable to local climate changes.

Second, ecological stress would not be caused by temperature rise alone. Changes in global temperature patterns would trigger widespread alterations in rainfall patterns (Hansen et al., 1981; Kellogg & Schware, 1981; Manabe et al.,

1981), and we know that for many species precipitation is a more important determinant of survival than temperature *per se.* Indeed, except at treeline, rainfall is the primary determinant of vegetation structure, trees occurring only where annual precipitation is in excess of 300 mm (Woodward, 1991). Because of global warming, some regions would see dramatic increases in rainfall, and others would lose their present vegetation because of drought. For example, the U.S. Environmental Protection Agency (1988) concluded, based on several studies, that a long-term drying trend is likely in the midlatitude, interior continental regions during the summer. Specifically, Kellogg & Schware (1981), based upon rainfall patterns during past warming periods, projected that substantial decreases in rainfall in North America's Great Plains are possible – perhaps as much as 40% by the early decades of the next century.

Other environmental factors important in determining vegetation type and health would change because of global warming. Soil chemistry would change (Kellison & Weir, 1987), as, for example, changes in storm patterns alter leaching and erosion rates (Harte, Torn, & Jensen, 1991). Increased carbon dioxide concentrations may accelerate the growth of some plants at the expense of others (NRC, 1983; Strain & Bazzaz, 1983; see discussion in Webb, 1991), possibly destabilizing natural ecosystems. And rises in sea level may inundate coastal biological communities (NRC, 1983; Hansen et al., 1981; Hoffman et al., 1983; Titus et al., 1984).

Estimates of how rapidly the sea will rise, given a certain amount of warming, have varied substantially, in part because of uncertainty about the contribution of Antarctic melting (see IPCC, 1990, for a table summarizing recent studies). The IPCC (1990) attempted to estimate how rapid this rise would be. It projected that for "business as usual," i.e., no significant reduction in greenhouse gas production, by 2030 sea level is likely to have risen 8–29 cm, with a best estimate of 18 cm. By 2070 it expects 21–71 cm, with a best estimate of 44 cm. Lest this amount of rise seem insignificant, it is important to realize that a relatively small increase in local sea level can translate into much increased rates of coastal erosion and height of storm surges. For example, Leatherman (1991) has recounted the fate of several Chesapeake Bay islands and associated towns that were washed away over the past 100 years as local sea level rose only 30 cm. Further, some regional or local increases will be substantially larger than the global average. For example, the IPCC (1990) cites Mikolajewicz et al. (1990), whose "dynamic ocean model showed regional differences of up to a factor of two relative to the global-mean value."

As mentioned, it is generally concluded by a variety of computer projections that warming will be relatively greater at higher latitudes (Hansen et al., 1987). This suggests that although tropical systems may be more diverse and are currently under great threat because of habitat destruction, temperate zone and arctic species may ultimately be in greater jeopardy from climate change, at least from temperature *per se* (see Hartshorn, 1991, for a discussion of precipitation effects on tropical forests). Arctic vegetation would experience widespread changes (Billings & Peterson, 1991; Woodward, 1991; Edlund, 1987). A recent attempt to map climate-induced changes in world biotic communities projects that high-latitude communities would be particularly stressed (Emanuel et al., 1985), and the boreal forest, for example, was projected to decrease by 37% in response to global warming of 3°C.

A final point, important in understanding species response to climate change, is that weather is variable, and extreme events, like droughts, floods, blizzards, and hot or cold spells, may have more effect on species distributions than average climate *per se* (e.g., Knopf & Sedgwick, 1987). For example, in northwestern forests, global warming is expected to increase fire frequency, leading to rapid alteration of forest character (Franklin et al., 1991).

Species' ranges shift in response to climate change

We know that when temperature and rainfall patterns change, species' ranges change. Not surprisingly, species tend to track their climatic optima, retracting their ranges where conditions become unsuitable while expanding them where conditions improve (Peters & Darling, 1985; Ford, 1982). Even very small temperature changes of less than one degree within this century have been observed to cause substantial range changes. For example, the white admiral butterfly (*Ladoga camilla*) and the comma butterfly (*Polygonia c-album*) greatly expanded their ranges in the British Isles during the past century as the climate warmed approximately 0.5°C (Ford, 1982). The birch (*Betula pubescens*) responded rapidly to warming during the first half of this century by expanding its range north into the Swedish tundra (Kullman, 1983).

On a larger ecological and temporal scale, entire vegetation types have shifted in response to past temperature changes no larger than those that may occur during the next 100 years or less (Baker, 1983; Bernabo & Webb, 1977; Butzer, 1980; Flohn, 1979; Muller, 1979; Van Devender & Spaulding, 1979). As the Earth warms, species tend to shift to higher latitudes and altitudes. From a simplified point of view, rising temperatures have caused species to colonize new habitats toward the poles, often while their ranges contracted away from the equator as conditions there became unsuitable.

During several Pleistocene interglacials, the temperature in North America was apparently 2–3°C higher than now. Sweet gum trees (*Liquidambar*) grew in southern Ontario (Wright, 1971); Osage oranges (*Maclura*) and papaws

(*Asimina*) grew near Toronto, several hundred kilometers north of their present distributions; manatees swam in New Jersey; and tapirs and peccaries foraged in North Carolina (Dorf, 1976). During the last of these interglacials, which ended more than 100,000 years ago, vegetation in northwestern Europe, which is now boreal, was predominantly temperate (Critchfield, 1980). Other significant changes in species' ranges have been caused by altered precipitation accompanying past global warming, including expansion of prairie in the American Midwest during a global warming episode approximately 7,000 years ago (Bernabo & Webb, 1977).

It should not be imagined, because species tend to shift in the same general direction, that existing biological communities move in synchrony. Conversely, because species shift at different rates in response to climate change, communities often disassociate into their component species (Fig. 6.1). Recent studies of fossil packrat (*Neotoma* spp.) middens in the southwestern United States show that during the wetter, moderate climate of 22,000–12,000 years ago there was not a concerted shift of plant communities. Instead, species responded individually to climatic change, forming stable, but by present-day standards, unusual assemblages of plants and animals (Van Devender & Spaulding, 1979). In eastern North America, too, postglacial communities were often ephemeral associations of species, changing as individual ranges changed (Davis, 1983; Graham, 1986, 1991).

A final aspect of species response is that species may shift altitudinally as well as latitudinally. When climate warms, species shift upward. Generally, a short climb in altitude corresponds to a major shift in latitude: a 3°C cooling of 500 m in elevation equals roughly 250 km in latitude (MacArthur, 1972). Thus, during the middle Holocene, when temperatures in eastern North America were 2°C warmer than at present, hemlock (*Tsuga canadensis*) and white pine (*Pinus strobus*) were found 350 m higher on mountains than they are today (Davis, 1983).

Because mountain peaks are smaller than bases, as species shift upward in response to warming, they typically occupy smaller and smaller areas, have smaller populations, and may thus become more vulnerable to genetic and environmental pressures (Murphy & Weiss, 1991). Species originally situated near mountaintops might have no habitat to move up to, and may be entirely replaced by the relatively thermophilous species moving up from below (Fig. 6.2). Examples of past extinctions attributed to upward shifting including alpine plants once living on mountains in Central and South America, where vegetation zones have shifted upward by 1000–1500 m since the last glacial maximum (Flenley, 1979; Heusser, 1974). See Murphy and Weiss (1991) for projections of some local extinction rates for butterflies, birds, and mammals.

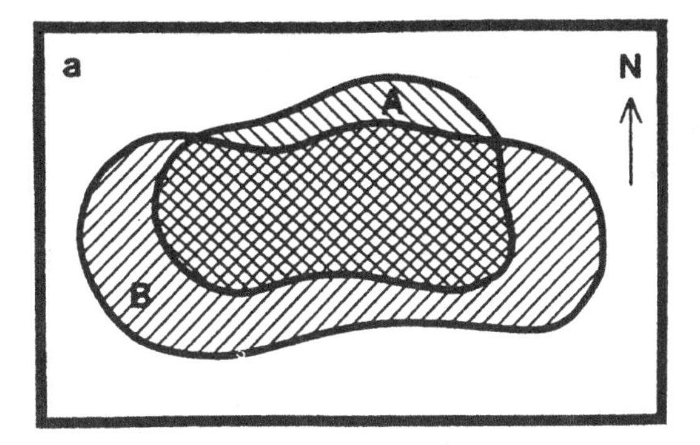

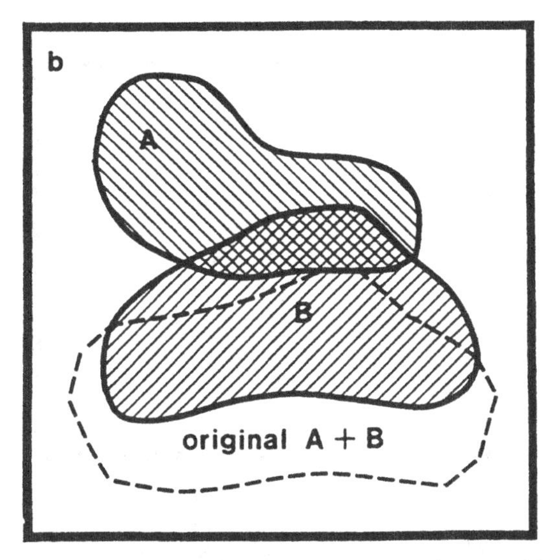

Fig. 6.1. (a) Initial distribution of two species, A and B, whose ranges largely overlap. (b) In response to climate change, latitudinal shifting occurs at species-specific rates and the ranges disassociate.

Magnitude of projected latitudinal shifts

If the proposed CO_2-induced warming occurs, species shifts similar to those in the Pleistocene would occur, and vegetation belts would move hundreds of kilometers toward the poles (Davis & Zabinski, 1991; Frye, 1983; Peters & Darling, 1985). Based on positions of vegetation zones during analogous warming periods in the past (Dorf, 1976; Furley et al., 1983), Peters and Darling (1985) conservatively estimated that 3°C of warming would cause a

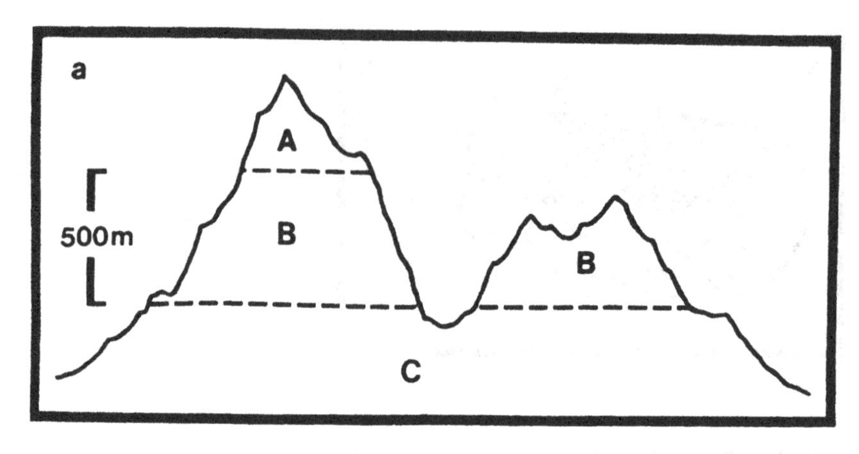

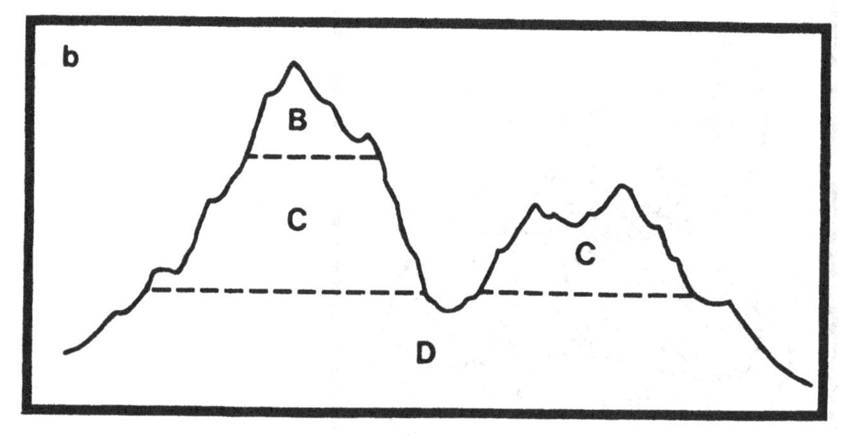

Fig. 6.2. (a) Present altitudinal distribution of three species, A, B, and C. (b) Species distribution after a 500 meter shift in altitude in response to a 3.25°C rise in temperature (based on Hopkin's bioclimatic law, MacArthur, 1972). Species A becomes locally extinct. Species B shifts upward and the total area it occupies decreases. Species C becomes fragmented and restricted to a smaller area, while species D successfully colonizes the lowest altitude habitats.

minimum 300 km shift in the temperate zone, which may have been an underestimate for many species. Additional confirmation that shifts of this magnitude or greater would occur comes from attempts to project future range shifts for some species by looking at their ecological requirements. For example, the forest industry is concerned about the future of commercially valuable North American species, like the loblolly pine (*Pinus taeda*). This species is limited on its southern border by moisture stress on seedlings. Based on its physiological requirements for temperature and moisture, Miller et al. (1987)

projected that the southern range limit of the species would shift approximately 350 km northward in response to a global warming of 3°C. Davis and Zabinski (1991) have projected possible northward range withdrawals among several North American tree species, including sugar maple (*Acer saccharum*) and beech (*Fagus grandifolia*), from 600 km to as much as 2,000 km in response to the warming caused by a doubled CO_2 concentration. Beech would be most responsive, withdrawing from its present southern extent along the Gulf Coast, and retreating into Canada.

Mechanisms underlying range shifts

The range shifts described above are the sum of many local processes of extinction and colonization that occur in response to climate-caused changes in suitability of habitats. These changes in habitat suitability are determined by both direct climate effects on physiology, including temperature and precipitation, and indirect effects secondarily caused by other species, themselves affected by temperature.

There are numerous examples of climate directly influencing survival and thereby distribution. In animals, the direct range-limiting effects of excessive warmth include lethality, as in corals (Glynn, 1984), and interference with reproduction, as in the large blue butterfly, *Maculinea arion* (Ford, 1982). For exothermic animals like insects, Rubenstein (1991) models how changes in temperature or humidity can alter substantially the rates of metabolism, fecundity, and feeding in insects, in turn determining the ranges of species. Tracy (1991) discusses how changing thermal regimes might change the composition of animal communities, favoring those species that are relatively adapted to high temperatures. Evidence of species loss from power plant cooling ponds, in which water temperatures are higher than normal, shows that, at least over the short-term, warming can decrease the total number of species within a community (Brisbin, 1974; Kolehmainen & Castro, 1974).

In plants, excessive heat and associated decreases in soil moisture may decrease survival and reproduction (Woodward, 1991). Coniferous seedlings, for example, are injured by soil temperatures over 45°C, although other types of plants can tolerate much higher temperatures (see Daubenmire, 1962). Many plants have their northern limits determined by minimum temperature isotherms below which some key physiological process does not occur. For instance, the grey hair grass (*Corynephorus canescens*) is largely unsuccessful at germinating seeds below 15°C and is bounded to the north by the 15°C July mean isotherm (Marshall, 1978). Moisture extremes exceeding physiological tolerances also determine species' distributions. Thus, the European range of the beech tree (*Fagus sylvatica*) ends to the south where rainfall is less than 600 mm annually (Seddon, 1971), and dog's mercury (*Mercurialis perennis*), an herb restricted to well-drained sites in Britain, cannot survive in soil where the water table reaches as high as 10 cm below the soil surface (Ford, 1982).

The physiological adaptations of most species to climate are conservative, and it is unlikely that most species could evolve significantly new tolerances in the time allotted to them by the coming warming trend. This is despite the fact that climate has strong effects on survivorship in many species (Holt, 1990). Indeed, the evolutionary conservatism in thermal tolerance of many plant and animal species – beetles, for example (Coope, 1977) – is the underlying assumption that allows us to infer past climates from faunal and plant assemblages.

Interspecific interactions altered by climate change will have a major role in determining new species distributions. Temperature can influence predation rates (Rubenstein, 1991; Rand, 1964), parasitism (Aho et al., 1976), and competitive interactions (Beauchamp & Ullyott, 1932). Tracy (1991) describes how groups of species, in particular a genus of darkling beetles (*Eleodes*), may partition the thermal habitat by time of day and season. Climate-induced changes in the ranges of tree pathogens and parasites may be important in determining future tree distributions (Winget, 1988). Soil moisture is a critical factor in mediating competitive interactions among plants, as is the case where the dog's mercury (*Mercurialis perennis*) excludes oxlip (*Primula elatior*) from dry sites (Ford, 1982).

Given the new associations of species that occur as climate changes, many species will face "exotic" competitors for the first time. Local extinctions may occur as climate change causes increased frequencies of droughts and fires, favoring invading species. One species that might spread, given such conditions, is *Melaleuca quinquenervia,* a bamboo-like Australian eucalypt. This species has already invaded the Florida Everglades, forming dense monotypic stands where drainage and frequent fires have dried the natural marsh community (Courtenay, 1978; Myers, 1983).

The preceding effects, both direct and indirect, may act in synergy, as when drought makes a tree more vulnerable to attack by insect pests.

Dispersal rates and barriers

The ability of species to adapt to changing conditions will depend to a large extent upon their ability to track shifting climatic optima by dispersing colonists. In the case of warming, a North American species, for example, would most likely need to establish colonies to the north or at higher elevations.

Survival of plant and animal species would therefore depend either on long-distance dispersal of colonists, such as seeds or migrating animals, or on rapid iterative colonization of nearby habitat until long-distance shifting results. A plant's intrinsic ability to colonize will depend upon its ecological characteristics, including fecundity, viability and growth characteristics of seeds, nature of the dispersal mechanism, and ability to tolerate selfing and inbreeding upon colonization. If a species' intrinsic colonization ability is low, or if barriers to dispersal are present, extinction may result if all of its present habitat becomes unsuitable.

There are many cases where complete or local extinction has occurred because species were unable to disperse rapidly enough when climate changed. For example, a large, diverse group of plant genera, including water-shield (*Brassenia*), sweet gum (*Liquidambar*), tulip tree (*Liriodendron*), magnolia (*Magnolia*), moonseed (*Menispermum*), hemlock (*Tsuga*), arbor vitae (*Thuja*), and white cedar (*Chamaecyparis*), had a circumpolar distribution in the Tertiary (Tralau, 1973). But during the Pleistocene ice ages, all went extinct in Europe while surviving in North America. Presumably, the east–west orientation of such barriers as the Pyrennes, Alps, and the Mediterranean, which blocked southward migration, was partly responsible for their extinction (Tralau, 1973).

Other species of plants and animals thrived in Europe during the cold periods, but could not survive conditions in postglacial forests. One such previously widespread dung beetle, *Aphodius hodereri*, is now extinct throughout the world except in the high Tibetan plateau where conditions remain cold enough for its survival (Cox & Moore, 1985). Other species, like the Norwegian mugwort (*Artemisia novegica*) and the springtail *Tetracanthella arctica*, now live primarily in the boreal zone but also survive in a few cold, mountaintop refugia in temperate Europe (Cox & Moore, 1985).

These natural changes were slow compared to predicted changes in the near future. Change to warmer conditions at the end of the last ice age spanned several thousand years, yet is considered rapid by geologic standards (Davis, 1983). We can deduce that if such a slow change was too fast for many species to adapt, the projected warming – possibly 40 times faster – will have more severe consequences. For widespread, abundant species, like the loblolly pine (modelled by Miller et al., 1987), even substantial range retraction might not threaten extinction; but rare, localized species, whose entire ranges might become unsuitable, would be threatened unless dispersal and colonization were successful. Even for widespread species, major loss of important ecotypes and associated germplasm is likely (Davis & Zabinski, 1991).

A key question is whether the dispersal capabilities of most species prepare them to cope with the coming rapid warming. If the climatic optima of temperate zone species do shift hundreds of kilometers toward the poles within the next 100 years, then these species would have to colonize new areas rapidly. To survive, a localized species whose present range becomes unsuitable might have to shift poleward at several hundred kilometers per century. Although some species, such as plants propagated by spores or "dust" seeds, may be able to match these rates (Perring, 1965), many species could not disperse fast enough to compensate for the expected climatic change without human assistance (Rapoport, 1982), particularly given the presence of dispersal barriers. Even wind-assisted dispersal may fall short of the mark for many species. In the case of the Engelmann spruce (*Picea engelmannii*), a tree with light, wind-dispersed seeds, fewer than 5% of seeds travel even 200 m downwind, leading to an estimated migration rate of 1–20 km per century (Seddon, 1971); this reconciles well with rates derived from fossil evidence for North American trees of between 10–45 km per century (Davis & Zabinski, 1991; Roberts, 1989). As described in the next section, many migration routes will likely be blocked by the cities, roads, and fields replacing natural habitat.

Although many animals may be physically capable of great mobility, the distribution of some is limited by the distributions of particular plants, i.e., suitable habitat; their dispersal rates therefore may be largely determined by those of cooccurring plants. Behavior may also restrict dispersal even of animals physically capable of large movements. Dispersal rates below 2.0 km/year have been measured for several species of deer (Rapoport, 1982), and many tropical deep-forest birds simply do not cross even very small unforested areas (Diamond, 1975). On the other hand, some highly mobile animals may shift rapidly, as have some European birds (Edgell, 1984).

Even if animals can disperse efficiently, suitable habitat may be reduced under changing climatic conditions. For example, it has been suggested that tundra nesting habitat for migratory shore birds might be reduced by high-arctic warming (Myers & Lester, 1991).

Synergy of habitat destruction and climate change

We know that even slow, natural climate change has caused species to become extinct. What is likely to happen given the environmental conditions of the coming century?

Some clear implications for conservation follow from the preceding discussion of dispersal rates. Any factor that would decrease the probability that a species could successfully colonize new habitat would increase the probability of extinction. Thus, as previously described, species are more likely to become extinct if there are physical barriers to colonization, such as oceans, mountains, and cities. Further, species are more likely to become extinct if their remaining populations are small. Smaller populations mean fewer colonists can be sent out and that the probability of successful colonization is smaller.

Species are more likely to become extinct if they occupy a small geographic range. It is less likely that some part will remain suitable when climate changes than if the ranges were larger. Also, if a species has lost much of its range because of some other factor, like clearing of the richer and moister soils for agriculture, it is possible that remaining populations are located in poor habitat and are therefore more susceptible to new stresses.

For many species all of these conditions will be met by human-caused habitat destruction, which, as discussed at the beginning of this chapter, increasingly confines the natural biota to small patches of original habitat, patches isolated by vast areas of human-dominated urban or agricultural lands.

Habitat destruction in conjunction with climate change sets the stage for an even larger wave of extinction than previously imagined, based upon consideration of human encroachment alone. Small, remnant populations of most species, surrounded by cities, roads, reservoirs, and farm land, would have little chance of reaching new habitat if climate change makes the old unsuitable. Few animals or plants would be able to cross Los Angeles on the way to new habitat. Figure 6.3 illustrates the combined effects of habitat loss and warming on a hypothetical reserve (see Harris & Cropper, 1991, for a discussion of how direct human effects and climate change will affect Florida's endangered fauna).

Amelioration and mitigation

Because of difficulty in predicting regional and local changes, conservationists and reserve managers must deal with increased uncertainty in making long-range plans. However, even given imprecise regional projections, informed guesses can be made at least about the general direction of change, specifically that most areas will tend to be hotter and that continental interiors, in particular, are likely to experience decreased soil moisture.

How might the threats posed by climatic change to natural communities be mitigated? One basic truth is that the less populations are reduced by development now, the more resilient they will be to climate change. Thus, sound conservation now, whereby we try to conserve more than just the minimum number of individuals of a species necessary for present survival, would be an excellent way to start planning for climate change.

In terms of responses specifically directed at the effects of climate change, the most environmentally conservative response would be to halt or slow global

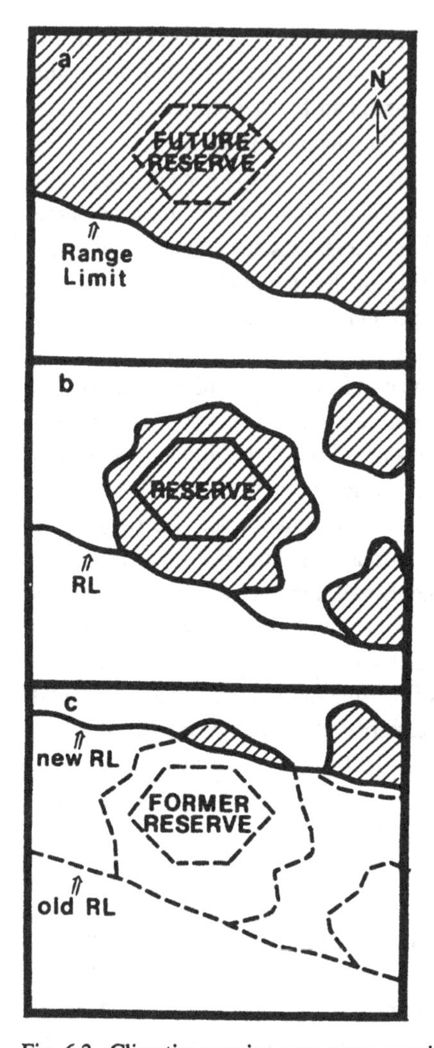

Fig. 6.3. Climatic warming may cause species within biological reserves to disappear. Hatching indicates: (a) species distribution before either human habitation or climate change (SL indicates southern limit of species range); (b) fragmented species distribution after human habitation but before climate change; (c) species distribution after human habitation and climate change.

warming by cutting back on production of fossil fuels, methane, and chlorofluorocarbons. Extensive planting of trees to capture carbon dioxide could help slow the rise in carbon dioxide concentrations (Sedjo, 1989; Woodwell, 1991). Nonetheless, even were the production of all greenhouse gases stopped today, it is very likely that there are now high enough concentrations in the air to cause ecologically significant warming anyway (Rind, 1989). Therefore, those concerned with the conservation of biological diversity must begin to plan mitigation activities now.

To make intelligent plans for siting and managing reserves, we need more knowledge. We must refine our ability to predict future conditions in reserves. We also need to know more about how temperature, precipitation, CO_2 concentrations, and interspecific interactions determine range limits (e.g., Picton, 1984; Randall, 1982), and, most important, how they can cause local extinctions.

Reserves that suffer from the stresses of altered climatic regimes will require carefully planned and increasingly intensive management to minimize species loss. For example, modifying conditions within reserves may be necessary to preserve some species; depending on new moisture patterns, irrigation or drainage may be needed. Because of changes in interspecific interactions, competitors and predators may need to be controlled and invading species weeded out. The goal would be to maintain suitable conditions for desired species or species assemblages, much as the habitat of Kirtland's warbler is periodically burned to maintain pine woods (Leopold, 1978; see Botkin & Nisbet, 1991). On the other hand, if native species die out because of physiological intolerance to climate change, despite management efforts, then some invading species might actually be encouraged as ecological replacements of those that have disappeared.

In attempting to understand how climatically stressed communities may respond, and how they might be managed to prevent the gradual pauperization of their constituents, restoration studies, or, more properly, "community creation" experiments can help. Communities may be created outside their normal climatic ranges to mimic the effects of climate change. One such "out of place" community is the Leopold Pines, at the University of Wisconsin Arboretum in Madison, where there is periodically less rainfall than in the normal pine range several hundred kilometers to the north (Jordan, 1985, 1988). Researchers have found that although the pines themselves do fairly well once established at the Madison site, many of the other species that would normally occur in a pine forest, especially the small shrubs and herbs, such as *Trientalis borealis,* the northern star flower, have not flourished, despite several attempts to introduce them.

If management measures are unsuccessful, and old reserves do not retain necessary thermal or moisture characteristics, individuals of disappearing species might be transferred to new reserves. For example, warmth-intolerant ecotypes or subspecies might be transplanted to reserves nearer the poles. Other species may have to be periodically reintroduced in reserves that experience occasional climate extremes severe enough to cause extinction, but where the climate would ordinarily allow the species to survive with minimal management. Such transplantations and reintroductions, particularly involving complexes of species, will be difficult. Many plants, for example, have their flowering times determined by photoperiod, and in such species southern strains flower later in the year than northern ones (McMillan, 1959). A southern strain transplanted to the north therefore might flower too late in the season for successful reproduction. Despite such difficulties, applicable restoration technologies are being developed for many species (Botkin, 1977; Jordan et al., 1988; Lovejoy, 1985).

To the extent that we can still establish reserves, pertinent information about changing climate and subsequent ecological response should be used in deciding how to design and locate them to minimize the effects of changing temperature and moisture. One implication is that more reserves may be needed. The existence of multiple reserves for a given species or community type increases the probability that, if one reserve becomes unsuitable for climatic reasons, the organisms may still be represented in another reserve.

Reserves should be heterogeneous with respect to topography and soil types, so that even given climatic change, remnant populations may be able to survive in suitable microclimatic areas. Species may survive better in reserves with wide variations in altitude, since, from a climatic point of view, a small altitudinal shift corresponds to a large latitudinal one. Thus, to compensate for a 2°C rise in temperature, a northern hemisphere species can achieve almost the same result by increasing its altitude only some 500 m as it would by moving 300 km to the north (MacArthur, 1972).

Corridors between reserves, important for other conservation reasons, would allow some natural migration of species to track climate shifting. Corridors along altitudinal gradients are likely to be most practical because they can be relatively short compared with the longer distances necessary to accommodate latitudinal shifting.

As climatic models become more refined, pertinent information should be taken into consideration in making decisions about where to site reserves in order to minimize the effects of temperature and moisture changes. In the northern hemisphere, for example, where a northward shift in climatic zones is likely, it makes sense to locate reserves as near the northern limit of a species' or community's range as possible, rather than farther south, where conditions are likely to become unsuitable more rapidly.

Maximizing the size of reserves will increase long-term persistence of species by increasing the probability that suitable microclimates exist, by increasing the probability of altitudinal variation, and by increasing the latitudinal distance available to shifting populations.

Flexible zoning around reserves may allow us to actually move reserves in the future to track climatic optima, as, for example, by trading present range land for reserve land. The success of this strategy, however, would depend on a highly developed restoration technology, capable of guaranteeing, in effect, the portability of species and whole communities.

Preparing for the future

What concrete steps should be taken now by agencies and organizations responsible for the management of reserves and natural resources? How can they act to preserve species that may soon be dying out over large portions of their ranges?

From the conservation manager's point of view, the changes will present many difficult practical and philosophical questions. Should the manager strive to preserve all the species within a reserve, given that climate change is causing some to disappear? Should management be used to conserve examples of community types, given that, on the time-scale of climate change, communities are temporary assemblages of species likely to break up as the earth warms? How should the recent evolution of a "let nature take its course" management philosophy within, for example, the U.S. National Park Service, be reconciled with the increasingly intensive management that will be necessary to conserve many species in a warming world?

Not only is this problem complex, but management response will be difficult because, although the changes will be rapid from an ecological point of view, they will be slow in relation to management's traditionally short-term planning horizon. Thus, to ensure rapid response, continuity, and adequate resources, it will be necessary for high level authority within conservation organizations, management agencies, and funding bodies to give this issue high priority. The following are some areas that conservation management should emphasize.

Begin monitoring

One of the most important steps is to begin the collection of baseline data on how species and communities respond to climate. Within a reserve, for example, abundances, ranges, and reproductive success of important species can be measured and analyzed in terms of ongoing climate measurements. Baseline information is necessary to identify the beginning of warming effects, to distinguish short-term from long-term changes, to help identify susceptible species and communities, to identify the nature of potential changes, such as the direction of climate-driven plant succession, and to provide the basis for identifying the relationships between changes in climatic variables and resultant changes in the biota.

Changes might be expected to show up first at high latitudes, in low-lying marine coastal environments, and generally at ecotones between vegetation types determined by both temperature and precipitation. Changeover from one vegetation type to another might first be identified where disturbance events create succession. Monitoring for climate change may be done at different locations than other sorts of ongoing monitoring. For example, transects in Merritt Island National Wildlife Refuge in Florida for studying effects of burning on vegetation are typically laid out in the center of a plant community, while a climate transect should more likely be at the interface between two communities.

Some programs already exist that could be focused on climate effects. For example, the National Park Service has ongoing ecological and species monitoring programs at several parks including monitoring and research into the declines of Fraser fir (*Abies fraseri*) and red spruce (*Picea rubens*) of Great Smoky Mountains National Park. Similarly, the integrated studies of small watersheds initiated at several parks as part of the National Acid Precipitation Assessment Program will provide useful information for assessing long-term trends in ecosystem processes associated with climate change. Coordination of monitoring at a variety of sites, along with development of remote sensing and geographical information systems, can yield information on important trends at the landscape level. Such efforts can provide the scientific basis for the flexible region-wide planning that will be needed to develop effective management responses.

Undertake ecological research

Monitoring should be backed up by specific experiments on species and community responses to climate variables. One of the common themes in the book *Global Warming and Biological Diversity* (Peters & Lovejoy, 1991) is the lack of good basic knowledge about how species react to climate. To take a single example, Whitford (1991) stresses the paucity of information about climate effects on soil biota, yet these biota will play a key role not only in determining the nature of future soils, but also in the severity of the greenhouse effect itself through their role in the carbon cycle. From the point of view of a reserve manager, autecological studies can demonstrate which species have their ranges within a park determined by climate. Species of particular interest, such as endangered species, could receive special attention as to the effect of climate on, for example, food supply. Paleoecological studies and dendrochronology within protected areas or other important sites can shed additional light on past climate change and biotic response within reserves.

Identify sensitive communities, species, and populations

The results of monitoring, research, and analysis based on present information should allow identification of species or communities of special concern, including those that are nationally or globally rare or endangered. Climate-sensitive species could be targeted for additional monitoring, research, and the development of management techniques.

Develop contingency plans

Long-term plans for endangered species, for species likely to become rare, and for protected areas should have provisions for climate change. Even though precise local or regional climate projections are not available, contingency plans could be developed, particularly for sensitive biota. For example, contingency plans could be made based upon assumptions of local average warmings of 2, 4, 6, and 8°C, or upon assumptions of various rainfall increases and decreases. Given that in many areas increased temperature will add to water stress, it would be reasonable to make long-term plans for dealing with lower water availability. This might include plans for mitigating effects on sensitive species, or political or legal maneuvers to ensure that biological resources receive adequate water in the face of future competition from other users, such as agriculture and development.

Develop regional plans that involve nonreserve habitat

As the location and abundance of habitat and critical resources change with climate, management of specific wildlife species will need to incorporate populations and resources lying outside protected areas. Therefore, nature reserves and other management units will increasingly be forced to become partners in planning and management that transcends the scale of protected areas. There is precedence for this: management of endangered species is already done as outlined in multiagency, multiinstitution endangered species recovery plans. The current efforts to provide a basin-wide conservation plan for the grizzly bear in the Greater Yellowstone Area might provide a model for such regional planning.

Develop management techniques

The increased disturbance likely to result from climate change demands a large increase in resources for the development and implementation of new management techniques, particularly those of restoration ecology. Despite the efforts previously mentioned (Botkin, 1977; Lovejoy, 1985), restoration and transplantation techniques are poorly developed and require extensive research (see Jordan et al., 1988).

Develop philosophical approaches to management

Conservationists need to begin the process of deciding philosophical questions that will affect management. Given the likelihood of community breakups, should efforts be expended on maintaining existing community types? As conditions become unsuitable for species existing within reserves today, should herculean efforts be expended to maintain them? Should reserves be used as transplantation sites for southern species in need of new habitat? At what point should efforts to maintain a particular species within a reserve be stopped and resources used elsewhere? How can reserves become integrated components of regional, national, and global strategies for conservation of species? Given stresses on the natural world, will the role of parks as refuges increase relative to their recreation role? At the moment, it is easier to ask these questions than to answer them.

Dedicate additional reserve lands

Global warming is a strong argument for the enlargement or creation of additional parks and other reserved lands. As mentioned above, multiple refuges provide additional chances that some protected habitat will remain suitable for a particular species as climate changes. Moreover, as parks become unable to provide adequate habitat and other resources for the species within, given climate change, enlargement may be necessary, as was done when Redwoods National Park was expanded during the 1970s to prevent external logging from threatening the parks' ecosystems.

Summary

In the geologic past, natural climate changes have caused large-scale geographic shifts in species' ranges, changes in the species composition of biological communities, and species extinctions. If the widely predicted greenhouse effect occurs, natural ecosystems will respond in ways similar to how they did in the past, but the effects will be more severe because of the very rapid rate of the projected change. Moreover, population reduction and habitat destruction due to human activities will prevent many species from colonizing new habitat when their old becomes unsuitable. The synergy between climate change and habitat destruction would threaten many more species than either factor alone.

These effects would be pronounced in temperate and arctic regions, where temperature increases are projected to be relatively large. It is unclear how affected the tropical biota would be by the relatively small temperature increases projected for the lower latitudes, because relatively little is known about the physiological tolerances of tropical species, but substantial disruption may occur due to precipitation changes. Throughout the world, geographically restricted species might face extinction while widespread species are likely to survive in some parts of their range. In the northern mid and high latitudes, new northward habitat will become suitable even as die-offs occur to the south. However, it may be difficult for many species to take advantage of this new habitat because dispersal rates for many species are slow relative to the rate of warming, and therefore ranges of even many widespread species are likely to show a net decrease during the next century. Range retractions will be proximally caused by temperature and precipitation changes, increases in fires, changes in the ranges and severity of pests and pathogens, changes in competitive interactions, and additional effects of nonclimatic stresses like acid rain and low-level ozone.

The best solutions to the ecological upheaval resulting from climatic change are not yet clear. In fact, little attention has been paid to the problem. What is clear, however, is that these climatological changes would have tremendous impact on communities and populations isolated by development and by the middle of the next century may dwarf any other consideration in planning for reserve management. The problem may seem overwhelming. One thing, however, is worth keeping in mind: If populations are fragmented and small, they are more vulnerable to the new stresses brought about by climate change. Thus, one of the best things that can be done in the short-term is to minimize further encroachment of development upon existing natural ecosystems. Further, we must refine climatological predictions and increase understanding of how climate affects species, both individually and in their interactions with each other. Such studies may allow us to identify those areas where communities will be most stressed, as well as alternate areas where they might best be saved. Meanwhile, efforts to improve techniques for managing communities and ecosystems under stress, and also for restoring them when necessary, must be carried forward energetically.

Acknowledgments

This chapter is being contemporaneously published in *Global Warming and Biological Diversity,* edited by Robert Peters and Thomas E. Lovejoy, and was previously published in somewhat different form in *Global Climate Change and Life on Earth,* edited by Richard Wyman, published by Chapman and Hall. Much of the text on global warming was also previously published as an article in *Forest Ecology and Management,* in a special 1989 volume containing the proceedings of the symposium on Conservation of Diversity in Forest Ecosystems, University of California at Davis, July 25, 1988. It draws heavily on other previously published versions, including those in *Endangered Species Update,* **5**(7), 1–8 and *Preparing for Climate Change: Proceedings of the First North American Conference on Preparing for Climate Change: A Cooperative Approach,* ed. J. C. Topping. Many of the ideas and all the figures derive from a paper by Peters and Darling, published in *Bioscience,* December, 1985. Please see Peters and Darling (1985) for a complete list of acknowledgments for help with this work.

References

Aho, J. M., Whitfield Gibbons, J., & Esch, G. W. (1976). Relationship between thermal loading and parasitism in the mosquitofish. In *Thermal Ecology II,* ed. G. W. Esch & R. W. McFarlane, pp. 213–218. Springfield, VA: Technical Information Center, Energy Research and Development Administration.

Alexander, V. (1991). Arctic marine ecosystems. *Global Warming and Biological Diversity,* ed. R. L. Peters & T. E. Lovejoy, pp. 221–232. New Haven: Yale University Press.

Baker, R. G. (1983). Holocene vegetational history of the western United States. In *Late-Quaternary Environments of the United States.* Vol. 2, ed. H. E. Wright, pp. 109–125. The Holocene, Minneapolis: University of Minnesota Press.

Barron, E. J. (1985). Explanations of the Tertiary global cooling trend. *Palaeogeography, Palaeoclimatology, and Palaeoecolog,* **50,** 17–40.

Beauchamp, R. S. A., & Ullyott, P. (1932). Competitive relationships between certain species of fresh-water triclads. *J. Ecol.* **20,** 200–208.

Bernabo, J. C., & Webb III, T. (1977). Changing patterns in the Holocene pollen record of northeastern North America: a mapped summary. *Quat. Res.* **8,** 64–96.

Billings, W. D., & Peterson, K. M. (1991). Some possible effects of climatic warming on arctic tundra ecosystems of the Alaskan North Slope. In *Global Warming and Biological Diversity,* ed. R. L. Peters & T. E. Lovejoy. New Haven: Yale University Press.

Botkin, D. B. (1977). Strategies for the reintroduction of species into damaged ecosystems. In *Recovery and Restoration of Damaged Ecosystems,* ed. J. Cairns, Jr., K. L. Dickson, and E. E. Herricks, pp. 241–260. Charlottesville: University Press of Virginia.

Botkin, D. B., & Nisbet, R. A. (1991). Projecting the effects of climate change on biological diversity in forests. In *Global Warming and Biological Diversity,* ed.

R. L. Peters & T. E. Lovejoy, pp. 277–296. New Haven: Yale University Press.

Brisbin, I. L. (1974). Abundance and diversity of waterfowl inhabiting heated and unheated portions of a reactor cooling reservoir. In *Thermal Ecology,* J. W. Gibbons & R. R. Sharitz, pp. 579–593. CONF-730505 U.S. Atomic Energy Commission.

Butzer, K. W. (1980). Adaptation to global environmental change. *Prof. Geogr.,* **32**(3), 269–278.

Caufield, C. (1985). *In the Rainforest.* New York: Knopf.

Coope, G. R. (1977). Fossil coleopteran assemblages as sensitive indicators of climatic changes during the Devensian (Last) cold stage. *Philos. Trans. R. Soc. Lond., B.,* **280,** 313–340.

Courtenay, W. R. Jr. (1978). The introduction of exotic organisms. In *Wildlife and America,* ed. H. P. Brokaw, pp. 237–252. Washington, D.C.: Council on Environmental Quality, U.S. Government Printing Office.

Cox, B. C., & Moore, P. D. (1985). *Biogeography: An Ecological And Evolutionary Approach.* Oxford: Blackwell.

Critchfield, W. B. (1980). Origins of the eastern deciduous forest. In Proceedings, *Dendrology in the Eastern Deciduous Forest Biome,* September 11–13, 1979, pp. 1–14. Virginia Polytech. Inst. and State Univ. School of Forestry and Wildlife Resources. Publ. FWS-2-80.

Daubenmire, R. F. (1962). *Plants and Environment: A Textbook of Plant Autecology.* New York: Wiley.

Davis, M. B. (1983). Holocene vegetational history of the eastern United States. In *Late-Quaternary Environments of the United States,* Volume 2, *The Holocene,* ed. H. E. Wright, Jr., pp. 166–181. Minneapolis: University of Minnesota Press.

Davis, M. B., & Zabinski, C. (1991). Changes in geographical range resulting from greenhouse warming effects on biodiversity in forests. In *Global Warming and Biological Diversity,* ed. R. L. Peters & T. E. Lovejoy. New Haven: Yale University Press.

Diamond, J. M. (1975). The island dilemma: lessons of modern biogeographic studies for the design of natural preserves. *Biol. Conser.,* **7,** 129–146.

Dorf, E. (1976). Climatic changes of the past and present. In *Paleobiogeography: Benchmark Papers in Geology 31,* ed. C. A. Ross, pp. 384–412. Stroudsburg, PA: Dowden, Hutchinson, and Ross.

Duffy, D. C. (1983). Environmental uncertainty and commercial fishing: effects on Peruvianguano birds. *Biol. Conserv.,* **26,** 227–238.

Edgell, M. C. R. (1984). Trans-hemispheric movements of Holarctic Anatidae: the Eurasian wigeon (*Anas penelope L.*) in North America. *J. Biogeogr.,* **11,** 27–39.

Edlund, S. A. (1987). Effects of climate change on diversity of vegetation in arctic Canada. In *Preparing for Climate Change: Proceedings of the First North American Conference on Preparing for Climate Change: A Cooperative Approach,* ed. J. C. Topping, pp. 186–193. Washington, D.C.: Government Institutes.

Ehrlich, P. R., & Mooney, H. A. (1983). Extinction, substitution, and ecosystem services. *Bioscience,* **33**(4), 248–254.

Emanuel, W. R., Shugart, H. H., & Stevenson, M. P. (1985). Response to comment: "Climatic change and the broad-scale distribution of terrestrial ecosystem complexes." *Clim. Change,* **7,** 457–460.

Environmental Protection Agency (EPA). (1988). *The Potential Effects of Global*

Climate Change on the United States: Draft Report to Congress, Vol. I. Washington, D.C.: Environmental Protection Agency.

Erwin, T. L. (1982). Tropical forests: their richness in Coleoptera and other arthropod species. *Coleopterists Bull.,* **36,** 74–75.

Erwin, T. L. (1983). Beetles and other insects of tropical forest canopies at Manaus, Brazil, sampled by insecticidal fogging. In *Tropical Rain Forest Ecology and Management,* ed. T. C. Whitmore & A. C. Chadwick, pp. 59–75. Oxford: Blackwell.

Flenley, J. R. (1979). *The Equatorial Rain Forest.* London: Butterworths.

Flohn, H. (1979). Can climate history repeat itself? Possible climatic warming and the case of paleoclimatic warm phases. In *Man's Impact on Climate,* ed. W. Bach, J. Pankrath, & W. W. Kellogg, pp. 15–28. Amsterdam: Elsevier Scientific Publishing.

Ford, M. J. (1982). *The Changing Climate.* London: George Allen and Unwin.

Franklin, J. F., Swanson, F. J., Harmon, M. E., Perry, D. A., Spies, T. A., Dale, V. H., McKee, A., Ferrell, W. K., Means, J. E., Gregory, S. V., Lattin, J. D., Schowalter, T. D., & Larsen, D. (1991). Effects of global climatic change on forests in northwestern North America. In *Global Warming and Biological Diversity,* ed. R. L. Peters & T. E. Lovejoy, pp. 79–90. New Haven: Yale University Press.

Frye, R. (1983). Climatic change and fisheries management. *Nat. Resources J.,* **23,** 77–96.

Furley, P. A., Newey, W. W., Kirby, R. P., & Hotson, J. McG. (1983). Geography of the Biosphere. London: Butterworths.

Glynn, P. (1984). Widespread coral mortality and the 1982–83 El Niño warming event. *Env. Conserv.,* **11**(2), 133–146.

Graham, R. W. (1986). Plant-animal interactions and Pleistocene Extinctions. In *Dynamics of Extinction,* ed. D. K. Elliott, pp. 131–154. New York: Wiley.

Graham, R. W. (1991). Using late Pleistocene faunal changes as a guide to understanding effects of greenhouse warming on the mammalian fauna of North America. In *Global Warming and Biological Diversity,* ed. R. L. Peters & T. E. Lovejoy, pp. 76–90. New Haven: Yale University Press.

Hamburg, S. P., & Cogbill, C. V. (1988). Historical decline of red spruce populations and climatic warming. *Nature,* **331,** 428–431.

Hansen, J., Johnson, D., Lacis, A., Lebedeff, S., Lee, P., Rind, D., & Russell, G. (1981). Climate impact of increasing atmospheric carbon dioxide. *Science,* **213,** 957–966.

Hansen, J., Lacis, A., Rind, D., Russell, G., Fung, I., & Lebedeff, S. (1987). Evidence for future warming: how large and when. In *The Greenhouse Effect, Climate Change, and U.S. Forests,* ed. W. E. Shands & J. S. Hoffman. Washington, D.C.: Conservation Foundation.

Hansen, J., Fung, I., Lacis, A., Lebedeff, S., Rind, D., Ruedy, R., & Russell, G. (1988a). Prediction of near-term climate evolution: what can we tell decision-makers now? In *Preparing for Climate Change: Proceedings of the First North American Conference on Preparing for Climate Change: A Cooperative Approach.* Washington, D.C.: Government Institutes.

Hansen, J., Fung, I., Lacis, A., Rind, D., Lebedeff, S., Ruedy, R., & Russell, G. (1988b). Global climate changes as forecast by Goddard Institute for Space Studies three-dimensional model. *J. Geophysical Res.,* **98**(D8), 9341–9364.

Harris, L. D., & Cropper, W. P. Jr. (1991). Between the Devil and the deep blue sea: implications of climate change for wildlife in Florida. In *Global Warming and Biological Diversity,* ed. R. L. Peters & T. E. Lovejoy, pp. 309–324. New Haven: Yale University Press.

Harte, J., Torn, M., & Jensen, D. (1991). The nature and consequences of indirect linkages between climate change and biological diversity. In *Global Warming and Biological Diversity,* ed. R. L. Peters & T. E. Lovejoy, pp. 325–343. New Haven: Yale University Press.

Hartshorn, G. S. (1991). Possible effects of global warming on the biological diversity in tropical forests. In *Global Warming and Biological Diversity,* ed. R. L. Peters & T. E. Lovejoy, pp. 137–146. New Haven: Yale University Press.

Heusser, C. J. (1974). Vegetation and climate of the southern Chilean lake district during and since the last interglaciation. *Quat. Res.,* 290–315.

Hoffman, J. S., Keyes, D., & Titus, J. G. (1983). *Projecting Future Sea Level Rise.* Washington, D.C.: U.S. Environmental Protection Agency.

Holt, R. D. (1990). The microevolutionary consequences of climate change. *Trends in Ecol. Envol.,* **5**(9), 311–315.

Houghton, J. T., Jenkins, G. J., & Ephraums, J. J. (eds.) (1990). *Climate Change: The IPCC Scientific Assessment.* New York: Cambridge Univ. Press.

Jordan III, W. R. (1985). Personal communication, University of Wisconsin, Madison.

Jordan III, W. R. (1988). Ecological restoration: Reflections on a half-century of experience at the University of Wisconsin–Madison Arboretum. In *Biodiversity,* ed. E. O. Wilson, pp. 311–316. Washington, D.C.: National Academy Press.

Jordan III, W. R., Peters, R. L., & Allen, E. B. (1988). Ecological restoration as a strategy for conserving biological diversity. *Env. Management,* **12**(1), 55–72.

Kellison, R. C., & Weir, R. J. (1987). Selection and breeding strategies in tree improvement programs for elevated atmospheric carbon dioxide levels. In *The Greenhouse Effect, Climate Change, and U.S. Forests,* ed. W. E. Shands & J. S. Hoffman. Washington, D.C.: Conservation Foundation.

Kellogg, W. W., & Schware, R. (1981). *Climate Change and Society: Consequences of Increasing Atmospheric Carbon Dioxide.* Boulder, CO: Westview Press.

Knopf, F. L., & Sedgwick, J. A. (1987). Latent population responses of summer birds to a catastrophic, climatological event. *The Condor,* **89,** 869–873.

Kolehmainen, T. M., & Castro, R. (1974). Mangrove-root communities in a thermally altered area in Guayanilla Bay, Puerto Rico. In *Thermal Ecology,* CONF-730505, ed., J. W. Gibbons & R. R. Sharitz, pp. 371–390. Washington, D.C.: U.S. Atomic Energy Commission.

Kullman, L. (1983). Past and present tree lines of different species in the Handolan Valley, Central Sweden. In *Tree Line Ecology,* ed. P. Morisset & S. Payette, pp. 25–42. Quebec: Centre d'etudes nordiques de l'Universite Laval.

Lanly, J. (1982). *Tropical Forest Resources.* FAO Forestry Paper Number 30. Rome: Food and Agriculture Organization of the United Nations.

Lashof, D. A. (1989). The dynamic greenhouse: feedback processes that may influence future concentrations of atmospheric trace gases and climatic change. *Climatic Change,* **14,** 213–242.

Leatherman, S. (1991, In prep.). Coastal land loss in the Chesapeake Bay region: an historical analog approach to global changes analysis and response.

Leopold, A. (1953). *Round River.* New York: Oxford Univ. Press.

Leopold, A. S. (1978). Wildlife and forest practice. In *Wildlife and America,* ed. H. P. Brokaw, pp. 108–120. Washington, D.C.: Council on Environmental Quality, U.S. Government Printing Office.

Lovejoy, T. E. (1980). A projection of species extinctions. In *The Global 2000 Re-*

port to the President: Entering the Twenty-First Century, pp. 328–331. Council on Environmental Quality and the Department of State. Washington, D.C.: U.S. Government Printing Office.

Lovejoy, T. E. (1985). *Rehabilitation of degraded tropical rainforest lands.* Commission on Ecology Occ. Pap. 5. Gland, Switzerland: International Union for the Conservation of Nature and Natural Resources.

MacArthur, R. H. (1972). *Geographical Ecology.* New York: Harper & Row.

Manabe, S., Wetherald, R. T., & Stouffer, R. J. (1981). Summer dryness due to an increase of atmospheric CO_2 concentration. *Clim. Change,* **3,** 347–386.

Marshall, J. K. (1978). Factors limiting the survival of *Corynephorus canescens* (L) Beauv. in Great Britain at the northern edge of its distribution. *Oikos,* **19,** 206–216.

McMillan, C. (1959). The role of ecotypic variation in the distribution of the central grassland of North America. *Ecol. Monog.,* **29,** 285–305.

Mikolajewicz, U., Santer, B., & Maier-Reimer, E. (1990). *Ocean response to greenhouse warming.* Max-Planck-Institut für Meterologie, Report 49.

Miller, W. F., Dougherty, P. M., & Switzer, G. L. (1987). Rising CO_2 and changing climate: major southern forest management implications. *The Greenhouse Effect, Climate Change, and U.S. Forests.* Washington, D.C.: Conservation Foundation.

Muller, H. (1979). Climatic changes during the last three interglacials. In *Man's Impact on Climate,* ed. W. Bach, J. Pankrath, & W. W. Kellogg, pp. 29–41. Amsterdam: Elsevier.

Murphy, D. D., & Weiss, S. B. (1991). Predicting effects of climate change on biological diversity in western North America: species losses and mechanisms. In *Global Warming and Biological Diversity,* ed. R. L. Peters & T. E. Lovejoy, pp. 355–368. New Haven: Yale University Press.

Myers, J. P., & Lester, R. T. (1991). Double jeopardy for migrating animals: multiple hits and resource asynchrony. In *Global Warming and Biological Diversity,* ed. R. L. Peters & T. E. Lovejoy, pp. 193–200. New Haven: Yale University Press.

Myers, R. L. (1983). Site susceptibility to invasion by the exotic tree *Melaleuca quinquenervia* in southern Florida. *J. Appl. Ecol.,* **20**(2), 645–658.

National Academy of Sciences (NAS). (1987). *Current Issues in Atmospheric Change.* Washington, D.C.: National Academy Press.

National Research Council (NRC). (1983). *Changing Climate.* Washington, D.C.: National Academy Press.

Ono, R. D., Williams, J. D., & Wagner, A. (1983). *Vanishing Fishes of North America.* Washington, D.C.: Stone Wall Press.

Perring, F. H. (1965). The advance and retreat of the British flora. In *The Biological Significance of Climatic Changes in Britain,* ed. C. J. Johnson & L. P. Smith, pp. 51–59. London: Academic Press.

Peters, R. L., & Darling, J. D. (1985). The greenhouse effect and nature reserves. *BioSci.,* **35**(11), 707–717.

Peters, R. L., & Lovejoy, T. E. (eds.) (1991). *Global Warming and Biological Diversity.* New Haven: Yale University Press.

Picton, H. D. (1984). Climate and the prediction of reproduction of three ungulate species. *J. Appl. Ecol.,* **21,** 869–879.

Rand, A. S. (1964). Inverse relationship between temperature and shyness in the lizard *Anolislineatopus. Ecology,* **45,** 863–864.

Randall, M. G. M. (1982). The dynamics of an insect population throughout its altitudinal distribution: *Coleophora alticolella* (Lepidoptera) in northern England. *J. Anim. Ecol.*, **51,** 993–1016.

Rapoport, E. H. (1982). *Areography: Geographical Strategies of Species.* New York: Pergamon Press.

Ray, G. C., Hayden, B. P., Bulger, Jr., A. J., & McCormick-Ray, M. G. (1991). Effects of global warming on the biodiversity of coastal marine zones. In *Global Warming and Biological Diversity,* ed. R. L. Peters & T. E. Lovejoy, pp. 91–104. New Haven: Yale University Press.

Rind, D. (1989). A character sketch of greenhouse. *EPA Journal,* **15**(1), 735–737.

Roberts, L. (1989). How fast can trees migrate? *Science,* **243,** 735–737.

Rubenstein, D. I. (1991). The greenhouse effect and changes in animal behavior: effects on social structure and life history strategies. In *Global Warming and Biological Diversity,* ed. R. L. Peters & T. E. Lovejoy, pp. 180–192. New Haven: Yale University Press.

Schneider, S. H., & Londer, R. (1984). *The Coevolution of Climate and Life.* San Francisco: Sierra Club Books.

Schneider, S. H. (1988). The greenhouse effect: What we can or should do about it. In *Preparing for Climate Change. Proceedings of the First North American Conference on Preparing for Climate Change: A Cooperative Approach,* pp. 19–34. Washington, D.C.: Government Institutes.

Schneider, S. H., Mearns, L., & Gleick, P. H. (1991). Climate-change scenarios for impact assessment. In *Global Warming and Biological Diversity,* ed. R. L. Peters & T. E. Lovejoy. New Haven: Yale University Press.

Seddon, B. (1971). *Introduction to Biogeography.* New York: Barnes and Noble.

Sedjo, R. A. (1989). Forests: a tool to moderate global warming? *Environment,* **31**(1), 14–20.

Schlesinger, M. E. (1989). Model projections of the climatic changes induced by increased atmospheric CO_2. In *Climate and Geosciences,* ed. A. Berger, S. H. Schneider & J. C. Duplessy, pp. 375–416. Dordrecht: Kluwer.

Simons, M. (1988). Vast Amazon fires, man-made, linked to global warming. *New York Times,* August 11, 1988.

Soulé, M. E. (1985). What is conservation biology? *Bioscience,* **35**(11), 727–734.

Strain, B. R., & Bazzaz, F. A. (1983). Terrestrial plant communities. In *CO_2* and Plants, ed. E. R. Lemon, pp. 177–222. Boulder, CO: Westview Press.

Titus, J. G., Henderson, T. R., & Teal, J. M. (1984). Sea level rise and wetlands loss in the United States. *National Wetlands Newsletter,* **6**(5), 3–6.

Tracy, C. R. (1991). Ecological responses of animals to climate. In *Global Warming and Biological Diversity,* ed. R. L. Peters & T. E. Lovejoy, pp. 171–179. New Haven: Yale University Press.

Tralau, H. (1973). Some quaternary plants. In *Atlas of Palaeobiogeography,* ed. A. Hallam, pp. 499–503. Amsterdam: Elsevier.

Van Devender, T. R., & Spaulding, W. G. (1979). Development of vegetation and climate in the southwestern United States. *Science,* **204,** 701–710.

Webb III, T. (1991). Past changes in vegetation and climate: lessons for the future. In *Global Warming and Biological Diversity,* ed. R. L. Peters & T. E. Lovejoy, pp. 59–75. New Haven: Yale University Press.

Whitford, W. G. (1991). Effects of climate change on soil biotic communities and soil processes. In *Global Warming and Biological Diversity,* ed. R. L. Peters & T. E. Lovejoy, pp. 124–136. New Haven: Yale University Press.

Winget, C. H. (1988). Forest management strategies to address climate change. In *Preparing for Climate Change: Proceedings of the First North American Con-*

ference on Preparing for Climate Change: A Cooperative Approach, pp. 328–333. Rockville, MD: Government Institutes.

Woodward, F. I. (1991). A review of the effects of climate on vegetation: ranges, competition and composition. In *Global Warming and Biological Diversity*, ed. R. L. Peters & T. E. Lovejoy, pp. 105–123. New Haven: Yale University Press.

Woodwell, G. M. (1991). How does the world work? In *Global Warming and Biological Diversity*, ed. R. L. Peters & T. E. Lovejoy, pp. 31–37. New Haven: Yale University Press.

World Meteorological Organization (WMO) (1982). *Report of the JSC/CAS: A Meeting of experts on Detection of Possible Climate Change (Moscow, October 1982)*. Geneva, Switzerland: Rep. WCP29.

Wright, H. E., Jr. (1971). Late Quaternary vegetational history of North America. In *The Late Cenozoic Glacial Ages*, ed. K. K. Turekian, pp. 425–464. New Haven: Yale University Press.

7

We do not want to become extinct: the question of human survival

NORMAN MYERS

Introduction

The situation facing humankind involves the collapse of any balance between our species and the rest of life on the planet. Paradoxically, at the time when we stand at the threshold of degeneration of the ecosystem and degradation of human quality of life, knowledge, science and technology are now in a position to provide both the human creativity and the technology needed to take remedial action and rediscover harmony between nature and humankind. Only the social and political will is lacking.

(UNESCO, 1989)

Among all the talk of a mass extinction of species underway, there is occasional mention of the thought that humankind itself could eventually become endangered if not extinct as a result of its reckless depletion of its habitats and life-support systems. While this prospect is extremely unlikely (though not impossible), there is every chance that if we continue to deplete the carrying capacity of our planet, civilized society as we know it will become destabilized and impoverished to a degree undreamed of by today's political leaders.

What would be the principal factors involved? What would be the main mechanisms at work? How far have the depletive processes advanced already? How readily can they be slowed and halted, even reversed? How far is the general public and its political leaders aware of the overall threat? These are some of the questions to be addressed in this chapter. Momentous as they are, they can receive no more than summary treatment in an assessment of this short scope. The aim is to identify and define the key issues, briefly expound their character and purview, and appraise some scenario outcomes.

The doomsday outlook has a long-standing pedigree. First advanced by Fairfield Osborne in his 1948 book *Our Plundered Planet,* it has been reiterated by Paul Ehrlich in *The Population Bomb* (1968), Dennis Meadows and his colleagues' *The Limits to Growth* (1972), Peter Raven's "We're Killing Our World: The Global Ecosystem in Crisis" (1987), Paul and Anne Ehrlich's *The*

Population Explosion (1990), and this writer's *The Gaia Atlas of Planet Management* (1985), among dozens of other books of similarly apocalyptic spirit.

While certain of these writings postulate a profound decline in the civilized perquisites of contemporary society, none of them envisages the ultimate debacle of humankind's extinction. Yet just such an outcome could now be on the environmental cards. Suppose global warming were to proceed until rapidly rising temperatures cause the die-off of many temperate-zone and boreal forests (Woodwell, 1986). The oxidizing biomass would release large amounts of carbon, thus aggravating the atmospheric buildup of the chief greenhouse gas, carbon dioxide. At the same time, there could be a number of other positive feedbacks to accelerate global warming; the IPCC Report (Houghton et al., 1990) listed 18 possible feedbacks with reinforcing capacity. The planetary ecosystem could soon start to experience the full rigors of a runaway greenhouse effect, whereby one compounded impact serves to multiplicatively amplify several other impacts. The process could quickly gather enough momentum to be beyond our capacity to slow it down except through efforts of altogether unprecedented scope, if then. All too readily the global climate could heat up until it reached levels beyond the tolerance of human life (Brown et al., 1990).

So the final catastrophe, from the standpoint of humankind at least, is no longer confined to the realm of science fiction. It has become entirely possible in the real world. We could reach a point of no return by the middle of next century, if not well before. It is 500,000 years since the emergence of *Homo sapiens* as a distinct species, and its future could be set at terminal risk within just 50 years, or one ten-thousandth as little time.

True, humankind's early extinction is a remote prospect. We shall surely learn to anticipate the final catastrophe, then move to prevent it. We must reckon that the ultimate downside outcome is entirely improbable, but entirely probable is the acute decline if not the collapse of civilization as we know it, due to the gross-scale degradation of our environmental-resource base. It is this dimension to our human predicament that will occupy the greater part of this chapter.

Gross-scale environmental degradation

Our biosphere is undergoing a series of environmental perturbations whose impacts are collectively exceeding all but the greatest geologic upheavals of the prehistoric past. The predominant driving force consists of two factors: growth in human numbers and growth in human consumerism.

Since the start of this century, human numbers have increased three-fold. At the same time, growth in human consumption of fossil-fuel energy has increased 12-fold, and growth in the global economy has increased 29-fold (Brown et al., 1989; World Resources Institute, 1989). The most rapid growth has occurred in just the past few decades. Of this century's economic growth, four-fifths has taken place since 1950; and the expansion of the global economy in 1988 and 1989, $600 billion, was as large as was the entire economy in the year 1900.

As a result of these various forms of growth in human enterprise, along with inappropriate technologies and deficient economic policies, the biosphere now suffers from a lengthening list of environmental assaults: widespread pollution (such as acid rain), soil erosion, tropical deforestation, spread of deserts, decline of water supplies, depletion of the ozone layer, and the greenhouse effect, to name but the main forms of environmental degradation and biospheric impoverishment.

To comprehend the full scope of our over-burdening of the biosphere, consider the capacity of humankind to feed its present population of 5.3 billion people. According to the World Hunger Project (Chen et al., 1990), the planetary ecosystem could, with present agrotechnologies and with equal distribution of food supplies, satisfactorily support 5.5 billion people if they all lived off a vegetarian diet. If they derived 15 percent of their calories from animal products, as do many people in South America, the total supportable population would decline to 3.7 billion. If they derived 25 percent of their calories from animal protein, as is the case with most people in North America, the Earth could support only 2.8 billion people.

True, these calculations reflect no more than today's food-production technologies. Certain observers (e.g., Simon, 1990) protest that such an analysis underestimates the scope for technological expertise to keep on expanding the Earth's carrying capacity. We can surely hope that many advances in agrotechnologies are still available to come on stream. But consider the population/food record over the past four decades. From 1950 to 1984, thanks largely to remarkable advances in Green Revolution agriculture, there was a 2.6-fold increase in world grain output. This achievement, representing an average increase of almost 3% per year, raised per-capita production by more than one third. But from 1985 to 1989 there was next to no increase, even though the period saw the world's farmers investing billions of dollars to increase output (fertilizer use alone expanded by 14%), supported by the incentive of rising grain prices and by the restoration to production of idled U.S. cropland. Crop yields had "plateaued"; it appeared that plant breeders and agronomists had exhausted the scope for technological innovation. So the 1989 harvest was hardly any higher than that of 1984. Meantime there were an extra 440 million people to feed. While world population increased by almost 8.5%, grain output per person declined by nearly 7% (Brown et al., 1990; U.S. Department of Agriculture, 1990).

Hence the present throngs of humankind press ever harder upon the planetary ecosystem. To summarize the situation through an alternative analysis, consider that humankind now co-opts some 40% of all net primary productivity, leaving only 60% for all the other 30 million species (Vitousek et al., 1986) – with all that implies for their life-support systems and hence for their ultimate survival.

Yet this is only the start of biospheric disruptions. The main population explosion is still to come. By the late 1990s humankind's numbers will expand by almost 100 million each year, in contrast to fewer than 90 million per year during the 1980s. Within the next 35–45 years human numbers are projected to double. Even more to the point, we must anticipate that consumption of food and fiber will triple, demand for energy will quadruple, and economic activity will quintuple (World Commission on Environment and Development, 1987). So the problem lies not only with sheer growth in human numbers. Remember that an average American utilizes 20 times as much energy as an average Bangladeshi, Ethiopian, or Bolivian, and a similarly disproportionate amount of other key raw materials.

Moreover it is pertinent to note, in a book dealing with biodiversity, that it is the activities of a single species that are precipitating the extinction of between 50 and 100 species per day right now, and could possibly trigger the demise of many millions of species within the foreseeable future.

Population growth and its environmental impacts

The crucial factor in environmental degradation is surely the sheer increase in human numbers. True, numbers in themselves are of scant significance. It is how people support themselves that counts. So the impact of population growth is expressed through the living levels (degree of affluence) enjoyed by human communities and the kinds of technologies they deploy to achieve their lifestyles. We can set out the relationships in the form of an equation

$$I = PAT$$

where P is population itself, I is environmental impact, A is per-capita consumption (determined by income and lifestyle), and T is environmentally harmful technology that supplies A (Ehrlich & Ehrlich, 1990; see also Davis et al., 1989; Myers, 1991; Pimentel et al., 1989; Shaw, 1989). The three factors, P, A, and T interact in multiplicative fashion, i.e., they compound each other's impacts. So whatever the size of *A* and *T*, the role of *P* is bound to be significant even when a population, and its growth rate too, are relatively small. For any type of technology, for any given level of consumption or waste, for any given level of poverty or inequality, we find that the more people there are, the greater is the overall impact on the environment.

This basic equation demonstrates why developing nations, with large populations but limited economic advancement, can generate a vast impact on the environment (hence on prospects for sustainable development), if only because the *P* multiplier on the *A* and *T* factors is so large. The equation also makes clear that developed nations also generate population impacts insofar as the *A* and *T* multipliers for each person are exceptionally large.

At the same time, a number of other factors are at work in addition to the three elements of the equation. They include socioeconomic inequities, cultural constraints, government policies, and the international economic order. Moreover these additional factors vary greatly throughout the global community of almost 200 nations, disparate as they are in agroclimatic zones, natural resource endowments, historical traditions and the like. But sooner or later all these additional factors operate through one or another of the equation's three variables.

To illustrate how the equation's interactions work, suppose that, by dint of exceptional effort, humankind managed to reduce the average per-capita consumption of environmental resources (*A* in the equation) by 5%, and to improve its technologies (*T*) so that they caused 5% less environmental injury, on average. This would reduce the total impact (*I*) of humanity by roughly 10%. But unless global population growth (*P*) were restrained at the same time, it would bring the total impact back to the previous level within less than 6 years (Ehrlich & Ehrlich, 1990).

Let us now use a variant of the basic equation above to examine the role of population growth on the part of a particular sector of humankind, the one billion people who live in absolute poverty. We can represent the linkages involved by relating environmental impact (*I*) to two further factors, population and poverty, in the form of another equation

$$I = \text{Pop}/\text{Pov}/E$$

where *I* is the environmental impact (confined to the developing world this time, by contrast with the global-scale impact in the first equation), Pop stands for the population factor, Pov stands for poverty, and *E* stands for the environmental resources available to support the impoverished multitudes. The interactions are again multiplicative, each one reinforcing the others' impacts.

At the same time, the communities in question tend to feature the highest population growth rates. Of the poorest one-fifth of developing-world households, 55–80% have eight or more members, whereas at national level the proportion is only 15–30% (Lipton, 1985). Moreover these people are unusually dependent for their survival upon the environmental-resource base of soil, water, forests, fisheries, and biotas that make up their main stocks of economic capital. In addition, they see scant alternative to exploiting their environmental-resource base at a rate they recognize is unsustainable: they experience unusually short "time preference" rates, meaning they feel obliged to misuse and overuse their resource stocks today even at cost to their prospects tomorrow. They thereby undercut their principal means of livelihood, thus entrenching their poverty. In turn, this appears to reinforce their motivation to have large families (Keyfitz, 1990). As a result, they face the prospect of ever-tightening constraints.

Note that these people's plight also reflects the failure of development in general. They have been bypassed by the usual forms of development, notably Green Revolution agriculture: they cannot afford costly inputs such as high-yielding seeds, fertilizer, irrigation, and farm machinery. So these people become "marginalized." They are marginalized, too, in that they generally lack economic, political, legal, or social status, meaning they can do little to remedy their plight. All too often this drives them to seek their livelihood in environments that are unsuitable for sustainable agriculture, being too wet, too dry, or too steep. Hence there is the phenomenon of the impoverished peasant who causes deforestation, desertification, and soil erosion on a wide scale: a case of the marginal person in marginal environments. In developing countries as a whole, these "bottom billion" people may often impose greater environmental injury than the other three billion of their fellow citizens.

In short, far from enjoying the development benefits that would ostensibly push them through a demographic transition to smaller families, these people are caught in a demographic trap. Given their severely constrained circumstances, population growth denies them the very inducements that could serve to reduce population growth (Keyfitz, 1990).

We should note a further factor, that of socioeconomic infrastructure. This refers to the capacity of governments to cope with the processes involved (population growth, increased consumerism, and technology expansion), i.e., to plan for them and otherwise accommodate them. In turn, this reflects a host of government activities: political responses, policy interventions, institutional initiatives, promotion of technology advances, and a host of measures to establish socioeconomic infrastructure. It is a planning challenge unprecedented in its character and extent. No societies in the past have had to cater for population growth at annual rates of 2% or more for decades on end, let alone the rates of 3–4% that have recently characterized a sizeable number of nations in sub-Saharan Africa and the Muslim world. Indeed it would tax the planning capacities of the most sophisticated and established societies. Yet it is a challenge that confronts nations that often have experience of only a few decades of nationhood and the modern state system. Moreover, many developing nations are further constrained by exogenous factors such as adverse trade relations, inadequate and often inequitable aid flows, and foreign debt. In these circumstances, it is remarkable that so many developing nations have managed to achieve so much in such a short period.

Ecological dislocations

Thus, we see that growth in human numbers, in conjunction with growth in human consumption and growth in environmentally adverse technology (the $I = PAT$ equation), serves to build up a situation that eventually generates an "overshoot" outcome. In turn, this outcome can precipitate a downturn in the capacity of environmental resources to sustain human communities at their former level. This amounts to a macrolevel reversal, made up of several components – a phenomenon well known in the environmental sphere. Designated as "jump effects" of ecological discontinuity, or threshold effects of irreversible injury, these occur when ecosystems have absorbed stresses over long periods without much outward sign of damage, then eventually reach a disruption level at which the cumulative consequences of stress finally reveal themselves in critical proportions. We can well anticipate that as human communities continue to expand in numbers, they will exert increasing pressures on ecosystems and natural-resource stocks, whereupon ecological discontinuities will surely become more common.

An example has arisen in the Philippines where the agricultural frontier closed in the lowlands during the 1970s. As a result, multitudes of landless people started to migrate into the uplands, leading to a buildup of human numbers at a rate far greater than that of national population growth. The uplands contain the country's main remaining stocks of forests, and they feature much sloping land. The result has been a marked increase in deforestation and a rapid spread of soil erosion (Myers, 1988). In other words, there occurred a "breakpoint" in patterns of human settlement and environmental degradation. As long as the lowlands were less than fully occupied, it made little difference to the uplands whether there was 50% or 10% space left. It was only when hardly any space at all was left that the situation altered radically. What had seemed acceptable became critical, and the profound shift occurred in a very short space of time.

Similarly, in Costa Rica agricultural expansion finally reached both oceans and both frontiers during the 1980s. For the first time in 400 years of their history, Costa Ricans (currently increasing in numbers at 2.5% per year) have no ready access to new land. Their predominantly agrarian society is having to adjust to a sudden change from land abundance to land scarcity (Augelli, 1984).

This problem of land shortages is becoming widespread in many if not most developing countries where land provides the livelihood for almost 60% of populations and where most of the most fertile and most accessible land has already been taken. During the 1970s, arable areas were expanding at roughly 0.5% per year. But during the 1980s the rate dropped to only half as much; and primarily because of population growth, the amount of per-capita arable land declined by 1.9% per year (United Nations Population Fund, 1990). Moreover, as far back as 1975 some 25 million km^2 of land already supported 1.2 billion people, yet only 563 million could be sustainably fed with the low-technology farming methods generally practiced. Most of these lands were in semiarid or montane zones, unusually susceptible to soil erosion. The population overloading served to aggravate the pace of land degradation (Food and Agriculture Organization, 1984).

Consider too an instance where a potentially renewable resource suddenly becomes overwhelmed by rapid population growth. Most people in the developing world derive their energy from fuelwood. As long as the number of wood collectors does not exceed the capacity of the tree stock to replenish itself through regrowth, the local community can exploit the resource indefinitely. They may keep on increasing in numbers for decades, indeed centuries, and all is well, provided they do not surpass a critical level of exploitation. But what if the number of collectors grows and grows until they finally exceed the self-renewing capacity of the trees – perhaps exceeding it by only a small amount? Quite suddenly a point is reached where the tree stock starts to decline. Season by season the self-renewing capacity becomes ever more depleted: the exploitation load remains the same, and so the resource keeps on dwindling more and more, meaning, in turn, that there is an ever-increasing overloading of the resource. The vicious circle is set up, and it proceeds to tighten, when once the level of exploitation becomes nonlinear. Note in particular that this scenario applies *even if* the number of collectors stops growing. The damage is done. But if the number of collectors continues to expand through population growth, the double degree of overloading (derived from an ever-dwindling stock exploited by ever more collectors) becomes compounded. There ensues a positive feedback process that leads to fuelwood scarcity, and then all too quickly the stock is depleted to zero. It is a process that occurs all the more rapidly as the stock is progressively depleted.

The essence of the situation is that the pace of critical change can be rapid indeed. As soon as a factor of absolute scale comes into play, the self-sustaining equilibrium becomes disrupted. A situation that seemed as if it could persist into the indefinite future suddenly moves on to an altogether different status. It is as if two lines on a graph approach each other with seeming indifference to each other, then when once they cross the situation is radically transformed.

We encounter this nonlinear relationship between resource exploitation and population growth with respect to many other natural resource stocks, notably forests, soil cover, fisheries, water supplies, and pollution-absorbing services of the atmosphere. Whereas resource exploitation may have been growing gradually for very long periods without any great harm, the switch in scale of exploitation induced through a phase of unusually rapid population growth can readily result in a slight initial exceeding of the sustainable yield, whereupon the debacle of resource depletion is precipitated with surprising rapidity.

Population growth and energy

We should also bear in mind that much environmental degradation is due to the industrialized nations – nations with very little though still significant population growth, and for which the imperatives of sustainable development are just as significant as for developing nations. North America and Europe are responsible for nearly three-quarters of the carbon dioxide emissions (almost all from burning of fossil fuels) that account for half of global warming, while containing only 8% of the world's population. Developing nations, with 77% of the world's population, contribute only 7% of industrial emissions of carbon dioxide. True, the twin factors of population growth and economic advancement in developing nations are projected to lift their share of carbon dioxide emissions a good deal higher during the next few decades. Meantime, the population of Bangladesh, now 115 million people and with a growth rate of 2.5%, is projected to expand in 1990 by 2.9 million; that of the United States, with 251 million people and a growth rate of 0.8%, by 2 million. But each Bangladeshi consumes commercial energy equivalent to only three barrels of oil per year, each American 55 barrels. So the population-derived increase in Bangladesh's consumption of oil equivalent will be 8.7 million barrels, that of the United States 110 million barrels. In terms of global carrying capacity, the United States plainly bears a large responsibility since its population growth rate of "only" 0.8% per year is allied with an exceptional technological and consumerist capacity to utilize polluting materials.

This is not to disregard the fact that population growth has played a prominent part in increased use of energy worldwide. Between 1970 and 1990 per-capita consumption of commercial energy expanded from 2.04 to 2.30 kW, a 13% increase. But because population expanded from 3.6 to 5.3 billion, a 47% increase, total energy consumption rose from 8.36 to 13.73 TW, a 64% increase. Moreover, the population of developing countries is projected to expand from 4.1 billion people in 1990 to 7.1 billion in 2025, a 73% increase. Even if their per-capita energy consumption increases during the same period only from 1.1 to 2.2 kW (compare 7.5 kW for developed-country citizens today), for a 100% increase, developing-world consumption of energy will soar from 4.5 TW in 1990 to 15.0 TW in 2025, a 232% increase (Holdren, 1990).

North–south relations

Certain north/south relationships serve to exacerbate problems of population and environment in developing countries (Agarwal, 1990; Alonso, 1987; Ramphal, 1990; South Commission, 1990). They thus act as ultimate as opposed to proximate factors – a somewhat covert dimension that is often overlooked (Shaw, 1989). Consider two such relationships, external debt and international trade.

Developing countries now owe some $1.3 trillion of outstanding loans from developed countries. This debt burden weighs heavily on developing-country prospects for development generally, and on population and environment concerns in particular. It has obliged many poorer countries to cut back on their government spending on health and family planning activities, and has thus contributed to the slowing in fertility-rate declines in Philippines, India, Tunisia, Morocco, Colombia, and Costa Rica (United Nations Population Fund, 1990). It can even be demonstrated to be causing the deaths of half a million children each year through general slowing or even reversal of development processes in developing countries generally (UNICEF, 1990), with all that implies for population planning prospects.

Next, consider international trade and its adverse impact on environmental concerns. Agricultural trade protectionism on the part of developed countries in North America and Western Europe results in the outlay of $200 billion a year to protect domestic agriculture in these two regions; Japan supports rice production up to eight times the world market price. These policies militate against agricultural exports from developing countries, depriving them of trade revenues worth $30 billion a year (Winglee, 1989; Zietz and Valdes, 1986). In turn, these direct losses reduce developing-country farmers' profits, leaving them less to invest in upgraded agriculture and perpetuating poverty. In turn again, they ultimately induce poor farmers to overload their croplands and to cultivate marginal lands, and they likewise foster mass migration of the rural impoverished into cities where they aggravate environmental degradation (Shaw, 1989).

The key concept of carrying capacity

Carrying capacity is a critical and controversial issue. Certain observers (often ecologists) tend to assert that it is not only a key constraint to population growth, but that it can readily become an absolute factor. Other observers (often economists) tend to assert that it is such a flexible affair, subject to endless expansion through technology and policy interventions, that it soon ceases to have much operational value at all.

Carrying capacity can be defined as "the number of people that the planet can support without irreversibly reducing its capacity to support people in the future" (Ehrlich et al., 1989; see also Daly and Cobb, 1989; Mahar, 1985; Pimentel and Pimentel, 1989). While this is a global-level definition, it applies at the national level too, albeit with many qualifications as concerns international relationships of trade, investment, etc. In fact, it is a highly complex affair, being a function of factors that reflect food and energy supplies, ecosystem services (such as provision of fresh water and recycling of nutrients), human capital, people's lifestyles, social institutions, political structures, and cultural constraints, among many other factors, all of which interact with each other. Particularly important are two points: carrying capacity is ultimately determined by the component that yields the lowest carrying capacity; and human communities must learn to live off the "interest" of environmental resources rather than off the "principle" (Ehrlich et al., 1989).

Thus, the concept of carrying capacity is closely tied in with the concept of sustainable development. There is now evidence that human numbers with their consumption of resources, plus the technologies deployed to supply that consumption, are often already exceeding carrying capacity. In many parts of the world, the three principal and essential stocks of renewable resources – forests, grasslands, and fisheries – are being utilized faster by humans than their rate of natural replenishment (Brown et al., 1989).

To consider the situation through an illustrative example, consider the case of sub-Saharan Africa. This is the only region of the developing world where population growth rates are still rising. Almost half of the increase in humanity's numbers during the foreseeable future is projected to occur in this region alone, which is projected to feature a four-times increase in numbers (Acsadi and Acsadi, 1990; Caldwell and Caldwell, 1990; Ohadike, 1990). Yet there are signs that the region already suffers from severe population pressures (Myers, 1989; World Bank, 1989), notably the environmental decline that is overtaking much of the region, especially with respect to the natural-resource base that supports food production (Harrison, 1987; Timberlake, 1985).

Almost 200 million people in sub-Saharan Africa are reckoned to receive less than 90% of the minimum of 2200 calories a day needed to support an active working life – that is, they are chronically undernourished (Chen et al., 1990; Mellor et al., 1987; World Bank, 1988). In much if not most of the region, per-capita food production has been declining for a full two decades. At least 62% of the entire populace endures absolute poverty. As much as 80% of croplands and 90% of stock-raising lands are affected to some degree by land degradation of various forms (World Bank, 1989). Yet today's population of 543 million people, with an annual growth rate of 3.0%, is projected to reach 733 million by the year 2000, and 1,268 million by 2020. This ultrarapid rate of growth operates in conjunction with the adverse climatic conditions that have obtained in much of the region for the past two decades.

Of the population of almost 450 million people in 1985, 250 million were chronically malnourished, 150 million were subject to acute food deficits, and 30 million were actually starving (Independent Commission on International Humanitarian Issues, 1985). In recent years there has been some respite, thanks to better rains. But because of unpromising baseline conditions generally, and particularly in respect to harsh climate (Falkenmark, 1989) together with widespread environmental degradation and poor agricultural policies, the return of only moderately adverse weather conditions could quickly trigger a renewed onset of broadscale famine. If these adverse conditions persist [they could well be aggravated by the climatic vicissitudes entrained by the greenhouse effect (Farmer and Wigley, 1985; Glantz, 1987)], the enfamished throngs that totalled 30 million in 1985 could well increase to 60 million by the early 1990s, and to 130 million by the year 2000 (McNamara, 1985). This means that the proportion of starving people would expand from less than 7% of the region's population in 1985 to 18% by the year 2000.

Even while sub-Saharan Africa has continued to feature a fertility rate higher than anywhere else in the world, since 1960 the region has been growing poorer and hungrier in absolute as well as relative terms. Today's average per-capita income of roughly $250 is only 95% of real income in 1960; of the 36 poorest countries in the world, 29 are in Africa (Chidzero, 1988; DeCuellar, 1988). Worse, average per-capita agricultural production has declined by an average of 2 percent annually since 1970, and the World Bank (1988) estimates that production is unlikely to grow at more than 2.5% per year for at least the next two decades, even while population growth remains at 3% or more per year. As a result, food output per head, which has declined by 20% since 1970, is scheduled to decline by a further 30% during the next 25 years (Dow, 1985;

United Nations World Food Council, 1988; World Hunger Program, 1989). Thus the region serves as a prime example of an "adverse outlook" scenario (Brown and Wolf, 1985; Goliber, 1989; Independent Commission on International Humanitarian Issues, 1985; Marcum, 1989; Sai, 1984; World Bank, 1984, 1986, 1989).

Environmental refugees

As a measure of the extent to which human society is in danger of disintegration in certain of its manifestations, consider the problem of environmental refugees, these being people who are obliged to leave their homelands because there is no longer any prospect of their deriving even the most meager livelihood there (Jacobson, 1988; Myers, 1991). The phenomenon derives from both population growth and environmental deterioration, leading to an extreme degree of human deprivation. To anticipate the eventual numbers of these refugees is to engage in an exercise of "creative speculation": it is not possible to determine the ultimate scope of the phenomenon. But since it will profoundly affect multitudes of people, a cursory assessment is appropriate for this section on quality of life.

Already there are at least 10 million of these environmental refugees, or several million more than total refugees at the end of World war II. For future prospects, consider a not unlikely outlook for Bangladesh, a Florida- or England-sized country with 115 million people today, projected to exceed 200 million by 2020 and 350 million by 2050. In terms of scenarios for sea level rise, the result could be the destruction of the homes and holdings of some 120 million Bangladeshis (Milliman et al., 1989). The victims could look for little help from an already poor nation that would have lost a sizeable share of its economic base within the inundated zone.

In the Nile delta, too, sea level rise is projected to lead to the elimination of almost one fifth of Egypt's habitable land – this being an area where the population today exists at twice the density of Bangladesh. Flooding would likely cause the displacement of around 20 million people (Milliman et al., 1989). The scenario, moreover, is cautious and conservative: there would surely be additional problems, such as the intrusion of saltwater up the foreshortened Nile, which would further reduce the irrigated lands that make up virtually the whole of Egypt's agriculture.

Deltas are unusually vulnerable to even a moderate amount of sea level rise. Yet these are precisely the areas that feature some of the densest human settlements and most intensive agriculture on Earth. Among such "severe risk" areas are the estuaries of the rivers Hwang Ho, Yangtze, Mekong, Chao Phraya,

Salween, Irrawaddy, Indus, Tigris/Euphrates, Zambezi, Niger, Gambia, Senegal, Courantyne and Maruni (in the Guyanas), and La Plata. Given present populations and their growth rates, at least 100 million people will find themselves flooded out or suffering related troubles (storm surges, etc.), plus the related effects such as acute congestion in the new coastal zones (El-Hinnawi, 1985).

Still other communities could be threatened in low-lying coastal territories, notably the mega-metropolises of Jakarta, Madras, Bombay, Karachi, Lagos, and Rio de Janeiro. If only half of their projected populations eventually become displaced, that will add 40 million people to the refugee total. At the same time a number of developed-world cities will be threatened as well, notably Rotterdam, Venice, New York, Miami and New Orleans. But developed nations will have the engineering skills and the finances to hold back the sea by dykes, after the manner of Netherlands, or they could go some way toward moving their cities inland. Neither of these options is open to developing nations.

In yet broader terms, note that a 1 m rise in sea level would threaten almost 5 million km^2 of coastal lands in total. While this is roughly equivalent to the United States west of the Mississippi River, it amounts to only 3% of Earth's land surface. Yet is encompasses a full third of global croplands and is home to well over one billion people already, projected to rise to well over two billion within a few decades. Since it includes the lands already listed above, we can reasonably suppose that nonlisted areas could well supply another 50 million refugees. Already 30 million Chinese live in coastal lands a mere half meter above sea level.

Then there is the prospect of other greenhouse effects, such as continuous droughts and dislocation of monsoon systems. According to Daily and Ehrlich (1990), and under an entirely plausible greenhouse scenario for the first part of next century, there could be a 10% reduction in the global grain harvest on average three times a decade (the 1988 droughts in just the United States, Canada, and China resulted in almost a 5% decline). Given that world's food reserves have dwindled almost to nothing today as a result of late-1980s droughts, it is not unrealistic to reckon that each such grain-harvest shortfall would result in the starvation deaths of between 50 and 400 million people. Catastrophes of this order would trigger mass migrations of people from famine-affected areas; it is not possible to say how many. But for the sake of "getting a handle" on the prospect, we can hazard a best-judgment guess – nothing more, but also nothing less – of 50 million refugees.

Of course these are very much a case of "best judgment" guesstimates – exploratory at most. Some of them could be inaccurate by tens of millions either way. But the combined total according to these calculations is 380 million. An extremely rough and ready reckoning, it supplies an initial insight into the scale of the upcoming problem of environmental refugees.

Conclusion

These, then, are some perspectives on the environmental predicament of human society. Acute as they already are, they appear set to become supersevere within the foreseeable future unless we mobilize society's preventive capacities in an entirely unprecedented effort. In my judgment, a response of this sort is little likely insofar as political leaders and the general public alike seem disinclined to do their environmental homework and learn what ultimately makes the world go round. But despair is not only counterproductive, it is occasionally shown to be unrealistic. Who would have supposed in mid-1989 that in just a few months' time the Berlin wall would come crashing down? What further walls – the barriers of ignorance, indifference, and sheer inertia as concerns our environmental outlook – may not be brought crashing through some unexpected breakthrough of public awareness or political leadership? The clock shows we still have a few minutes to midnight!

References

Acsadi, G.T., & Acsadi, G. J. (eds.). (1990). *Population Growth and Reproduction in Sub-Saharan Africa.* Washington, D.C.: The World Bank.

Agarwal, A. (1990). The North–South Perspective: Alienation or Interdependence? *Ambio,* **19,** 94–96.

Alonso, W. (ed.). (1987). *Population in an Interacting World.* Cambridge, Massachusetts: Harvard University Press.

Augelli, J. P. (1984). Costa Rica: Transition to Land Hunger and Potential Instability. *1984 Yearbook of Conference of Latin Americanist Geographers,* **10,** 48–61.

Brown, L. R., & Wolf, E. C. (1985). *Reversing Africa's Decline.* Washington, D.C.: Worldwatch Institute.

Brown, L. R. et al. (1989). *State of the World 1989.* New York: W. W. Norton.

Brown, L. H. et al. (1990). *State of the World 1990.* New York: W. W. Norton.

Caldwell, J. C., & Caldwell, P. (1990). High Fertility in Sub-Saharan Africa. *Scientific American,* **262,** 82–89.

Chen, R. S., Bender, W. H., Kates, R. W., Messer, E., & Millman, S. R. (1990). *The Hunger report: 1990.* Providence, Rhode Island: World Hunger Program, Brown University.

Chidzero, B. T. (1988). Africa and the World Economy. *World Futures,* **25,** 157–162.

Daily, G. C., & Ehrlich, P. R. (1990). *An Exploratory Model of the Impact of Rapid Climate Change on the World Food Situation.* Stanford, California: The Morrison Institute for Population and Resource Studies, Stanford University.

Daly, H. E., & Cobb, J. B. Jr. (1989). *For the Common Good.* Boston, Massachusetts: Beacon Press.
Davis, K., Berstam, M. S., & Sellers, H. M. (1989). *Population and Resources in a Changing World.* Stanford, California: Morrison Institute for Population and Resource Studies, Stanford University.
DeCuellar, P. (1988). *Africa's Economic Situation.* Office of the Secretary General. New York: United Nations.
Dow, M. M. (1985). Food and Security. *Bulletin of the Atomic Scientists,* **41,** 21–26.
Ehrlich, P. (1968). *The Population Bomb.* New York: Ballantine.
Ehrlich, P. R., & Ehrlich, A. H. (1990). *The Population Explosion.* New York: Simon and Schuster.
Ehrlich, P. R., Daily, G. C., Ehrlich, A. H., Matson, P., & Vitousek, P. (1989). *Global Change and Carrying Capacity: Implications for Life on Earth.* Stanford, California: Institute for Population and Resources Studies, Stanford University.
El Hinnawi, E. (1985). *Environmental Refugees.* Nairobi, Kenya: United Nations Environment Programme.
Falkenmark, M. (1989). The Massive Water Scarcity Now Threatening Africa – Why Isn't it Being Addressed? *Ambio,* **18,** 112–118.
Farmer, G., & Wigley, T. M. L. (1985). *Climatic Trends for Tropical Africa.* Norwich, U.K.: Climatic Research Unit, University of East Anglia.
Food and Agriculture Organization. (1984). *Potential Population Supporting Capacities of Lands in the Developing World.* Rome, Italy: Food and Agriculture Organization.
Glantz, M. H. (ed.). (1987). *Drought and Hunger in Africa.* Cambridge, U.K.: Cambridge University Press.
Goliber, T. J. (1989). *Africa's Expanding Population: Old Problems, New Policies.* Washington, D.C.: Population Reference Bureau.
Harrison, P. (1987). *The Greening of Africa.* London, U.K.: Paladin Books.
Holdren, J. P. (1990). Energy in Transition. *Scientific American,* **263**(3), 109–115.
Houghton, J. T., Jenkins, G. J., & Ephraums, J. J. (1990). *Climate Change: The IPCC Scientific Assessment.* Cambridge, U.K.: Cambridge University Press.
Independent Commission on International Humanitarian Issues. (1985). *Famine: A Man-Made Disaster.* London, U.K.: Pan Books.
Jacobson, J. L. (1988). *Environmental Refugees: A Yardstick of Habitability.* Washington, D.C.: Worldwatch Institute.
Keyfitz, N. (1990). *Population Growth Can Prevent the Development that Would Slow Population Growth.* Laxemburg, Austria: Population Programme, International Institute for Applied Systems Analysis.
Lipton, M. (1985). *The Poor and the Poorest: Some Interim Findings.* Washington, D.C.: The World Bank.
Mahar, D. (1985). *Rapid Population Growth and Human Carrying Capacity.* Washington, D.C.: World Bank.
Marcum, J. A. (1989). Africa: A Continent Adrift. *Foreign Affairs,* **68,** 159–179.
McNamara, R. S. (1985). The Challenges for Sub-Saharan Africa. Sir John Crawford Memorial Lecture, Washington, D.C. November 1st, 1985.
Meadows, D. H., Meadows, D. L., Randers, J., & Behrens III, W. (1972). *The Limits to Growth.* New York: Universe Books.
Mellor, J. W., Delgado, C. L., & Blackie, M. J. (eds.). (1987). *Accelerating Food Production in Sub-Saharan Africa.* Baltimore, Maryland: Johns Hopkins University Press.
Milliman, J. D., Broadus, J. M., & Gable, F. (1989). Environment and Economic

Implications of Rising Sea Level and Subsiding Deltas: The Nile and Bengal Examples. *Ambio,* **18,** 340–345.
Myers, N. (1985). *The Gaia Atlas of Planet Management.* New York: Doubleday.
Myers, N. (1988). Environmental Degradation and Some Economic Consequences in the Philippines. *Environmental Conservation,* **15,** 205–214.
Myers, N. (1989). Population Growth, Environmental Decline and Security Issues in Sub-Saharan Africa. In *Ecology and Politics: Environmental Stress and Security in Africa,* ed. A. Hjort af Ornas & M. A. Mohamed Salih, pp. 211–231. Upsala, Sweden: Scandinavian Institute of African Studies.
Myers, N. (1991). *Population and the Environment: Issues, Prospects and Policies.* New York: United Nations Fund for Population Activities.
Ohadike, P. O. (ed.). (1990). *The State of African Demography.* Liege, Belgium: International Union for the Scientific Study of Population.
Osborn, F. (1948). *Our Plundered Planet.* New York: Little, Brown.
Pimentel, D., & Pimentel, M. (1989). Land, Energy and Water: The Constraints Governing Ideal U.S. Population Size. Teaneck, New Jersey: *The NPG Forum,* Negative Population Growth, Inc.
Pimental, D., Fredrickson, L. M., Johnson, D. B., McShane, J. H., & Yuan, H.-W. (1989). Environment and Population: Crises and Policies. In *Food and Natural Resources,* ed. D. Pimentel & C. W. Hall, pp. 363–389. New York: Academic Press.
Ramphal, S. (1990). Third World Grievances. *U.S. Environmental Protection Agency Journal,* **16**(4), 39–43.
Raven, P. H. (1987). *We're Killing Our World: The Global Ecosystem in Crisis.* MacArthur Foundation Occasional Paper, Chicago, Illinois.
Sai, F. T. (1984). Population Factor in Africa's Development Dilemma. *Science,* **226,** 801–805.
Shaw, R. P. (1989). Rapid Population Growth and Environmental Degradation: Ultimate *versus* Proximate Factors. *Environmental Conservation,* **16,** 199–208.
Simon, J. L. (1990). *Population Matters: People, Resources, Environment and Immigration.* New Brunswick, New Jersey: Transaction Publishers.
South Commission (The). (1990). *The Challenge to the South: The Report of the South Commission.* Oxford, U.K.: Oxford University Press.
Timberlake, L. (1985). *Africa in Crisis: The Causes, the Cures of Environmental Bankruptcy.* London, U.K.: Earthscan Publications Ltd.
UNESCO. (1989). *Vancouver Declaration of Symposium "Science and Culture for the 21st Century: Agenda for Survival,"* Paris: UNESCO.
UNICEF. (1990). *State of the World's Children.* New York: UNICEF.
United Nations Population Fund. (1990). *1990 Revision of Global Demographic Estimates and Projections.* New York: United Nations Population Fund.
United Nations World Food Council. (1988). *The Global State of Hunger and Malnutrition: 1988 Report.* New York: United Nations World Food Council.
U.S. Department of Agriculture. (1990). *World Agricultural Production.* Washington, D.C.: Foreign Agricultural Service, U.S. Department of Agriculture.
Vitousek, P. M., Ehrlich, P. R., Ehrlich, A. H., & Matson, P. M. (1986). Human Appropriation of the Products of Photosynthesis. *BioScience,* **36,** 368–373.
Winglee, P. (1989). Agricultural Trade Policies of Industrial Countries. *Finance and Development,* **26**(1), 9–12.
Woodwell, G. M. (1986). Forests and Climate: Surprises in Store. *Oceanus,* **29,** 71–75.
World Bank. (1984). *Toward Sustained Development in Sub-Saharan Africa: A Joint Program of Action.* Washington, D.C.: World Bank.

World Bank. (1986). *World Development Report 1986.* Washington, D.C.: The World Bank.
World Bank. (1988). *The Challenge of Hunger in Africa: A Call to Action.* Washington, D.C.: World Bank.
World Bank. (1989). *Sub-Saharan Africa: From Crisis to Sustainable Development.* Washington, D.C.: World Bank.
World Commission on Environment and Development. (1987). *Our Common Future.* Oxford, U.K.: Oxford University Press.
World Hunger Program. (1989). *Beyond Hunger: An African Vision of the 21st Century.* Providence, Rhode Island: World Hunger Program, Brown University.
World Resources Institute. (1989). *World Resources 1989.* Washington, D.C.: World Resources Institute.
Zietz, J., & Valdes, A. (1986). *Costs of Protectionism to Developing Countries.* Washington, D.C.: The World Bank.

8

Germplasm conservation and agriculture

GARRISON WILKES

Introduction

Agriculture is a prehistory technology that still has a far-reaching impact on the planet. The quest for an assured food supply has done more to decrease biodiversity and physically alter the environment than any other activity in which we engage. Approximately 60% of the human population directly or indirectly makes their living from agriculture. Tragically, food production is population driven. As we produce more food, the human population becomes larger and the demand for increased yield creates an open spiral of greater impact on the land. Before the advent of agriculture, we lived like any other animal in the sense that we hunted and gathered our food daily, and on the days we were not successful we went hungry. Our population density did not exceed 1 person per 25 km^2, and we were sustained on our forage territory. Today our density exceeds 25 persons per km^2, and in the urban zones, such as the one between Boston and Washington, D.C., our density approximates 600 persons per km^2 (Wilkes, 1989). Quite literally we are absolutely dependent on domesticated plants and animals and there is no turning back to hunting and gathering. Cultivated plants and domestic animals provide an assured food supply that liberates us from the daily quest for food so we can be free to engage in such human activities as the arts and learning and/or live at high densities in large metropolitan centers, but at what cost!

It is difficult to visualize a challenge more profound in its implications yet less appreciated by the general public than our food production systems, which are dependent on sun, soil, water, and genetic resources (IUCN, 1980; WWI, 1992). Food security and biodiversity will be the most obvious challenges in this decade. How can we supply adequate nutrition to the 5.4 billion humans that exist now and the more than 6 billion that will exist by the year 2000 (UN,

1989). To put the demands in perspective, in the first twenty years of the next century one out of every ten humans that ever existed will be sitting down to dinner. The food production of a single year will equal that of the century 1850–1950, and in the two decades 2000–2020, we will produce and consume as much food as we have since the beginning of agriculture 10–12 thousand years ago (Wilkes, 1989)! Most of this population increase will take place in the developing countries where demand for food and agricultural products will double. The problems of a precarious food supply and rural poverty is expected to aggravate pressure on scarce land for arable farming, deforestation for new agricultural lands will increase, and pushing systems beyond sustainable limits will result in increased habitat destruction. Our human population will grow with short term food increases and not be sustained by long term "real" reasons (Walsh, 1991), so that, by FAO estimates, by the year 2000 there will be at least 600–650 million undernourished people on the planet, (FAO, 1981) most of them children.

For the developed nations, population increases can be accommodated by eating lower on the trophic level and consuming grains directly, but the developing world is already doing that. Since new arable land in the developing world will become steadily more scarce as the natural systems they replace become more valued, higher yields mean using more fertilizer, plowing, and water-lifting energy and improved plant material (Walsh, 1991). All but the last are agricultural inputs, which compete for meager resources available in developing nations. Therefore, breeding for better crop plants will be the focal point around which all strategies to increase crop yields will develop. Human knowledge to grow and maintain crop plants was central to the origin of agriculture and so it will be in the high density human population sustainable agriculture of the 21st century. It will be the positive genetic response of crop plant seed and vegetatively propagated plants to soil, water, fertilizer, pest, and social institutions that will determine the success of our attempt to feed ourselves (Plucknett et al., 1987).

The limited number of crop plants

The actual number of plant species that has historically fed the human population is only about 5000. This small number is a fraction of 1% of the flora of the world. The number that have entered agriculture is only about 1500 and as the human population has grown in number over the last two thousand years, and especially so since the development of the science of genetics, we have depended increasingly on the shorter list of the most productive and most easily stored/shipped plants. Today only about 150 plant species with about a quarter million local races are important in meeting the calorie needs of humans. Urban dwellers, who are two-thirds of the world's population, are essentially dependent on three cereals (rice, wheat, maize) for subsistence (Prescott-Allen & Prescott-Allen, 1990).

Evolution under domestication of crop plants

The human population has exceeded the carrying capacity for long term sustainable agriculture without energy inputs in many parts of the world (Ehrenfield, 1978). Agriculture cannot continue on the treadmill of always producing more. In 400 human generations we have come from wild food plants to high yielding domesticated ones, but we are at or near the limits. Two generalizations can characterize the present world condition: 1) we now are in a state of diminished resources on a per-capita basis; 2) farming is no longer a subsistence activity where the cultivator eats and depends on the actual harvest, but one in which money is generated by the sale of the crop. With subsistence farming there are negative feedback loops that limit production, but cash is a neutral (universal) commodity and cropping for cash has the tendency to be an open, ever-expanding, feedback system (Vasey, 1992).

We have not reached present levels of agricultural productivity without cost to the environment. The Agricultural Revolution has:

1. Replaced native vegetation with open plowed fields (opened the soil system for possible deterioration). Agricultural systems are generally less mature ecosystems with high productivity/biomass ratios.
2. Changed water relationships by opening the habitat to sunlight for cultivated heliophytes and enhanced the physical conditions for more extreme temperature fluctuations and increased water stress.
3. Replaced natural selection with artificial selection, which in its ultimate form leads to pure stands (monocultures) and warfare with competitors (insects with insecticides, weeds with herbicides, and fungi with fungicides).

Charles Darwin called artificial selection telescoped evolution because what was normally a drawn out affair was collapsed to a short period of time (Darwin, 1897). Plants which are potential candidates for telescoped evolution or domestication must meet three criteria. They must be:

1. capable of thriving in full sun in human made environments (heliophytes);
2. productive in terms of land and labor;
3. genetically plastic – will hybridize and mutate, i.e., have a depth of genetic resources.

The last criterion is especially important, because it is the one area where we can still make substantial improvements in crop plants (Harlan, 1975).

The gene pool of crop plants and their wild relatives can be thought of as a bucket holding a fluid. Under natural selection (normalizing selection mode) the gene pool is comparable to a bucket sitting on its bottom. The surface expresses the phenotype and the volume the gene pool. The top surface is small (wild type genes) when expressed as a ratio to the total volume below (recessive multiple alleles). Under artificial selection the bucket is tipped, and some of the fluid is poured off. The top surface is larger, and the ratio of top area to volume is also greater. Many traits selected under artificial selection are recessive (nonbrittle rachis in cereals for example), and once they are fixed, the allelic diversity of the gene pool narrows to one. Artificial selection increases the phenotypic diversity but decreases the allelic diversity. It is only on hybridization of two diverse forms that the allelic diversity is again broadened in the artificial selection process.

The story of agriculture over the last 10,000 years has been essentially the same in the four independent sites where it began: Central Mexico/Guatemala, Andean South America, the Fertile Crescent in the Near East, and Mid-northern China (Heiser, 1990; see Fig. 8.1). First there was human selection for unique traits and a proliferation of phenotypic variance as hidden recessive genes came to the surface through artificial selection and close inbreeding enforced by physical isolation and small population size (when compared to the widespread wild populations). Next, the environment was restructured or rearranged, making it possible for the survival of genetic lethals (mutations) and, more importantly, semilethals. The seed was carried into new zones beyond its native distribution, and alleles deep in the volume of the bucket rose to the surface, because they conveyed adaptation factors for the new conditions. Also, these plants were now suddenly growing alongside wild relatives with whom they were not genetically protected by isolating mechanisms, and, on hybridization, a wealth of variation was generated that has driven the selection process ever since (Simmonds, 1976). Examples are the hexaploid wheats (6x) with their three genomes. Bread wheats represent three plants in one because of the polyploid state. Maize has hybridized with teosinte and kept the same chromosome number, but through introgressive hybridization part of the teosinte genome is in maize and the reciprocal is also true. Because of introgressive hybridization, teosinte is a "little second plant" in the maize gene pool (a plant within a plant). Almost all major crop plants have developed a mechanism of one form or another to expand the gene base, which has countered the tendency of artificial selection to narrow it (Wilkes, 1989).

I emphasize the importance of germplasm for crop plants because it is the

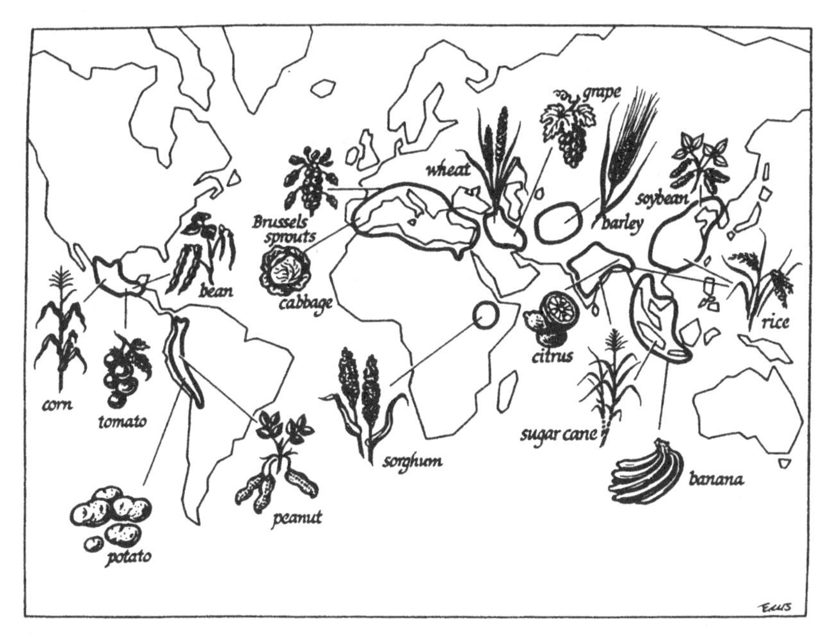

Fig. 8.1. Map of major Vavilov centers. The places where plant domestication of our major food plants has taken place are not distributed at random on the planet but are highly localized in areas of ancient civilization and long human settlement. These regions of crop plant diversity are called Vavilov centers, after the Russian plant breeder and geneticist N. I. Vavilov, who made lucently clear the existence of zones from which many of the land races and valuable genes used by plant breeders in this century have come.

area where I expect the greatest advances in the next two decades, and we stand to lose its great potential if management of plant genetic resources does not improve. For genetic advance of any crop there are two resources: (1) the gene or genes that control specific traits and (2) the knowledge of how to exploit those traits in the total genetic background of the finished variety for farmers' fields. This loss of knowledge disturbs me because most of the talk about germplasm loss centers around genetic erosion and genetic wipeout (Altieri et al., 1987; Marshall, 1990), but equally important is the decrease in the number of plant breeders relative to the sheer numbers that must be fed round the world (Collins & Phillips, 1991). Compounding the trend is the privatization of plant breeding and the rush to biotechnology, which is drawing funds away from germplasm management in the public sector.

History of plant breeding

Plant breeding as a human activity can be viewed as developing historically through three phases; we are currently on the threshold of the fourth. The earliest domesticated crops were probably not much more productive than their wild progenitors, but the act of cultivation was a radical break with the past, as artificial selection was applied to small isolated populations. This was the first stage of plant breeding or human control over crop plant evolution. All the important world food crops (wheat, corn, rice, barley, potatoes, beans, and soybeans) were developed in this first stage (Vavilov, 1935; Wilkes, 1988).

The second stage of plant breeding came with the discovery of the New World and trade that followed the circumnavigation of the world. This rapid diffusion of crops, livestock, and farming techniques coupled with emigration brought about tremendous genetic recombination as distinct and far-flung land races were brought together and hybridized in farmers' fields. This era also saw the development of colonial empires in the tropics and cash crops such as coffee, cotton, rubber, and sugarcane. Many of these cash crops, such as coffee, were moved from the Old World to the New, where, free of their diseases, they became more productive. Or the reverse took place when crops such as rubber were moved from the New World to the Old. These plantation crops were the first cash crops. Now all field crops, with the possible exception of kitchen gardens, are cash crops (Purseglove, 1968, 1972). It is my contention that when the threshold was crossed from crops for subsistence to crops for sale, the selection forces for nutritional quality and respect for local habitat carrying capacity were shed, and many of the problems we now experience began. The crossing of this threshold brought the imperative to anticipate the future and better manage the crop-plant gene pool, because the mechanisms that once worked to create and maintain this diversity were no longer in operation (FAO, 1989; CGIAR, 1990).

The third stage of plant breeding and improvement began with the rediscovery of Gregor Mendel's classic experiments on the heredity of garden peas at the beginning of this century. For the first time the plant breeding community had a set of principles by which to proceed with the crop improvement process. Products of this era are hybrid corn, changes in the photoperiod response of soybeans, and the dwarf-stature wheat and rice from CIMMYT and IRRI, respectively. These Green Revolution cereals and the genes they hold now enter the food supply of two billion people and are directly responsible for feeding over 800 million people with their increased yields (Walsh, 1991).

The fourth era of plant breeding, genetic engineering, promises to have an impact equal to the magnitude of computers in the way we go about managing and structuring the productivity of the world around us. Traditional plant breeding is, in fact, genetic engineering, but this term is now being limited to biotechnologies such as *in vitro* cell culture and recombinant DNA techniques, where genetic material is introduced directly into cell cultures by gene splicing, completely side-stepping the usual sexual process of meiosis, fertilization, and seed production, or when cell cytoplasm is altered, as in protoplast fusion. And, of course, all of these changes depend on cloning technologies, where hundreds of identical plants are grown from units as small as a single cell.

The advent of these technologies has ramifications that affect our current state in regard to plant germplasm. Because these advances are produced using unnatural means, they are potentially protected by the laws of ownership. The spectre of patent protection has polarized the public views of germplasm as either public good (common heritage) or private good (rewards for value added). The developed world has the skills and resources (manpower, institutions, and finances) to enter the biotechnology game on a broad scale, and the developing world lacks them and is fearful of what these new technologies hold for it in the future (Juma, 1989). Much of this fear is reflected in the FAO resolution known as "The Undertaking" to bring genetic resources under the central control of the FAO (Witt, 1985). This has been good because it has focused attention on why plant genetic resources have been undervalued, but unfortunately the solutions sought have been more political and are not advancing with the more important strategies necessary to deal with the major issues of crop plant genetic diversity, its preservation and utilization (Keystone, 1991; Williams, 1989, 1991).

Genetic erosion

Ten thousand years ago the earliest domesticated crops were probably not much more productive than their wild progenitors and less productive than their weedy forms today. But the act of cultivation was a radical break with the normalizing selection of wild plants. This favoring of certain genotypes and restructuring of the environment (open soil and changed water relationships) set in motion numerous interlocking forces, many not at the time consciously intended, that have directed the evolution of the crops and the societies that attended them. The process of artificial selection is a dynamic one of genetic selection in response to changing diseases, pests, planting densities, agricultural techniques, and human use. Starting about 5000 years ago the human population increased and the growing of crops expanded into many different environments. As a result an enormous wealth of genetic variation was created and preserved over the centuries in locally adapted races (Wilkes, 1977). Only a small fraction of this variation has been sampled in breeders' collections and included in the present leading crop varieties. Approximately 75–90% of the variation of four major crops (rice, wheat, corn, and potatoes) and less than 50% for many minor crops is found in gene banks (Plucknett et al., 1987). The largest gene pool is found in the silently shrinking land races and folk varieties of indigenous and peasant agriculture (Altieri et al., 1987; Brush, 1991). Increasingly, the centers of genetic diversity for crop plants have become the mega-gene bank seed storage facilities and not the countryside (Wilkes, 1988; Kloppenburg, 1988).

The technological bind of improved varieties is that they eliminate the resource upon which they are based. Current elite varieties yield better than the varieties they displace. Once the displaced variety is no longer planted, its genes are lost to future generations. Elite varieties also have a second force: they create market expectations. Once a highly uniform variety captures a large fraction of the market share, other varieties are bred to mimic or have the same attributes as this leading variety (Duvick, 1984, 1986). This further eliminates diversity within the crop and even across crops. The marketplace has the potential to influence and limit the genetic diversity of crops. These market forces of "volume sales" and specialized handling of specific varieties has actually promoted the decrease in the total number of crop plants that enter commerce. Long-distance transport focuses on a limited number of crops and on only certain varieties, with the result that local producers and small unique-variety suppliers are forced out. In developing and developed countries the deep well for the gene pool in land races and in folk- or farmer-maintained varieties, which has been the foundation of the plant breeding process, is disappearing (Wilkes, 1977, Walsh, 1991). When a tree falls in the forest there is a crash; when the seed of a unique variety is no longer planted in peasant agriculture, it is a silent loss. Like soil erosion, it is gone with no drama of disappearing, leaving a void and a diminished gene pool.

To be stored safely for a long time, living seeds and the genes they hold must be at a low moisture content. A large number of seed crops may be stored for up to 25 years at refrigerated temperatures just above the freezing point. This is called "active storage conditions." Longer storage periods are obtained at subfreezing temperatures. At temperatures of −18–20°C safe storage for over fifty years is the norm. This is called "base storage conditions." At the temperatures of liquid nitrogen (super dry: approximately 6–7 % moisture and in some cases 3%) seed is expected to remain viable for up to 500 years, and certainly periods of a hundred years is a reasonable time frame (Hawkes, 1983; CIMMYT, 1988; Marshall, 1990).

The number of national and institutional gene banks with refrigerated active collection storage conditions exceeds one hundred. The number of gene banks with base collection storage facilities is approximately thirty (Wilkes, 1989). The number of gene banks that are actually regenerating their collections with rejuvenated fresh seed is less than ten. Creating gene banks does not correct the problem. It is only when genes increase the diversity in farmers' fields through the characterization, evaluation, and prebreeding process of growing the seed and searching for useful traits that we can say the technology of gene banks really works to replace the older system of farmer-held seed (Brown et al., 1989; Holden & Williams, 1984; Cohen et al., 1991). The concept of gene banks looks good on paper, but so far only the banks of the international agricultural research centers have a proven record of moving the genes back into peasant agriculture (Table 8.1).

Genetic wipeout

The wholesale loss of plant genetic resources is called "genetic wipeout" by some (Harlan, 1975). Genetic erosion is a slow, gradual process based on the independent individual decisions of farmers while genetic wipeout is the rapid

Table 8.1. *Germplasm – long term preservation and evaluation of crop-plant genetic resources.**

Ex Situ	*In Situ*
Breeders collections	Passive gene banking (most banks)
Seed depositories	Farmers' fields
Active – short-term storage	Wilderness Areas
Base – long-term storage	National Parks
No management	Biosphere reserves
	Inventoried areas
Management of germplasm	Active gene banking (few examples)
Gene bank with storage plus:	Monitored
Regeneration	Managed
Characterization	Researched sustainable systems
Evaluation	
Prebreeding or enhancement which links to active breeding programs, yield trials, etc. (making gene banks useful)	

*Worldwide there are about 700 significant germplasm collections for crop plants, approximately 100 seed depositories with short-term storage conditions, approximately 30 with long-term storage conditions, and less than 10 that meet the prebreeding criteria of being a fully managed gene bank. Worldwide there is a definite need to cross over the dividing line to the lower half of the chart in this decade or we will have failed at the conservation of genetic resources for crop plants.

and one-stroke destruction of genetic resources, usually by institutional failure. Social disruptions such as political instability or crop failure and famine can eliminate genetic resources rapidly. Quite literally, the genetic heritage of a millennium in a particular valley can disappear in a single bowl of porridge if the seeds are cooked and eaten instead of saved as seed stock (Wilkes, 1977). Equally dramatic is the discarding of a genetic collection because a curator retires or the collection is no longer of use to the institution. The processes of genetic erosion and genetic wipeout are not mutually exclusive but are, in fact, two ends of a spectrum interlocked by the demands of an increasing human population under which biodiversity impoverishment increases daily.

Genetic conservation

In the face of genetic erosion and genetic wipeout how can we preserve the genetic foundation of our food supply and the raw material for the plant-breeding arts of hybridization and selection? The conservation for the future of plant genetic resources can take three forms (Oldfield, 1989).

Entire biomes preservation. Vast tracts are preserved, with their populations of animals and plants. This level of preservation will be extremely important in slowing the species-extinction rate but will have little impact on genetic resources of cultivated plants. An exception might be the Man in the Biosphere (MAB) Reserves such as the one in the Sierra de Manatlan, Jalisco, Mexico which protects wild relatives of both maize and beans and the fields of the indigenous Indian cultivators (Guzmán & Iltis, 1991).

In situ preservation. Land races and wild relatives are preserved in selected small areas, along with the genetic diversity generated where the two hybridize. These are evolutionary systems that are difficult for plant breeders to simulate and these systems should not be knowingly destroyed. Their preservation may not be possible, but the slowing of their disappearance by monitoring and population intervention, such as by fencing, will give us more time to better understand how these systems evolved. Considerable potential for creative institutional arrangements exist for village-level small area *in situ* preservation, especially in developing countries (Alcorn, 1991; Brush, 1991; Oldfield & Alcorn, 1991).

Ex situ preservation. Seed or *in vitro* cell lines are stores in gene banks under appropriate conditions for active or base storage. *Ex situ* gene banks have three distinct functions regarding the preservation of seed: 1) exploration of under-collected zones to maintain representative holding in the gene bank, 2) evaluation and documentation to maintain a useful data base on the holdings, and, lastly, 3) prebreeding and early breeding for enhancement so that gene bank materials will be in a useful form for the plant-breeding community (Brown et al., 1989; Williams, 1989; CGIAR, 1989, Cohen et al., 1991).

This is the mode of preservation for most plant genetic resources. It is generally the easiest to establish and does not come to a cost or management crisis until the seed needs to be grown out for regeneration of fresh seed and genetic evaluation.

Gene banks have certain disadvantages. Essentially they store seed, and in that sense they draw genes out of circulation. For the seed to be useful requires documentation and evaluation, activities in which very few gene banks engage in. Information management with *ex situ* gene banks is as important as the physical arrangements for safekeeping of the seed. Gene banks slow down crop-plant evolution, so both hybridizing and segregating populations of the breeding process become a necessary part of the well-functioning gene bank. Gene banks pass on the genes; they don't breed finished varieties. Until we improve our expectations and funding of gene banks they will continue to function short of being truly useful to the plant-breeding process (Goodman, 1990; Kloppenburg, 1988).

Genetic vulnerability

Genetic vulnerability is the potentially dangerous condition of "thin ice" – having a narrow genetic base. Never before have there been such widespread monocultures (dense, uniform stands of billions of identical or nearly identical plants) covering thousands of acres. The narrowness of the genetic base is responsible for, on the one hand, the predictability of higher yields, but, on the other hand, the greater risk of crop failure. The Irish potato famine of the 1840s is a classic example of genetic vulnerability. More recently, there has been the southern corn blight of 1970 on T cytoplasm maize.

The plant-breeding community is the primary user of germplasm, yet if breeders don't supply varieties that find favor with farmers, they have no product. Therefore, to satisfy farmers, these two concerns are always addressed by the breeder in any released variety: (1) yield and (2) predictability. Selection for superior genotype is primary in creating high-yielding varieties (Duvick, 1986; Goodman, 1990). Yield has many facets. There are the obvious tons/acre, or farm-gate value, and cost of inputs, but also the very important "quality" (nutritional content, taste, texture, etc.) and long term sustainability with current farming practices. Predictability means that the heritability must be high, and to achieve this standard there is considerable pressure for genetic uniformity. Genetic uniformity can be either in the form of homozygosity (often recessive) or the F1 hybrid of inbred parents (as in hybrid corn). The kinds of uniformity desired in the crop are: (1) rapid and uniform germination of seeds, (2) nearly simultaneous flowering, (3) nearly simultaneous maturation of the harvest, (4) stature that promotes mechanical harvest (ear height), (5) product uniformity for taste, flavor, and chemical composition, and, of course, (6) year-to-year stability of yield (Wilkes, 1977). This last point demands special mention, because in the last 25 years plant breeders have developed selection techniques for buffered systems that yield well over a range of conditions across a wide range of habitats (Duvick, 1984, 1986). This push for genetic uniformity over vast acreage has increased the potential for genetic vulnerability as plant breeding has become an international effort (Frankel, 1989), first at the International Agricultural Research Centers (IARCs) such as Centro Internacional Mejoramento de Maiz y Trigo (CIMMYT) and International Rice Research Institution (IRRI) (Chang, 1985), but now at private seed companies that sell seed around the world. The potential now for a crop failure hinged to genetic vulnerability is considerable, and there is significant reason for concern.

Genetic vulnerability is the risk of high-input agriculture with commercial food crop varieties, typically in developed nations but now extending into developing nations. In fact, for developing nations this extension of commercial seed is pushing the genetic erosion of locally adapted land races and thus decreasing the gene base (Brennan & Byerlee, 1991).

There is nothing biologically unsound about breeding for high yields and using a narrow genetic base as a plant-breeding expediency to insure predictability of yield, but the price of this expediency must be constant vigilance and backup of the plant-breeding system (genes in reserve; see Duvick, 1986). Throughout the world there exists an unstable evolutionary truce between crop plants and their pathogens. Genetic changes, either mutations or new recombinations, are always taking place in populations of pathogens (bacteria, fungi, insects, nematodes, etc.), and if a new genetic combination of an individual suddenly grows successfully on a previously resistant plant host, it will spread across the entire host population if the latter is genetically uniform. With land races, such genetic uniformity seldom extends beyond the fields of a single farmer or the fields of a village or district. With commercial seed, the uniformity often extends across nations, to fill a continent and beyond.

If genetic vulnerability is so obvious, why hasn't more attention been paid to it? Following the U.S. corn blight, the U.S. National Academy of Science Committee on Genetic Vulnerability of Major Crops (NAS, 1972) looked at the genetic diversity of American crops and found them dangerously narrow. For example 96% of the garden pea crop was planted to only two pea types, 95% of the peanut crop to only nine varieties, and over half of our two largest harvests, corn and soybeans, was based on less than six seed sources. These

Academy findings of vulnerability in the dominant crops of 1967 are essentially true over two decades later. There is little public awareness of the problem and a complete absence of institutional arrangements, either nationally or internationally, for monitoring genetic vulnerability to mitigate the impact of genetic disasters.

The NAS Genetic Vulnerability study in 1972 identified the major forces contributing to genetic uniformity in the U.S. seed industry and major crops so well that it is worth repeating sections of the report here. The chapter on the Challenges of Genetic Vulnerability begins,

> Two points are clear: (a) vulnerability stems from genetic uniformity; and (b) some American crops are on this basis highly vulnerable. This disturbing uniformity is not due to chance alone. The forces that produce it are powerful and they are varied. They pose a severe dilemma for the science that society holds responsible for its agriculture.
>
> The severity of the dilemma will stimulate some intensive thought by the scientists and policy makers concerned. There will be shifts in philosophy, shifts in the allocation of present resources and hopefully the allocation of additional resources.

This has not happened: the U.S. is on average as prone to vulnerability now as when the report was written more than two decades ago. The allocation of public resources for crop diversity in real dollars has actually decreased (NAS, 1991).Why is this so? The answer is not simple but relates to the nature of the issue, which is a technical one little appreciated by the public, and to the fact that no major epidemic has taken place in the developed nations since the U.S. maize blight in 1970 to shock society into action.

Food stability: a discussion

The existence of the food supply stability we enjoy is underappreciated by modern society, which has not experienced a food shortage in recent memory. In most developed countries a food shortfall is almost beyond public comprehension. Yet the size of the problem is proportional to the human population. The sheer number of humans to feed has made the seriousness of the potential for genetic vulnerability a major concern to the few who recognize the dangers. The fact that the full impact of genetic erosion of land races replaced by highly uniform commercial seed is not recognized by the decision makers and society at large does not lessen our current state of genetic vulnerability or reduce the problem of food-production stability.

Genetic vulnerability is always more clearly recognized after the fact than before, although the potential exists to predict to an extent an occurrence of vulnerability in some crops. Historically our reaction to pest and pathogen epidemics on crop plants has been a crisis response. Yet there has not been long term planning for either 1) building up germplasm collections and reserve pools of germplasm to back up existing varieties or 2) well developed alternative strategies to monocultures, such as crop mixtures or multilines, which, interestingly, were often found in the older agriculture with land races. Clearly our national priorities with regard to genetic vulnerability and food stability strategies are deficient or nonexistent; their absence endangers the future welfare of the nation. The same can be said internationally.

The developing countries are now a significant factor in genetic vulnerability. Both hybrids and high yielding varieties have come to dominate within the last two decades (Plucknett, 1990). The immediate effect of these developments has been: 1) production has increased and native land races have been replaced by elite germplasm in farmers' fields, and 2) essentially the same genetic backgrounds can now be found in the agricultures of both developed and developing countries (Brennan & Byerlee, 1991).

As a few elite materials have come to dominate the basic cereal and legume crops worldwide, increasing vulnerability is almost certain to emerge. Genetic vulnerability can no longer be considered in the national context alone, as it was in 1972, but in the international context of today's agriculture with elite germplasm, such as dwarf-stature wheat and rice. What we are doing currently is promoting a carpet of dwarf-stature wheat and rice varieties across the grain belts of the world. The magnitude of the potential becomes clear when the fact is known that most of the rice planted in Asia for the last decade shared the same cytoplasm (common female parentage) or that most of the high yielding bread wheats are presently based on only three maternal cytoplasms.

The burden of genetic vulnerability has, up until now, been placed primarily on the shoulders of plant breeders because there are elements in the technology of plant breeding that can be designed to minimize its impact (Marshall, 1990). Genetic vulnerability is, on the other hand, only one concern affecting the total array of uses of plant genetic resources. Farmers have always been alert to variations in a crop and have taken advantage of unplanned, fortuitous genetic variants that appeared long before these evolutionary processes had the names of mutation, introgression, and polyploidy. Genetic selection is a powerful tool both in controlled crossing or peasant farmer opportunism with land races. It can be employed to overcome the expression of genetic susceptibility once it appears and is recognized. The trouble with genetic vulnerability is that uniform monocultures are so widespread and the impact so magnified that it doesn't matter if the plant breeders can correct the problem. That it has happened at all is sufficient to disrupt the world food supply. Collectively, plant breeders have a very important influence on the amount of genetic uniformity to be found in commercial crop varieties, but demands of the industry and limitations of research resources ultimately define how diverse these varieties really are. Genetic uniformity in and of itself is not necessarily undesirable (if deployed wisely) and is necessary for much of our present agricultural markets (Plucknett et al., 1987), but to disperse a uniform variety very extensively is to spread a wide net for pests and pathogens. The price of the emphasis on maximum productivity is vigilance, based on a thorough knowledge of current varieties and the ability to trace any new infection with great care. Ultimately, varieties under development should originate from parentage wider than the varieties they displace. For most breeders, however, there are few incentives to go any further for breeding parents than the very small number of elite materials at the top of the pyramid (Duvick, 1986). Individual breeders watch with vigilance for outbreaks of pests and pathogens, but there is no overall management strategy to quickly bring genes from the lower levels of "genes in reserve" for most of the major crops, and there is no strategy at all for keeping the reserve ranks filled with planned diversity of genotypes for resistance to pests, weather, or soil problems.

Nowhere is the impact more serious than the extension of the high yielding grain belt germplasm into microhabitats and marginal lands of the tropics (CIMMYT, 1987). This has cast a larger net to capture pests and pathogens. These habitats are often without the pronounced dieback seasons of the major grain belts where severe cold (as in the U.S. corn belt), severe drought (as for Punjab wheat), or wet flooding (as in southeast Asian rice) push back the pests and pathogen populations. Presently, with wheat being harvested over most of the year somewhere in the world, the potential to capture pests and pathogens has increased.

Instead of efforts to develop wheat for the mountainous microhabitat zones of Thailand, there should be a strategy of patchiness to keep wheat in the grain belt of the Punjab and develop minor crops for the mountains of the tropics. Instead of promoting commercial seed in Iran and Turkey to form a continuous carpet from the Punjab into Europe, or commercial seed for Nepal and Thailand to form a continuous carpet from Punjab to China, there ought to be a mosaic of cultivated species so there will not be a bridge of single crop genetic uniformity to promote the march of a pest or pathogen. An integrated crop management strategy is needed to mitigate unwanted increases in genetic vulnerability of widely grown elite plant materials.

Future needs and priorities

If plant breeders breed from only adapted elite material, as many commercial companies tend to do, and aim only for the short run, narrow germplasm bases result, thus increasing the potential for genetic vulnerability. A certain amount of long-range breeding aimed at increasing genetic diversity must always be present. But it is difficult to determine the minimum number of plant breeders necessary to insure this goal of providing sufficient genetic diversity in reserve. If the total number of individuals committed to breeding decreases, the amount of germplasm actively used by breeders also may decrease unless some way is devised of increasing output per breeder, particularly with regard to the enhancement type of breeding for broadening the useful germplasm base. Much potential germplasm is lost due to lack of manpower. An inordinate amount of public responsibility must be assumed by a very few individuals who prepare gene bank or land race materials for further breeding work (Wilkes, 1989; CIMMYT, 1989). The degree to which these few can prebreed and effectively create, evaluate, release, and store valuable germplasm is limited. Further, as breeders retire and administrators change their emphasis, valuable breeding pools and germplasm collections often are discarded, a tragic loss for the public at large.

Changing patterns of susceptibility of major crops to diseases will always result in surprises, but the measure of success is how appropriate our preparedness and timely our response are to the vulnerability. Investing by the IARCs in improved minor crops for the zones between major grain belts should be a priority to protect their yield advancement with corn, wheat, and rice. Improvement in the evaluation of germplasm in gene banks and the development of alternate maternal parents (cytoplasm) should broaden the backup varieties to current elite materials. To do this we need plant breeders not only at the IARCs, but at the regional level dealing with local adaptation. Improvement of crop varieties depends on new and diverse germplasm resources, but also the human skill to see promise in segregating populations early on in the breeding process (CIMMYT, 1988). There exists an obvious need for more plant breeders worldwide to hybridize crops and generate genetic diversity than exist presently. The return to a more diverse genetic mosaic of species will be an important strategy to reduce genetic vulnerability within and between crops. Following the yield leaps of the Green Revolution of the late 60s, most of the breeding at the IARCs have been maintenance breeding to protect the crops from pests and pathogens and to keep the yield close to the earlier achieved high levels (CIMMYT, 1989).

Agriculture, germplasm conservation, and strategies for the future

The challenge as we approach the next century will be to design agricultural management strategies and institutional arrangements that can successfully ameliorate the negative environmental effects of agricultural and urban intensification primarily in the grain belts. The long range goal should be an agricultural development that does not come at the expense of any geographic region or later human generation, nor threaten the remaining noncultivated biodiversity of the planet (CGIAR, 1989, 1990; WWI, 1992). This is not an impossible task but one we won't achieve with only a half-hearted commitment.

The uniformity of major crops and the fact that they are displacing indigenous land races is a given of the present agricultural-consumer context. Subsistence farmers are a decreasing minority. Most farmers sell some or all of their harvest for cash and are, therefore, yield-over-cost-of-inputs driven. They need to use the best varieties available to them. Genetic vulnerability is inherent in the use of uniform elite germplasm. New varieties are continually brought up from breeders' plots to replace older ones. However, the lack of a broad-scale management strategy to anticipate expected problems and minimize their impact exhibits tunnel vision about how biological systems organize themselves and an insensitivity to the plant-breeding establishment's public responsibility. Positive actions to deal with genetic erosion, wipeout, and genetic vulnerability for both national and international levels must involve the following initiatives:

a) *New programs* to assess genetic erosion and vulnerability – variety surveys and collaboration in international efforts to monitor the use and geographic distribution of elite germplasm.
b) The *development* (especially in the public sector) of appropriate gene pools introgressed by exotic derived germplasm to support the commercial breeder with useful materials.
c) *Increase the training* in the maintenance of plant genetic resources both nationally and internationally. Active gene banks should be engaged in prebreeding and in moving genes from the collectors into enhanced materials but not finished varieties. The future will be best protected from genetic vulnerability if a worldwide cooperative network of crop-specific prebreeding develops.
d) *Conduct basic and applied research* in order to more efficiently and effectively: (1) measure genetic distance between varieties, and (2) improve evaluation techniques, especially relating to the nature of resistance to pests and pathogens and pest–host interaction, this in conjunction with early detection of changes in the virulence of pest and pathogens as well as shifts in the varietal picture.
e) *Educate and inform* the food industry, farmers, seedmen, plant breeders, and

others about the relationship of genetic and spatial diversity with regard to the potential for crop vulnerability so they can develop effective management strategies. Also, current trends to individualize food products (such as high-fructose maize for the sweetener industry) or specialty items might be exploited to help increase genetic diversity of farm crops.

1. Develop alternative management strategies to genetic uniformity such as: parallel breeding, genes in reserve, enhancement breeding for gene pools, wide crosses, and biotechnology.
2. Develop alternative management strategies to monocultures such as: crop rotation, tillage practices, crop mixtures, multilines for resistance, pyramided resistance factors, manipulations of pest parasites, pest trap crops, insecticides/fungicides, and better monitoring.

Conclusion

More than ever before, international efforts are required to help slow genetic erosion, establish and encourage active gene banks, and help prevent epidemics in the developing countries where the greatest threats of germplasm erosion and genetic vulnerability now exist. Because of its major role in ensuring global food stability, the IARCs and the plant breeding community can no longer continue the casual neglect of the tissues posed by genetic vulnerability.

References

Alcorn, J. B. (1991). Ethics, Economies, and Conservation. In *Biodiversity: Culture, Conservation and Ecodevelopment*, ed. M. Oldfield & J. Alcorn, pp. 317–349. Boulder, Colorado: Westview Press.

Altieri, M., Anderson, M. K., & Merrick, L. (1987). Peasant agriculture and the conservation of crop and wild plant resources. *Conservation Biology,* **1,** 49–88.

Brennan, J. P., & Byerlee, D. (1991). The rate of crop varietal replacement on farms: measures and empirical results for wheat. *Plant Varieties & Seeds,* **4,** 99–106.

Brown, A. H. D., Frankel, O. H., Marshall, D. R., & Williams, J. T. (eds.). (1989). *The Use of Plant Genetic Resources.* Cambridge, U.K.: Cambridge University Press.

Brush, S. (1991). A farmer-based approach to conserving crop germplasm. *Economic Botany,* **45,** 153–165.

Chang, T. T. (1985). Crop history and genetic conservation: Rice – A case study. *Iowa State Journal of Research,* **59,** 425–460.

CGIAR. (1990). Partners in conservation: Plant genetic resources and the CGIAR system. Washington, D.C.: Consultative Group on International Agriculture Research.

CIMMYT. (1987). The Future Development of Maize and Wheat in the Third World. Mexico, DF: CIMMYT.

CIMMYT. (1988). Recent Advances in the Conservation and Utilization of Genetic Resources: Proceedings of the Global Maize Germplasm Workshop. Mexico, DF: CIMMYT.

CIMMYT. (1989). CIMMYT 1988 Annual Report Delivering Diversity. Mexico, DF: CIMMYT.

Cohen, J., Williams, J. T., Plucknett, D. L., & Shands, H. (1991). Ex situ conservation of plant genetic resources: global development and environmental concerns. *Science,* **253,** 866–872.

Collins, W., & Phillips, R. (1991). Plant breeding training in public institutions in the United States: A survey conducted by the national plant genetic resources board. *Diversity,* **7,** 28–32.

Darwin, C. (1897). The Variation of Animals and Plants Under Domestication. Vol. II. 2nd ed. New York: D. Appleton & Co.

Duvick, D. N. (1984). Genetic diversity in major farm crops on the farm and in reserve. *Economic Botany,* **38,** 161–178.

Duvick, D. N. (1986). Plant breeding: Past achievements and expectations for the future. *Economic Botany,* **40,** 289–297.

Ehrenfeld, D. (1978). *The Arrogance of Humanism.* Oxford: Oxford University Press.

FAO. (1981). *Agriculture: Towards 2000.* Rome: Food and Agriculture Organization.

FAO. (1989). Plant Genetic Resources. Rome: Food and Agriculture Organization of the United Nations.

Frankel, O. (1989). Point of view: Perspectives on genetic resources. In CIMMYT 1988. Annual Report (International Maize and Wheat Improvement Center): *Delivering Diversity,* pp. 10–17. Mexico, DF: CIMMYT.

Goodman, M. (1990). Genetic and germplasm stocks worth conserving. *J. Heredity,* **81,** 11–16.

Harlan, J. R. (1975). *Crops and Man.* Madison, Wisconsin: American Society of Agronomy.

Hawkes, J. G. (1983). *The Diversity of Crop Plants.* Cambridge, Massachusetts: Harvard University Press.

Heiser, C. B. (1990). *Seed to Civilization: The Story of Food.* Cambridge, Massachusetts: Harvard University Press.

Holden, J., & Williams, J. T. (1984). Crop Genetic Resources: Conservation and Evaluation. London: Allen and Unwin.

IUCN. (1980). *World Conservation Strategy.* Gland, Switzerland: International Union for the Conservation of Nature and Natural Resources.

Juma, C. (1989). *The Gene Hunters: Biotechnology and the Scramble for Seeds.* Princeton, NJ: Princeton University Press.

Keystone. (1991). Oslo Report: Third Plenary Session Keystone International Dialogue Series on Plant Genetic Resources. Washington, D.C.: Genetic Resources Communication System.

Kloppenburg, J. R. (ed.). (1988). *Seeds and Sovereignity: The Use and Control of Plant Genetic Resources.* Durham, NC: Duke University Press.

Marshall, D. R. (1990). Crop genetic resources: current and emerging issues. In *Plant population genetics, breeding and genetic resources,* ed. A. H. D. Brown, M. Clegg, A. Kahler, & B. Weir. Sunderland, Mass.: Sinauer Associates.

NAS. (1972). *Genetic Vulnerability of Major Crops.* Washington, D.C.: National Academy of Science.

NAS. (1991). *Managing Global Genetic Resources: The U.S. National Plant Germplasm System.* Washington, D.C.: National Academy Press.

Oldfield, M. (1989). *The Value of Conserving Genetic Resources.* Sunderland, Massachusetts: Sinauer Associates.

Oldfield, M. L., & Alcorn, J. B. (eds.). (1991). *Biodiversity: Culture, Conservation and Ecodevelopment.* Boulder, Colorado: Westview Press.

Plucknett, D. L., Smith, N. J. H., Williams, J. T., & Anishelty, N. M. (1987). *Genebanks and the World's Food.* Princeton, NJ: Princeton University Press.

Prescott-Allen, R., & Prescott-Allen, C. (1990). How many plants feed the world? *Conservation Biology,* **4,** 365–374.

Purseglove, J. W. (1968). *Tropical Crops: Dicotyledons.* London: Longman.

Purseglove, J. W. (1972). *Tropical Crops: Monocotyledons.* Harlow, U.K.: Longman.

Simmonds, N. W. (ed.). (1976). Evolution of Crop Plants. London: Longman.

U.N. (1989). Department of International Economic and Social Affairs: World Population Prospects 1988. (ST/ESA/SERA/106), 1989.

Vasey, D. E. (1992). *An Ecological History of Agriculture.* Ames, Iowa: Iowa State University Press.

Vavilov, N. T. (1935). The phytogropgraphiç basis of plant breeding. In *The Origin, Variation, Immunity and Breeding of Cultivated Plant: Selected Writing of N.T. Vavilov* (translated from the Russian by K. Stan Chester), pp. 13–54. Chron. Bot. Bol. 13 1949/50. Waltham, Massachusetts: Chronica Botanica.

Walsh, J. (1991). *Preserving the Option: Food Productivity and Sustainability.* Issues in Agriculture, No. 2. Washington, D.C.: Consultative Group on International Agriculture Research.

Wilkes, G. (1977). The world's crop plant germplasm: an endangered resource. *Bull. Atom Scientist,* **33,** 8–16.

Wilkes, G. (1988). Plant genetic resources over ten thousand years: From a handful of seed to the crop-specific mega-genebank. In *Seeds and Sovereignity,* ed. J. Kloppenburg, pp. 68–89. Durham, NC: Duke University Press.

Wilkes, G. (1989). Germplasm preservation: objectives and needs. In *Biotic Diversity and Germplasm Preservation: Global Imperatives,* ed. L. Knotson & A. K. Stoner, pp. 13–41. Dordrecht: Kluwer.

Williams, J. T. (1989). Practical consideration relevant to effective evaluation. In *The Use of Plant Genetic Resources,* ed. A. H. D. Brown, O. H. Frankel, D. R. Marshall, & J. T. Williams, pp. 235–244. Cambridge: Cambridge University Press.

Williams, J. T. (1991). Plant Genetic Resources: Some New Directions. *Adv. in Agronomy,* **45,** 61–91.

Witt, S. C. (1985). *Brief Book: Biotechnology and Genetic Diversity.* San Francisco: Center for Science Information.

WWI World Resources Institute. (1992). *Global Biodiversity Strategy.* Washington, D.C.: IUCN & UNEP.

Part four

Management of biodiversity and landscapes

9

The paradox of humanity: two views of biodiversity and landscapes

EUGENE C. HARGROVE

Biodiversity is usually defended on the ground that humans are part of nature – that they are tied to nature instrumentally. This argument is about *prudence,* not *ethics.* It is designed to counter the arrogant claim that because humans technologically and culturally are no longer part of nature, nonhuman nature is expendable. In contrast, an ethical argument depends on the view that the special characteristics of humans that separate them from other parts of nature provide the foundations for their ethical obligations to protect it. Both kinds of arguments (prudential and ethical) and both views of the human–nature relationship (that humans are part and not part of nature) are needed if biodiversity is to be preserved. Moreover, the ethical argument needs to be supported by an aesthetic argument analogous to arguments for preserving artistic beauty.

Humans as part of nature

In terms of the general arguments given for and against the preservation of nature and biodiversity, there is considerable disagreement about the relationship of humans and nature – whether humans are part of nature or not. In the modern period, pride about human achievements has frequently turned into an arrogant disregard of nature. Many humans have come to think of themselves as godlike beings who have gone beyond nature and no longer have any need for it. They are confident that, whatever might occur, science and technology will find solutions and provide artificial environments for them. In short, these humans have come to the conclusion that they are no longer part of nature and that nature itself is an unnecessary extravagance. To counter this position, conservation biologists and other environmentalists have taken the opposite position – that humans are part of nature and unable to survive without it. These arguments are generally agricultural and medical, and focus, often implausibly, on the possible instrumental value of nature to humans.

Reliance on these "bread-and-medicine" arguments are not without a price, for they encourage defenders of nature and biodiversity to ignore, and frequently deny, other arguments and values that are more consistent with their basic motivations. For example, when conservation biologists and other environmentalists argue that rain forests ought to be preserved because of their possible instrumental benefits to humans – in order to find a cure for cancer or to provide genetic reserves for agricultural purposes – they are vulnerable to the counterargument that these reasons do not reflect their real motives: that they just like to study rain forests and want to have them preserved and available for study whether they possess any instrumental benefits to humans or not.

This counterargument is especially damaging because it is usually true. The real reasons Westerners from the temperate zones of the Northern Hemisphere want to preserve rain forests are normally scientific and aesthetic, and the values involved are not instrumental but intrinsic (see Hargrove, 1989, chap. 3). Ultimately, they want to preserve rain forests for their own sake, in a non-instrumental sense, not because they might be useful in an instrumental sense someday in the future. When conservation biologists point out in the standard rain forest argument that species are being lost without humans *knowing* that they ever existed, they are not simply lamenting that humans have not yet had a chance to study their possible medical and agricultural benefits, for quite clearly after they have studied them thoroughly they will *still* want them to continue to exist even if they know positively, beyond any doubt, that they have no agricultural or medical value at all. They want knowledge of these species because they consider such knowledge to be of intrinsic value, not because it has possible instrumental value. They want these species and systems to continue to exist because a world with them is intrinsically more interesting and worthwhile than a world without them.

Defenders of nature and biodiversity, nevertheless, avoid intrinsic value arguments. They believe that we live in a peculiar century in which value talk has fallen into disrepute. They think people want facts, not values, on the ground that values, with the possible exception of instrumental ones, are arbitrary, subjective expressions of emotion. It is not a fact, however, that such values are worthless. It is merely a claim that was made by an early twentieth-century group of philosophers, called logical positivists, that has trickled down to the level of the general public and transformed itself into a kind of twentieth-century folk belief. Moreover, it is not a fact that the only values worth thinking about at all are instrumental values. This is merely a vulgarization of another claim made by another group of late nineteenth and early twentieth-century philosophers, called pragmatists, that has also trickled down and achieved folk-belief status. Such beliefs are not, in fact, supported by the previous twenty-five centuries of philosophy and value theory. Nor, luckily, are they supported by the basic intuitions of people who are not environmental policy makers. We live in a world today that *believes* and *acts* as if values are not worthy of consideration, but *wishes* that they were. The call for rights for nature, for example, is one attempt to reinstate the instrumental/intrinsic value distinction. Seen in this way, the situation is not hopeless because the solution is simple – all we need to do is reinstate a more balanced value system and marry it up with a more complex conception of the human–nature relationship.

Four positions of the human–nature relationship

At first glance, there are only two positions concerning the relationship of humans to nature. Either humans are part of nature or they are not. The possibilities increase, however, when we try to transform the claim, one way or the other, into the beginning of an argument about human concern for the environment, for there are two possible arguments associated with each of these claims, and in each case one argument is proenvironment and the other is not.

If humans are considered part of nature, there is, of course, the standard argument that the continued life and welfare of humans depends entirely on the continued collective health and welfare of the ecosystems around the planet in which humans are said to be embedded. There is, nevertheless, an unpleasant counterargument: that because humans are part of nature, whatever they do is natural. If they destroy their environment, this is just another natural event in the evolutionary history of the planet and is not, therefore, a matter of concern: what will be will be.

Turning to the claim that humans are not part of nature, we have already discussed the argument that there is no need to worry about the environment because humans have separated themselves from it, and simply by employing science and technology can survive without it. It can, nevertheless, also be argued that the emergence of humans out of nature provides the very foundation for concern for nature. The argument is based on two senses in which humans are no longer part of nature. The first is a sense in which humans have emerged as self-conscious beings who are aware of the effects of their actions and through conscious, rational planning are frequently able to determine these consequences in advance. The second is the sense in which science and technology have given humans the power to have unnatural impacts on nature – that is, unnatural geological impacts in terms of human rather than geological time scales and on entire systems and groups of systems rather than short-term localized natural biological impacts within systems. Putting the two of these together, it can be said that although the new status of humans is the cause of the problem – because humans are no longer limited to natural influences within systems – it is, nevertheless, also a solution – because humans, unlike other animals, are aware of their actions and have the ability to alter their behavior.

There are curious interrelationships between these arguments. Note that the response to the argument that humans must protect nature because they are instrumentally dependent on nature (which is grounded in the claim that humans are part of nature) is answered by the counterargument that science and technology have removed this dependency (which is grounded in the claim that humans are no longer part of nature). Likewise, the argument that no action is required because everything that humans do is natural (because they are part of nature) is countered best by the argument that it is our special attributes (which set us apart from nature) that generate the ethical basis for concern about nature and provide us with the ability to act.

These relationships actually make it almost impossible for someone to maintain an ideologically pure stance with regard to the human–nature relationship whether he or she happens to be proenvironmental or antienvironmental, although I have encountered environmentalists who are willing to support both the pro- and antipreservationist arguments associated with the claim that humans are part of nature in order to avoid accepting a more complex human–nature relationship – for example, by arguing that all-out nuclear war is natural. If environmentalists are more careful, they can probably argue in terms of the idea that humans are not in any sense whatsoever apart from nature without having to defend the propriety of the total destruction of all life on Earth through nuclear war. Nevertheless, there are still other unpleasant consequences that cannot be avoided. First, the "in nature only" position produces an exclusively instrumental, anthropocentric defense of nature: nature should be protected in order to protect human beings. This argument is not completely satisfactory because the arguments that it is attacking are frequently said to be wrong because they are instrumental and anthropocentric. It is dissatisfaction with these kinds of arguments, once again, that frequently prompts calls for rights for nature. To be sure, there is at least one way out. Deep ecologists, for example, go on to argue that humans can become one with nature through a mystical process called self-realization, so that instead of protecting nature to protect oneself, one is now protecting nature to protect one's Self (see Reed, 1989; Callicott, 1985). Although this maneuver in a curious way elevates the status of nature so that it is no longer merely instrumental, by letting it become one with the human self (or the human self become one with it) so that it can share the intrinsic value of the individual human with which it is united, it is doubtful that such a mystical union will ever become widely accepted by the general public in this country and around the world (see Guha, 1989).

Second, the "in nature only" position does not produce an ethical argument. As Immanuel Kant pointed out two centuries ago, actions undertaken for personal benefit are usually not moral (Kant, 1785). If they are undertaken for human benefit, individual or collective, they are at worst immoral and at best prudent – that is, morally neutral. They are immoral if their motive involves selfishness. They are prudent if they are motivated by justifiable self-interest. They are moral, in contrast, if their motive is to do the right thing independent of selfishness and self-interest. As Kant also noted, actions that are prudent in terms of self-interest may also be moral on independent grounds; however, whenever self-interest is involved, it may be difficult to tell which is the correct justification, given that humans are frequently not able to sort out their real motivations even with the help of years of psychological analysis. Because humans usually have no control over what happens to be in their self-interest, there is nothing wrong with appeals to self-interest or prudence as long as they do not supplant all moral considerations. It is just this, however, that the instrumental, anthropocentric argument based on the "in nature only" position tries to do, even, I believe, in its self-realizing variant. The argument is intended as a shortcut that will appeal to the lowest common denominators, selfishness and self-interest, and will avoid ethical arguments which are more problematic because immoral people may not find them attractive.

More sympathetically, it could be argued that the instrument, anthropocentric appeal to selfishness and self-interest is required not simply because it is likely to be more effective with immoral people, but also because it is necessary, in order to be consistent, to have one and only one ultimate argument for the defense of nature. To have two arguments, one based on the idea that humans are part of nature and one based on the idea that they are distinct and independent of nature, would be inconsistent and indeed contradictory. Although this position is frequently encountered, especially in relation to moral issues, it is nevertheless an academic approach to argumentation that does not recognize practical realities. In the real world, having more than one argument for one's position, especially if they are based on independent grounds, is better than just having one. In defending nature, it seems reasonable to conclude that having one argument for selfish people and another for altruistic people is preferable to just having one for one group and none for the other. It is, after all, just possible that people who are moral might also want and need an argument of their own. Moreover, it is not contradictory to assert in one breath that humans are both part and not part of nature, for they are part and not part in different senses – the former referring to their biochemical dependency and their evolutionary origins and the latter to their unnatural (geological, rather than biological) impact on natural systems.

Christopher Stone's *Earth and Other Ethics* (1987) is an explication and defense of a position called moral pluralism, which is currently causing a good deal of controversy in theoretical environmental ethics literature. It is a position I support and wish to recommend. While it might be more elegant and intellectually tidy to have a single argument, from a single set of premises that are closely tied together, it is not necessary. Moreover, there is no way to tell at this point if and when such an argument will ever be developed, it is not prudent to wait, and it is not wise, if the object is to win the debate, to ignore arguments that might be effective among some other groups of humans – for example, moral people.

Consider for a moment the standard conservation biology slide-show defense of rain forests. As the audience listens to speculations about the possible instrumental value of rain forests as a cure for cancer or as a genetic reserve for agriculture, it watches a long succession of breathtaking slides displaying the natural beauty of rain forests. Never, however, does the audience hear an argument for protecting these systems because they are beautiful or because they have intrinsic value. Although the slides are obviously intended to be a nonverbal aesthetic argument, without a verbal reinforcement it is just as likely that the effect will be negative as positive. Because the rain forest defender never stops to point out that the beauty and scientific interest of rain forests are also good reasons to preserve them, the audience can easily conclude that only instrumental arguments are worth talking about and therefore suitable reasons for rain forest protection. When this happens, people who were previously open to intrinsic-value arguments have been converted to the simpler, twentieth-century belief that only the life, health, and welfare of human beings matter.

Although aesthetic arguments for the preservation of nature are seldom presented today, there are many people ready and willing to hear them. I learned this from my own personal experience when I was trying to protect a cave in Missouri from water pollution. Although I argued for the protection of the cave in terms of public health (the cave stream flowed past a picnic area in a state park) my opponents turned out, much to my own surprise, to be willing to ask about and listen to an aesthetic argument. Upon reflection it occurred to me that I should not have been surprised at all. The aesthetic appreciation of nature is a three-hundred-year-old Western tradition that developed out of aesthetic interest in the sublime in the mid-1600s and the picturesque in the early 1700s. Moreover, this aesthetic tradition came to North America with the first colonists and by the early 1800s, with the help of a multitude of American landscape painters and poets, had been translated into a national pride in the beauty of natural landscapes (Hargrove, 1989, chap. 3). Note that this tradition is nearly two centuries older than the economic and political considerations that currently dominate the form and content of official environmental policy and decision making.

Humans apart

If environmentalists abandon their exclusive reliance on anthropocentric, instrumental "bread-and-medicine" arguments and expand their repertoire to include a defense of natural beauty in terms of its intrinsic value, this change will require them also to adopt a more complex position generally. They will not only become moral pluralists, but also proponents of a more complex human–nature relationship in which humans in various senses are both part and not part of nature. In order to make the argument that humans must protect nature in order to protect themselves, they will still claim that humans are part of nature biochemically, but, in order to move successfully from prudence to ethics, they will also hold that there are differences between humans and the rest of nature, which set them apart from nature.

This change is, of course, likely to be a matter of concern initially because the "apart from nature" position has been so strongly associated with excessive human pride and arrogance in the modern period. These human character traits and the abusive behavior toward nature that has come with them, however, are only accidentally or historically associated with the view that humans have in some sense emerged out of and gone beyond nature. They are not even based on the idea that humans have become masters of nature. Rather, they are based on the belief that humans have become masters of something that has no value in its own right. The question, in other words, is not "Are humans completely independent of nature?" or "Are humans masters of nature?" but simply "Does nature have value independent of its instrumental value to humans?"

Art and natural beauty

This independent value could be, but need not be, independent of human judgment. A good model for this kind of value is the value we place on works of art. We do not go to art museums to obtain instrumental value from paintings. The object of aesthetic appreciation is not simply the feelings of pleasure derived from viewing works of art. These feelings accompany the act of aesthetic appreciation, to be sure, but they are not viewed as being the instrumentally derived object of the act. If the object was the instrumental generation of pleasure, these feelings could be induced more quickly and reliably, for example, by taking drugs. Rather the focus in the act of aesthetic appreciation of an art object is on the object itself, not because it makes us feel this way or that, but simply because it is wonderful that it exists. We delight in the fact that it is beautiful and in the fact that such beautiful things exist. We do not merely delight in our delight. Our delight is in the value of the artwork for its own sake, in its intrinsic value, not in its instrumental value as a pleasure generator.

At the beginning of this century, because of the prominence of American pragmatism at the philosophical level, because of Pinchot's doctrine of land value as use in environmental affairs, and because of the emergence of modern economic theory out of utilitarianism and positivism (Hargrove, 1989, p. 209), the aesthetic appreciation of natural landscapes and of natural beauty in general came to be viewed in instrumental terms. This conversion of natural beauty into an instrumental value has produced a great deal of confusion in environmental policy and decision making because the official valuational frameworks as they have developed in this century run counter to our basic (traditional) intuitions about what is valuable, how things are valuable, and which things are most valuable. When the aesthetic appreciation of nature is conceived as a use of natural resources, the use is the production of aesthetic feelings within humans, and when the production of those feelings is then weighed against the instrumental value of the landscape as a source of raw materials for industry, the aesthetic value appears to be marginal, trivial, frivolous, eccentric, and sometimes even ludicrous and undemocratic. It can be asked why the generation of feelings of pleasure in a few tourists, backpackers, and naturalists should be allowed to stand in the way of the economic growth and well-being of the United States and countries around the world and the life, health, and welfare of human beings.

This kind of question arises, however, only because the shift from the intrinsic to the instrumental perspective is a shift away from a valuational approach focused on the value of natural landscapes for their own sake to one that is centered on the value of fleeting feelings of pleasure. When this shift is accepted, as it nearly always is, natural beauty and natural landscapes are compromised and demeaned, for although natural beauty is a higher value, since intrinsic values are traditionally higher than instrumental values, feelings, on the other hand, have always ranked very low on instrumental priority lists, and probably always will. Even when natural beauty is valued instrumentally on its own, without comparison to higher instrumental values, counterintuitive policies and decisions usually result. For example, when tourist visitation begins to damage a natural landscape, all of the options permitted by an instrumental-value approach usually involve the continued destruction of the landscape through use. Because the value of a natural area is its use and the value is therefore lost if it is not used, all policy options have to be formulated in levels of use, which means that the alternatives are frequently little more than levels of aesthetic consumption that will be judged in terms of the length of time that feelings of pleasure can be efficiently generated before the pleasure-generating qualities are depleted.

The solution to this problem is not to abandon instrumental aesthetic arguments and try to substitute better instrumental nonaesthetic arguments, but rather to abandon the instrumental framework in favor of aesthetic arguments reformulated in terms of their intrinsic value. Because works of art are valued intrinsically and are not viewed as instrumental pleasure generators, the idea of efficiently consuming a deteriorating art object by finding a way to generate the most feelings of pleasure over the longest finite period of time does not arise. Because museum curators and art critics remained uninfluenced by the instrumental turn at the beginning of this century and did not convert to an instrumental-value framework, works of art that are being damaged by visitation are speedily removed from view until the problem is solved. Because this approach is uncontroversial and has widespread public support, the best way to proceed in developing more appropriate aesthetic arguments in defense of nature is to follow common practice with regard to the treatment of art objects, deviating from this practice only as required by significant differences between natural and artistic beauty.

At this point, it could be objected that basing aesthetic defenses of nature on such an analogy is perilous because artistic beauty is superior to natural beauty. First, artistic beauty is a conscious, planned act of creation whereas natural beauty is the unplanned accident of random biological and geological processes. Second, the creation of artistic beauty is akin to a creative act of God, whereas the emergence of natural beauty is simply blind chance that does not involve an act of any kind, creative or otherwise.

When confronted with these arguments, environmentalists have a number of ways to respond that need not put them on the defensive. There are very significant differences between natural and artistic beauty (Hargrove, 1989, chap. 6). Artistic beauty is produced in a very short period of time in accordance with a predetermined plan. The end product is a static entity, which can admit no additional change and which must be continuously protected from deterioration. Natural beauty in contrast is the product of complex interactions that have been taking place over millions of years. There is no static final product. Natural beauty changes in a seasonable cycle and slowly evolves into new forms over longer geological periods of time. While natural beauty can sometimes mimic the static nature of art, the reverse is usually not true. No human artist has ever created an art object that continually changes and evolves and is beautiful at every moment during its various transformations.

If humans could duplicate natural beauty, but nature could not duplicate art, there would be grounds for claiming that artistic beauty is superior to natural beauty. In reality, however, the reverse is the case: nature produces a kind of beauty that humans cannot match. Art is generally reductionist in character and is aimed at the generation of simple forms or at the construction of the appearance of complexity out of simple forms. Natural beauty in contrast involves a degree of complexity that is frequently beyond the comprehension and imagination of humans and cannot be duplicated by humans within human time scales. Although humans can produce landscapes of many kinds, they cannot produce a natural landscape, only a carefully manicured representation of a natural landscape, but still with the feel of a landscape garden. To produce the natural landscape, they must abandon it for a hundred years or more, let natural process reassert itself, and hope for the best. Nature employs methods that humans are unable and unwilling to employ. To act, a human must first make a plan. The act, in accordance with the plan, then structures the natural objects involved in such a way that they are no longer natural, whether they look natural to the untrained eye or not.

Turning to the second issue, is artistic beauty the result of creative activity and natural beauty not? Is artistic beauty the product of a creative process akin to the creative power of God and natural beauty merely the product of accidental and chaotic forces? Curiously, if one attends to the mainstream analysis of God's creative activity in late medieval philosophy and theology, creativity in nature is closer to the creativity of God than is the creativity of humans. According to late medieval and early modern philosophers, God, in order to protect his omnipotence, created the world indifferently, without a plan. To have followed a plan, these scholars maintained, would have required God to have recognized independent standards, which as limitations would have impinged on God's all-powerfulness and prevented him from being God. As a result, God did not make the world in accordance with any standards of beauty and goodness. Rather he just made it, and the result became beautiful and good because of his creative act. Although the indifferent creativity of nature is completely compatible with this conception of divine creativity, the carefully planned imaginative creativity of humans is at odds with it. The results of indifferent creativity in nature is natural beauty. In contrast, when humans try to act indifferently, imitating nature, the result is, to put it bluntly, not beautiful.

Creative imagination is frequently cited as the element in human creativity that makes artistic beauty superior to natural beauty. I would contend in response that it is a poor substitute for the creative indifference of nature and is actually derivative from it. Ultimately, the forms created by human imagination are, as David Hume pointed out long ago, simply the mixing and rearranging of the forms found in and taken from nature by human artists (Hume, 1748). Such creativity is really not very creative at all, for it is usually either direct borrowing or a simple negation of the natural. Following the contrasts established at the beginnings of modern aesthetic theory, natural beauty is complex and artistic beauty simple; natural beauty is big and artistic beauty small; natural beauty is geometrically irregular and artistic beauty geometrically regular and simple; natural beauty is rough and artistic beauty smooth; natural beauty is impermanent and artistic beauty permanent and unchanging. To be sure, these connections are blurred today. For example, a mountain that was terrifying in the 1600s may project a reassuring feeling of calm and harmony today. Such changes, however, in no way alter the basic facts: art is derived from and defines itself in contrast to natural beauty, and these lines of dependency hardly bespeak superiority.

In defending natural beauty, and biodiversity, it is essential that the arguments be developed in terms of a human–nature relationship in which humans are not part of nature, in which nature is viewed as an other. Natural beauty arises out of the unfolding of natural history through the creative indifference of uniformitarian geological and evolutionary biological processes, independent of human involvement. Our aesthetic admiration and appreciation for natural beauty is an appreciation of the achievement of complex form that is entirely unplanned. It is in fact *because* it is unplanned and independent of human involvement that the achievement is so amazing, wonderful, and delightful. This kind of admiration and appreciation cannot be accounted for in terms of a position in which humans are irrevocably part of nature, for that position leads at its worst to efforts to perfect nature and at its best to artificial natural landscapes, based on the romanticism of the eighteenth and nineteenth centuries – careful approximations of the wild in nature which, no matter how hard the gardeners may try, remain representations.

Nature as other

It may be objected that separating humans from nature to permit the indifferent creativity of nature to continue will also keep the door open for the arrogant contempt for nature that is characteristic of the modern period, which the "in nature only" view is supposed to counter. I do not see that the ability to dominate nature must lead to domination and a contemptuous dominating attitude. Counterexamples are possible. In J. R. R. Tolkien's *Lord of the Rings*, for example, early in the first volume, a very unusual fellow, named Tom

Bombadil, makes a brief cameo appearance. Bombadil does find instrumental value in the forest. In addition, he is incredibly powerful and has very considerable (ecological-like) power and control over the forest. This mastery, however, does not translate into domination or possessiveness or excessive pride. In a letter, written in 1954 to a reader of the trilogy (Carpenter, 1981), Tolkien elaborates in some detail on Tom Bombadil's view of nature, explaining that he is an example of pure natural science that wants knowledge of, and takes delight in, other things, in terms of their history and nature, because they are "other," completely independent of the mind. He compares this knowledge with zoology and botany and contrasts it with cattle-breeding or agriculture (Tolkien, 1981, p. 192). Tom is in this characterization a scientist who simply desires knowledge of other things. Although he can use this knowledge instrumentally, that is not his primary aim. He wants knowledge for its own sake – he wants knowledge that is intrinsically valuable. He is not seeking this knowledge to come to understand himself better (by finding himself in nature), but to understand that which is not himself, that which is other. According to Tolkien, this view of nature is a morally significant alternative to the arrogant modern view of nature as something to be dominated and controlled – a view incidentally shared by biologists studying species in rain forests.

Conclusion

It does no good for environmentalists to argue in terms of a simplified conception of the human–nature relationship that humans should consider themselves exclusively part of nature and that they should value nature exclusively in terms of its instrumental value to the life, health, and welfare of humans. To be sure, humans are complex creatures who crave simplicity; yet, paradoxically, once they have made things simple for themselves, they quickly feel that something is missing. Although they like the convenience of having one conception of nature and one conception of value, the framework is not adequate to permit them to account for all of the strange and complex facets of their human–nature relationship or to express all the ways in which they wish to value nature. They need to be able to value nature not just instrumentally, insofar as it benefits humans, but also intrinsically, as an other with a sake of its own. They need to be able to express their relationship to nature not just prudently in terms of economics, self-interest, survival, and at the far extreme arrogant self-indulgence, but also ethically and aesthetically in terms of concern for, delight in, wonder at, and admiration and love for nature as an other. The problem, as I see it, is not to find a way for humans and nature to become one, but rather to find a way for nature to retain an independent identity so that natural history can continue in some sense and not be subsumed, humanized, or civilized into a part of an all-encompassing conception of human history.

In his famous essay, "The Land Ethic," Aldo Leopold wrote that the environmental movement had made itself trivial by trying to make itself too easy, suggesting that it is for this reason that philosophy has not yet paid attention to it (Leopold, 1949, pp. 209–210). The situation today is very different from the time when Leopold wrote these words. Philosophers have now heard of the environmental movement and for nearly twenty years they have been trying to provide the arguments that Leopold and others have called for. Curiously, however, most environmentalists have not heard of environmental ethics, or if they have heard of it, have chosen to ignore it because it is too hard to understand, preferring to continue to talk about rights for nature and instrumental anthropocentric values instead.

There is a serious danger in ignoring environmental ethics literature in favor of simpler argumentation. There have been three periods in the history of Western civilization: the ancient, medieval, and modern periods. It is generally believed that we are at the beginning of a new period – tentatively called the postmodern period, until a better name is invented for it. We do not yet know what the new period is like. If the postmodern period and beyond brings about changes as dramatic as those between the medieval and modern periods, environmentalism may not simply become trivial, in Leopold's words, but irrelevant, or even inconceivable. Supposedly, the next age will be based on ecology. But can we depend on it? Only in developing a sound theoretical foundation for environmentalism and presenting arguments to the public at the level of complexity that this foundation requires is there likely to be any reasonable chance that nature preservation, and biodiversity, will still mean something important in the next century and beyond.

References

Callicott, J. B. (1985). Intrinsic Value, Quantum Theory, and Environmental Ethics. *Environmental Ethics,* **7,** 257–275.

Carpenter, H. (ed.). (1981). *The Letters of J. R. R. Tolkien.* Boston: Houghton Mifflin.

Guha, R. (1989). Radical American Environmentalism and Wilderness Preservation: A Third World Critique. *Environmental Ethics,* **11,** 71–83.

Hargrove, E. C. (1989). *Foundations of Environmental Ethics.* Englewood Cliff, NJ: Prentice-Hall.

Hume, D. (1748). *An Enquiry Concerning the Human Understanding.*

Kant, I. (1785). *Foundations of the Metaphysics of Morals.*

Leopold, A. (1949). *A Sand County Almanac and Sketches Here and There.* New York: Oxford University Press.

Reed, P. (1989). Man Apart: An Alternative to the Self-Realization Approach. *Environmental Ethics,* **11,** 53–69.
Stone, C. D. (1987). *Earth and Other Ethics: The Case for Moral Pluralism.* New York: Harper & Row.
Tolkien, J. R. R. (1965). *The Lord of the Rings,* vol. 1: *The Fellowship of the Ring.* Boston: Houghton Mifflin.

10
Biodiversity and landscape management

ZEV NAVEH

Introduction

In this book the problems of biodiversity and landscapes are rightly approached within a broad interdisciplinary context, transcending the realms of biology and natural science. One of the main objects of this chapter is to show that landscapes, as the total natural and human living space, can be fully comprehended only by such a transdisciplinary approach, and that their biological diversity is closely related to their cultural diversity.

In landscape ecology, as the scientific basis for landscape study, management, and conservation, biodiversity is considered an integral part of the broader concept of landscape heterogeneity. This has recently become a central issue in landscape ecology (Merriam, 1988) and a special symposium has been devoted to this subject in the United States (Turner, 1987).

First, some of the major premises of landscape ecology as related to biodiversity and landscape heterogeneity will be introduced and then the problems of conservation management of open landscapes will be dealt with. Although this discussion will be restricted chiefly to Europe and the Mediterranean, its implications are much more far reaching, especially for industrialized countries and regions with similar temperate and Mediterranean climates, and especially the eastern United States and California.

Holistic approaches to landscapes and landscape heterogeneity

In landscape ecology, the problems of the spatial, temporal, and functional landscape heterogeneity and its management are addressed on scales of a few meters to kilometers, and changes from the distant past to the present and predicted changes for the future are considered. Landscape heterogeneity can be recognized visually in the three-dimensional integration of physical, biological, and noospheric phenomena from the geosphere, the biosphere, and the noosphere (the sphere of the human mind and consciousness) into horizontal–chorological patterns of landscape units and vertical–topological patterns of landscape attributes. A further, fourth dimension is added by changes through time (Zonneveld, 1990).

Most landscape ecologists in the United States are concerned chiefly with the large-scale heterogeneity of the expansive so-called "natural" – but in reality only seminatural – landscapes of North America. Many worthwhile efforts are devoted to the study of the interrelations between spatial landscape patterns, patch dynamics, natural and human disturbances, and landscape heterogeneity along different spatial and temporal scales (Turner, 1987; Dale et al., 1989). The quantitative methods applied for this purpose have been discussed recently by Turner and Gardner (1990). However, in most of these studies on landscape heterogeneity, only the bioecological aspects – not the human ecological, cultural, and perceptional aspects – are considered. Their methods are based mostly on sophisticated mathematical models and are highly technical. Their practical value as guidelines for the conservation of biological and ecological landscape diversity in the decision making process of land use planning and management has still to be shown.

In Europe, on the other hand, and more recently also in Canada (Moss, 1988), more holistic problem-solving oriented approaches to landscape ecology and management have evolved. More attention is also paid to the smaller scale, closely interwoven natural and cultural dimensions. In Europe, landscape ecologists are aware of the fact that in many thousands of years of human intervention, first the paleolithic food-gatherer hunter and then the prehistoric and historic agropastoralist opened up, with the help of fire, the dense pristine forests and gradually converted all natural landscapes into fine-grained mosaic-like open cultural landscapes of seminatural forests, woodlands, and grasslands and agricultural fields and plantations, interspersed with hedgerows, terraces, roads, and settlements. Therefore, humans are not considered as external disturbance factors but as interacting coevolutionary ecosystem components (Naveh, 1984; Haber 1990a,b).

Most landscape ecologists use the "ecotope" as the smallest landscape subunit and not the vaguely defined "ecosystem." This is actually the tangible site of its ecosystem. Its boundaries are pragmatically defined according to the requirements of the study. More-or-less similar ecotopes with recurring properties can be aggregated into ecotope-types and all of these in turn, can be mapped together as regional landscape units of different spatial scales. Their ecotope-type diversity can be treated in ways similar to species diversity in ecosystems (Haber, 1990a).

As in Naveh and Lieberman (1983), landscapes should be conceived as integrated ordered wholes and interacting systems. To their physical, geographical space, *Homo sapiens* have added throughout their cultural evolution also conceptual-noospheric space and new, emerging structural and functional qualities. Our present landscapes are therefore complex natural and cultural gestalt systems, which contain more than the tangible, measurable, and quantifiable parameters of our space–time dimensions. They represent thereby a higher order of complexity in the ecological hierarchy than the natural bioecosystems. Following Egler (1964), we suggest calling this most complex ecological level the "total human ecosystem."

However, together with these new qualities of the open landscapes, a new type of artificial, human-created technological ecosystem or "technoecosystem" also evolved in the built-up landscape. Since the industrial revolution these have formed the rapidly growing technosphere and its urban–industrial complexes. All these open and build-up landscapes constitute the global ecosphere landscapes of our total human ecosystem.

Biodiversity and the total human ecosystem

The biological impoverishment of our total human ecosystem can now be viewed in a more general way as a gradual shift from biosphere to technosphere landscapes and their technoecosystems and ecotopes.

In Figure 10.1 a hierarchy of landscape ecotopes is presented with decreasing biospheric inputs of solar energy, natural material and organisms, and regulating biophysical information from the biosphere, and inversely, of increasing technospheric inputs of fossil energy, technologically produced or converted material and noospheric, controlling, and regulating information. In seminatural landscapes, the level of biodiversity – its enrichment or impoverishment – is determined by the kind and intensity of human modifications and their effect on the self-organizing and stabilizing capacities of these systems. Therefore, a clear understanding of the underlying mechanisms of these closely interwoven natural and cultural ecological processes is of great importance for their conservation management (Rickleffs et al., 1984). In intensively managed semiagricultural ecotopes of noncultivated pastures and forests, biodiversity is reduced considerably. As will be shown below, our modern agroindustrial ecotopes are replacing and degrading the remaining open landscapes and impoverishing biodiversity at alarming rates.

This neotechnological landscape degradation is the result of the accelerating and mostly uncontrolled expansion of the technosphere and its entropy and pollution-producing waste products. It is driven by mutually amplifying and

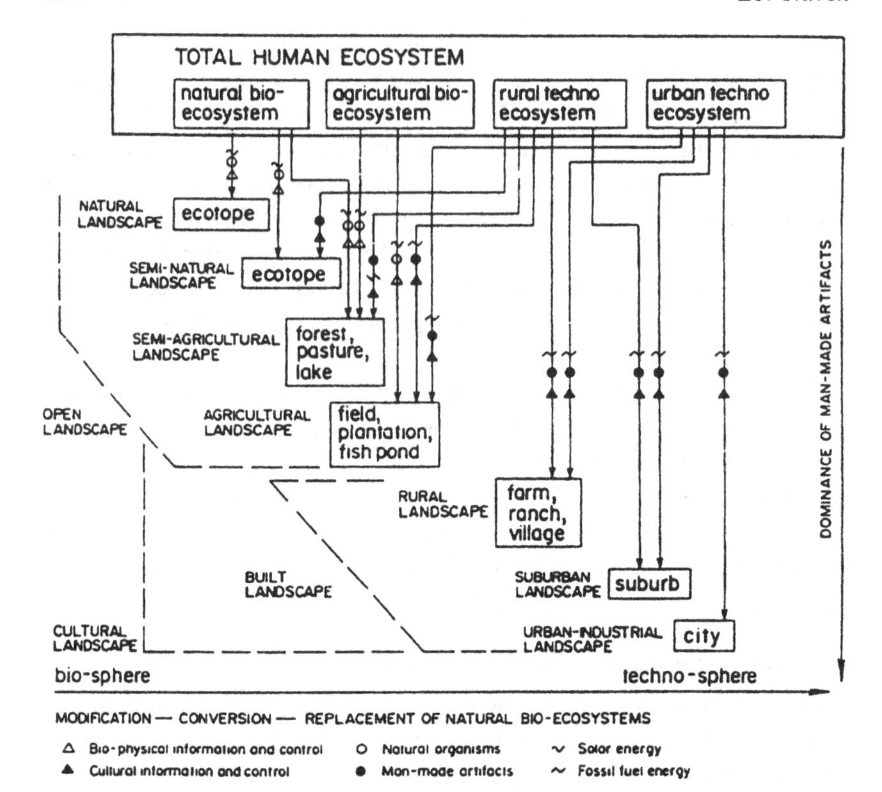

Fig. 10.1. Hierarchy of landscape ecotopes according to energy, matter, and information inputs from bio- and techno-ecosystems (Naveh & Lieberman, 1984).

destabilizing "runaway" feedback loops between unrestrained growth of population, energy, and consumption, and technological and political power. With the removal of most of the regulating and restraining natural and cultural feedback loops, our open landscapes are losing their biological and cultural richness, diversity, and ecological stability. Many of them are turned thereby into vulnerable monotonous agroindustrial steppes and man-made technological deserts, endangering our global total human ecosystem (Naveh, 1987). These threatening trends can only be reversed by more powerful negative inputs – restraining and controlling noospheric feedbacks of scientific, educational, and political information. In this process landscape ecology and other conservation-oriented sciences and information-providing activities, such as the conference from which this book resulted, can play an active role. An excellent example for such restraining negative feedbacks of scientific information has been provided by landscape ecologists in Germany, using innovative "ecological balancing methods" (Schaller & Haber, 1988) and nature conservation planning and management strategies (Bruns, 1988) to rescue biologically rich habitats and biotopes, which otherwise would have been lost by land consolidation and homogenization of agricultural landscapes.

New methods for the study of landscape biodiversity and connectivity

An important first step in this direction is the development and application of comprehensive methods for the assessment of biodiversity on the landscape scale and its relationship to ecotope diversity and heterogeneity.

For this purpose great amounts of data from many different sources on relevant chorographical and topological landscape attributes of the study area have to be collected, analyzed, and synthesized with the help of remote sensing and computerized Geographical Information Systems (GIS). These systems allow not only the storage, retrieval, and output of this information in the form of maps and tables, but also cartographic modeling methods, based on the conceptual understanding of the landscape processes and patterns and the transformation of these source data into practical information. They can also be used for dynamic evaluation of different scenarios and therefore are of great value for dynamic biodiversity conservation oriented landscape management (Bourrough, 1984).

Davis et al. (1990) have recently proposed such a comprehensive GIS model to organize existing data on biodiversity and to improve spatial aspects of the assessment for a U.S. National Biodiversity Center for the United States and to identify gaps in the network of nature reserves for California. The taxonomical and ecological data required for assessing the status of biodiversity are arranged along multileveled spatial scales from the local site to the regional scale and to the biogeographic scale. In each spatial level the specific cultural features of land use and their effect on environmental quality and landscape structure are also taken into consideration. The model shows clearly the complexity of biodiversity assessments for management and conservation purposes and is meant to facilitate the very much needed cooperation and coordination across administrative and political borders.

Other landscape ecologists in the U.S.A. have transformed the nonspatial Shannon-Wiener and Simpson diversity indices into spatially explicit measures of biotic landscape diversity (Romme, 1982; Hoover & Parker, 1991; Gardner & O'Neill, 1991).

Probably of greatest relevance for conservation management on the landscape scale is the determination of "connectivity" (Merriam, 1984), as a functional parameter of landscape connectedness by structural links or corridors between landscape units, described from mappable elements, such as woodland patches, hedgerows, roads, etc. It measures the processes by which subpopulations of organisms are interconnected into a functional demographic unit and into metapopulations and applies also to the interconnection of other functionally related ecological processes such as subunits of nutrient pools, interconnected by fluxes into landscape nutrient pools (Baudry & Merriam, 1988) (Fig. 10.2). In the special symposium devoted to connectivity, many examples of its application in landscape ecological studies have been provided (Schreiber, 1988).

As explained in detail by Merriam et al. (1989) connectivity has already lead to new insights into the effect of natural and man-made corridors and barriers on metapopulation demography and genetics, down to the subcellular and DNA levels. These insights should become the major guidelines for planning and management in heterogeneous environments. These should be approached on the landscape level by recognizing the heterogeneity of the land unit and addressed at the scales of size and quality used by the species of interest. The planning unit should be the minimum land unit necessary to understand demographic and genetic survival of the species, also taking into consideration temporal changes to landscapes and populations, as well as global climatic changes. This requires the integration of all the landscape and regional processes that will be operating on a relatively long-term basis. Retention of the array of required habitats, and connection of vital patches and landscape features are fundamental safeguards for the integrity of large ecological management units, such as watersheds, political units, etc. Within each landscape type a nested hierarchy of landscape elements at several scales appropriate to groupings of ecological types of organisms should be managed by a holistic "top-to-bottom" framework (Noss & Harris, 1986; Noss, 1987).

Another promising approach is offered by broadening of the integrated assessment of landscape heterogeneity with the help of information theory into the measurement of organization, order, and disorder in landscapes. This is accomplished by assessing the degree of predictability (as a parameter of order) in the occurrence of a specific ecotope type, on basis of its ecological and cultural features, as a neighbor of another type, or by the "mutual information" or "redundancy" of two neighboring ecotopes (Kwakernaak, 1984; Phipps, 1984). This method has been further developed in an important study on vanishing landscapes in Tuscany by Vos and Stortelder (1988) and will be reported below.

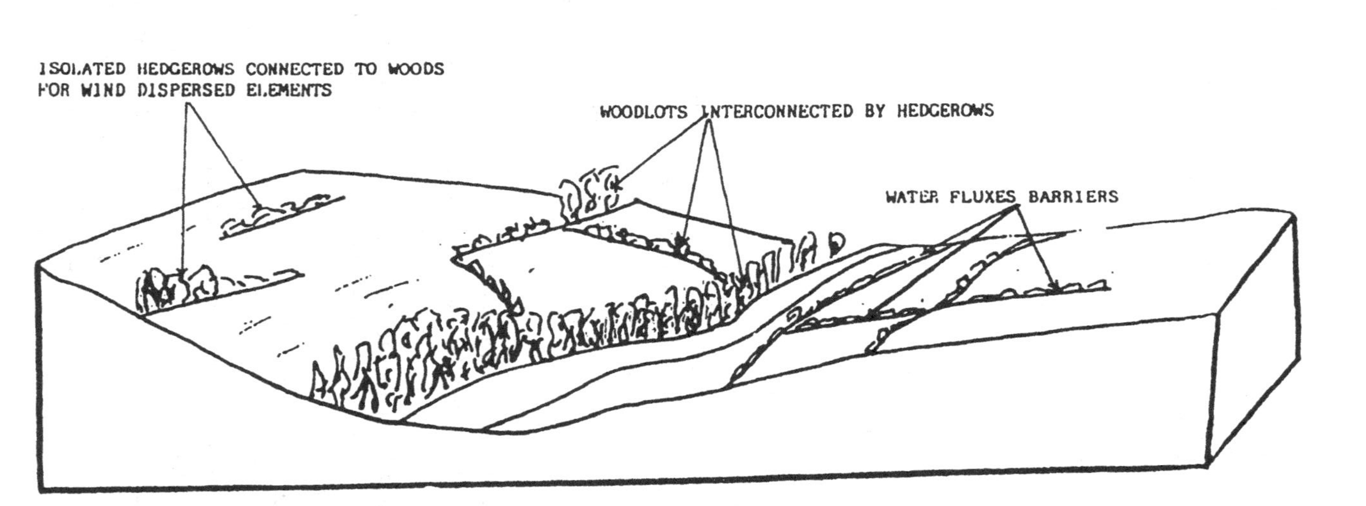

Fig. 10.2. Functional units in a hedgerow network landscape (Baudry & Merriam, 1988).

Biodiversity and landscape heterogeneity in European agricultural landscapes

Probably nowhere else has the recent neotechnological revolution caused more rapid and drastic detrimental changes to the open landscape than in Western and Central Europe. Landscape ecologists are very much aware of this and they are especially concerned with the fate of the small biosphere islands and connecting "ecolines" (Bridgewater, 1987) in the agricultural landscapes. By providing guidelines for their conservation and restoration to land planners and decision makers, they help to pave the way for richer and more stable post-industrial landscapes.

Of special importance in this respect are management-oriented connectivity studies in relation to hedgerows and to other small landscape elements and network patterns, reported in the above-mentioned special symposium on connectivity (Schreiber, 1988).

A good example is a study in eastern Denmark by Agger and Brandt (1988) on the dynamics of "patch biotopes," such as natural or artificial lakes and pools, bogs, marl pits, natural thickets and plantations for game, and "line biotopes" at least 10 m long, such as rivers, brooks, canals, dykes, drainage ditches, road verges, hedges, tree rows, and footpaths. As a result of the recent intensification of agricultural production, as well as road construction, marl pitting, and extraction of gravel, there is a clear tendency for their total decrease in area density, especially of smaller wet-patch biotopes. On the other hand, there is a rapid growth of other patch biotopes on small abandoned areas.

These small-scale landscape elements serve as the only source for biodiversity and are important for their amenities, cultural value, accessibility, recreation potentials, game production, and for marking field and estate borders. Protection against their removal, pollution, and disturbance on poorer land could be achieved by voluntary arrangements if the problem of public accessibility to private land could be solved. But in areas with better agricultural conditions and continued intensification there is need for stricter regulations. These can be enforced through the Danish Nature Conservation Act, requiring a permit from the nature conservation authorities for any changes in the beds of open watercourses, lakes, bogs, moors, heaths, salt meadows, and marshes. The authors proposed special management models for roadside verges and estate boundaries to form general barriers against further landscape fragmentation and to serve as skeletons for the creation of a new landscape as marginalization proceeds. These models indicate where marginalization of fields might satisfy recreational needs and require that all boundaries around boroughs, parishes, and estates shall carry some small biotopes. These developments will be hopefully supported by new legislation for the creation of uncultivated buffer zones along watercourses to limit the leaching of fertilizers, which has induced oxygen deficit in the Danish Sea, causing severe cases of fish death in the last years. However, if general price and market regulations continue to dominate agriculture policy within the European Economic Community (EEC) and agricultural development will not be integrated with landscape planning, such models are doomed to fail.

The same is also true in all other EEC countries in which agriculture policies were aimed exclusively towards an increase of production through increases use of chemical fertilizers, pesticides, and herbicides, the replacement of labor by heavy machinery, and the adaptation of the agriculture infrastructure to the latter by land consolidation.

Harms et al. (1987) summed up these developments in the Netherlands by stating that "agriculture has grown from controlling nature and landscape into their greatest threat" by distorting the strong relations between natural landscape elements and agricultural activities. The latter were restricted in time and space by highly diverse local conditions of water table and soil fertility levels to which the farmer had to adapt his land uses. The resulting removal of the dynamic equilibrium between nature, landscape, and man is threatening more than half of the Dutch flora and especially some of the most valuable wetlands and grasslands for breeding birds in Central Europe, as well as habitats of organisms that were adapted to lower soil fertility and lower water tables. The leveling process of land consolidation to greater parcel size of rectangular shape and the neglect of the seminatural landscape elements eliminated further the fine-grained patterns of ditches, hedges, and woods (Fig. 10.3), and many valuable cultural and historical elements and structures enriching the landscape, together with characteristic scenic features.

As noted by Green (1989), in Great Britain the small-scale historically diversified landscape patterns of seminatural and farmed and forested vegetation and of villages and small country towns, greatly cherished for their beauty and environmental quality, are also threatened by environmental agricultural impacts accompanying intensification and land consolidation, such as pollution from pastures by nitrogen fertilizers, loss of species and their habitats, and increasing incidence of soil erosion. Here, between 1947 and 1980, about 40% of broadleaved woodland, 110 miles of hedgerows, and 25% of the seminatural vegetation were lost. Ten species of flowering plants and several insect species have become extinct and many plant and animal species, especially birds, are endangered. Much of this can be attributed to scrub encroachment and secondary successions of woodlands in formerly grazed, marginal heathlands and downlands.

However, in recent years there is a growing concern about these environ-

Fig. 10.3. Small woods and hedges in 1950 and 1970 in a land consolidation area in "de Achterhoep" (east Netherlands) (Harms et al., 1987).

mental consequences and a tendency to shift from increased agricultural production into increasing multibeneficial productivity of the land including outdoor facilities, nature, and whole-landscape reserves. For the latter, the Dutch government proposed the establishment of more-or-less protected landscapes as "Natural Landscape Parks," with completely protected nature reserves, traditional agriculture and buildings and sites with historical and cultural importance on 200,000 ha, making up about 10% of the total agricultural area of the

Netherlands. In these, farmers may voluntarily manage their farms under restrictions for financial compensation. However, up to now only a few farmers are ready to join. There is concern that this may lead to the deterioration of the rest of the landscape. New conservation strategies are now considered to ensure a reasonable existence for farmers in harmony with a warranted management of nature and landscape, by focusing on production control.

In the Netherlands landscape ecological planning and management has already contributed much to dynamic nature conservation, environmental forestry, and land development in the new polders. Let us hope, therefore, that it will also help to implement these new strategies in the Dutch agricultural landscapes without further delay.

In Great Britain many farmers have already joined schemes for subsidized farming, restricting the use of agrochemicals and requiring the maintenance of hedgerows, walls, and other landscape amenities, as well as "Set Aside" schemes, in which marginal cropland is temporarily fallowed or converted into woods. There is a strong body of opinion that believes that the best way to protect the countryside is "for its own sake," and to adopt alternative strategies of returning to more sustainable, lower input/output farming systems.

In Germany Haber (1990b) proposed diversification in land-use intensity, based on gamma ecotope diversity, as guidelines for comprehensive planning and management strategies, including nature and biodiversity conservation and landscape stewardship and care. This should be achieved by preserving at least 10–15% of the land within each regional natural unit for natural and seminatural ecotopes, semiagricultural, unmanaged pastures, and forests managed by selective cutting only, to ensure a sufficient number of wild plants and animals that can coexist with human land uses. In order to promote a fine-scaled diversity of intensive land uses, these should form a more or less evenly distributed natural ecotope network, and agricultural field size units should not exceed 8–10 ha in densely populated regions. Presently the main value of this "Differentiated Land Use" strategy is to slow down the accelerating changes in land use structure wherever these changes have not yet gone too far.

However, in Europe, as well as in all other industrialized countries, the increasingly severe problems of soil pollution and eutrophication have shown convincingly that the short-sighted increase of economic production at all costs has not only reduced biodiversity and the protection and regulation functions of seminatural ecotopes and diminished the soil biota and their soil-fertility building and stabilizing functions in arable fields but has also distorted the flow of nutrients and spread them indiscriminately over the whole landscape through chemical fertilizers, manure, and sludge and by drift, run-off, and rain, together with toxic pesticides. According to Rose (1987), nitrogen inputs into European open-landscape ecosystems in rain are now in the range of 30–40 kg N/ha, as compared to the natural background of 1–3 kg N/ha. This is endangering not only the ecological diversity but also future economic productivity. Therefore no piecemeal solutions but only a radical shift from agroindustrial farming to ecologically sustainable, organic farming will be able to reverse these trends and restore the health of these rural landscapes. It is highly doubtful that this can be achieved by the creation of small biosphere islands and cultural open-door museums, surrounded by a biological and technological desert. Much more far-reaching efforts are required for innovative regional planning and management of the open and built-up landscape as a whole, guided by a comprehensive landscape-ecological determinism.

Neotechnological degradation of Mediterranean uplands and their dynamic conservation management

In the countries around the Mediterranean Basin the situation is different. Here, with the exception of the few remaining dunes and wetlands, the major refuge for wild plants and animals are the extensive uplands, with soils too shallow, steep, and/or rocky for mechanized, modern agriculture. Wherever the semi-natural grasslands, shrublands, woodlands, and forests have not been degraded into poor scrub and rock deserts by overuse or into dense, unpenetrable, and highly inflammable thickets by underuse or protection, they are distinguished by great biological richness, especially in herbaceous plants. Amongst these are many flowering composites and geophytes with great ornamental value, and grasses and legumes that served as progenitors for domestication and have great genetic potentials. Amongst the woody plants are many Labiatae with great value for medical and other economic uses. Their striking floristic and faunistic richness and mosaic-like patch dynamics of diverse regeneration patterns was the result of thousands of years of lasting and regularly repeated defoliation pressures by burning, grazing, cutting, and coppicing, superimposed on the great macro- and microsite heterogeneity of the rough and rocky terrain. The combination of natural, climatic, seasonal, and annual fluctuation and human perturbations created a metastable flow equilibrium between the herbaceous and woody components. Therefore, after the cessation of cultivation, woodcutting, grazing, and burning, aggressive tall grasses and thistles crowd out smaller herbs and a dense, species-poor and a highly combustible weed thicket establishes itself in open woodlands and grasslands. In maqui shrublands and forests this leads to the dominance of a few taller woody plants and the almost complete suppression of the herbaceous understory. This causes a heavy reduction of plant and animal diversity and the loss of the richest and most attractive, more open and lower grass and shrub "degradation" stages, including many light demanding, flowering geophytes and endemics.

Typical examples from protected nature reserves are the mixed *Quercus ilex* and *Pinus halepensis* forests at the unique island nature reserve of Lokrum in Croatia in the western Mediterranean (Ilijanic & Hecimovic, 1981), and in the Eastern Mediterranean, the *Q. calliprinos* and *P. halepensis* forests at the Mt. Carmel nature reserve in Israel. Here, as reported by Naveh and Whittaker (1979), in comparison with the disturbed, semiopen, multilayered communities, species richness is much lower (21 plant species in 1/10th ha as compared with 120 species) and the same is true also for species richness and abundance of birds, rodents, reptiles, and insects. Both have become stagnating and species-poor thickets and not rich "climax forests." The Mt. Carmel reserve has been heavily damaged recently by uncontrollable wildfires and the same can happen also to the heavily visited Lokrum Island.

Not only in the Mediterranean uplands but also in all other seminatural landscapes such a homeorhetic flow equilibrium (from the Green *homeorhesis,* preserving the flow) has been ensured by a combination of periodic natural and human-caused perturbations. They should be treated, therefore, as metastable perturbation dependent systems. These cannot be returned to a stationary state of homeostasis, like natural, undisturbed systems, but will go on to change in the same way as they have changed in the past, as long as these perturbations are continued with similar "optimum" intensities (Waddington, 1975). It would be, therefore, futile to attempt to restore the so-called climax and its mature vegetation by simply stopping these perturbations. On the contrary, here we have to conserve and to reestablish all ecological processes to which they have been adapted throughout their long history.

The threats of neotechnological landscape degradation were demonstrated in a very convincing way in a thorough landscape-ecological study of the changes that occurred in the agro-silvo-pastoral uplands of Tuscany, Italy, in the Solano Basin, in the last 50 years (Vos and Stortelder, 1988). With the help of the above-mentioned mutual information method, similar ecotopes were clustered spatially as recurring patterns of landscape units on maps of 1:50,000. These illustrated very lucidly the changes that have occurred in the last 50 years and which are apt to occur in the next 50 years if no conservation measures will be introduced.

As a result of large-scale emigration and agricultural intensification (encouraged by EEC policies and subsidized wheat prices), the rich mosaics of coppiced and managed oak and chestnut forests, mountain grasslands, and the terraced mixed vine/olive and crop cultures were replaced by intensive agriculture of monocultures of wheat and viniculture, planted canopy coniferous forests and secondary scrub on abandoned pastures and fields. The fine-grained patterns broke down into coarser patterns of fewer and floristic and faunistically much poorer ecotopes and land units. In addition, the soils of arable slopes, now cultivated by heavy machinery, are suffering from erosion and are losing their fertility. Within 50 years, together with the spread of coniferous forests and the progress of secondary successions of scrub and monotonous oak forests, five major landscape types will vanish altogether and the biological, cultural, and scenic richness will be further severely impoverished (Fig. 10.4).

In a recent study by the same team on the assessment of the impacts of the planned Farva River Barrage for the irrigation of the Grosseto Plain in Tuscany, these methods were expanded from spatial and topographic heterogeneity to the evaluation of biological and scenic landscape attributes (Pedroli et al., 1988). This comprehensive landscape ecological impact study prevented the implementation of an ill-conceived project that could have disastrous environmental consequences.

As discussed in detail by Naveh and Lieberman (1983), for the preservation and restoration of this biological, cultural, and scenic richness, there is urgent need for a new, holistic approach to the open landscape as a whole. It should be based on well-coordinated, simultaneous efforts for better public and professional conservation education, more comprehensive landscape conservation planning and management, and more interdisciplinary, integrated landscape ecological and ecosystem conservation oriented research. The main aim should be to relieve the present heavy and uncontrolled traditional and neotechnological pressures on these lands and to reconcile the conflicting demands by optimization of all relevant "soft" and "hard" landscape values for long-term and overall benefit by the establishment of closely interwoven networks of multibeneficial land-use patterns with great flexibility in space and time, ensuring the following functions:

1. *Bio-ecological functions* of natural and seminatural uplands and wetland ecotopes as last refuges for spontaneous organic evolution of the genetic potentials of wild plants and animals, as protective and regulative "life supporting" systems, buffer zones for watershed protection from floods, erosion, sedimentation, and environmental pollution for the densely populated and intensively cultivated lowlands and coastal regions.
2. *Socioecological, cultural, and psychohygienic functions* of scenic beauty, solitude, and wilderness and of preservation of unique natural and historical features for future generations, endangered now even in protected areas because of mass recreation and tourism pressures.

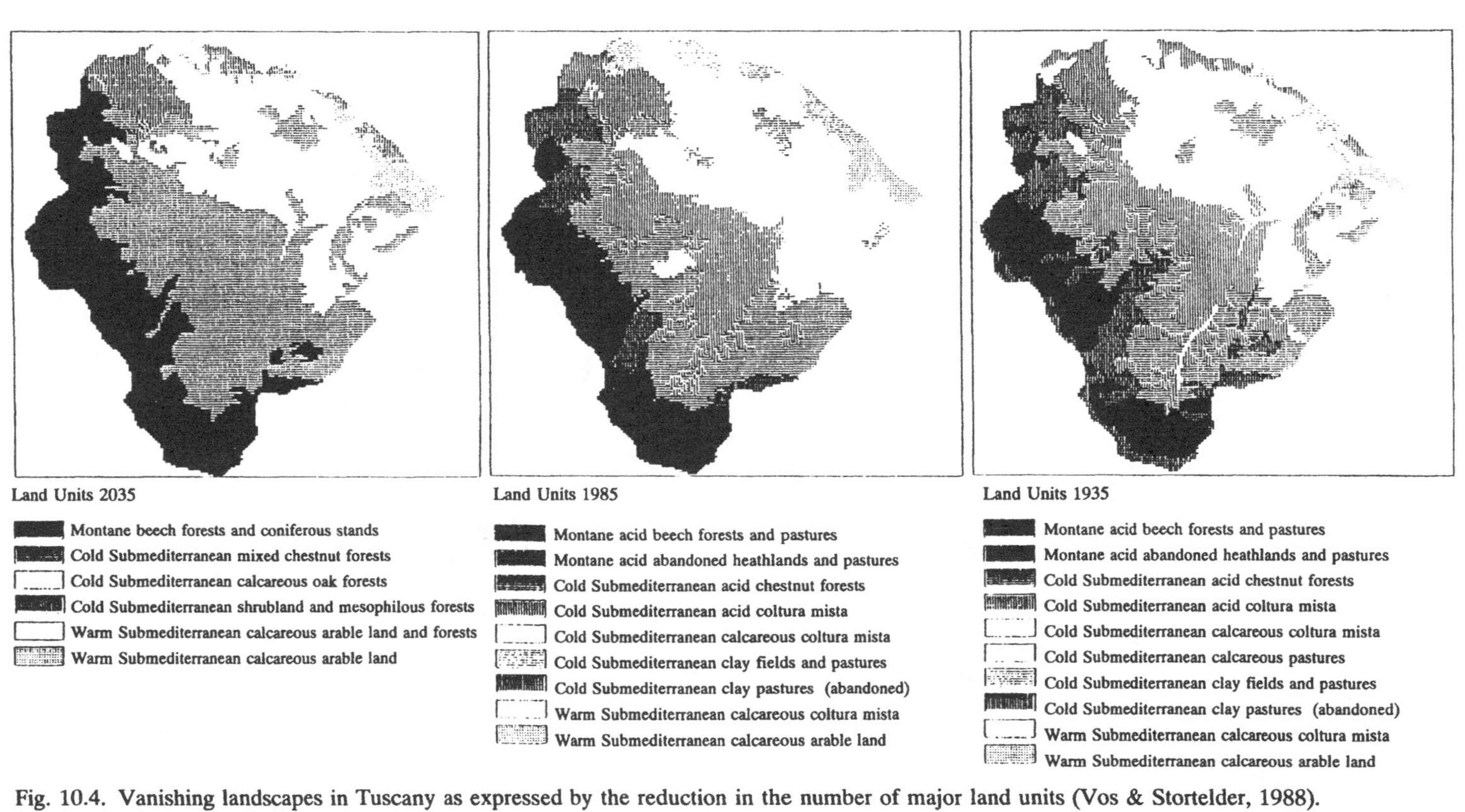

Fig. 10.4. Vanishing landscapes in Tuscany as expressed by the reduction in the number of major land units (Vos & Stortelder, 1988).

3. *Socioeconomic functions* as source of food and revenues from plant and animal products and tourism that can be ensured by rational management on a long-term sustained basis.

This can be achieved by applying two main "ecotechniques":

1. *Vegetation and ecotope management* by manipulation of the existing plant cover and the soil–plant–animal complex by controlled grazing, cutting, coppicing, and burning, and in improved natural pastures by fencing, reseeding, manuring or applying slowly released phosphate fertilizers, and selective weed control.

In nature reserves, the continuation of light to moderate defoliation pressures is essential for the achievement of maximum attainable biological productivity and diversity and ecological metastability. With exception of undisturbed "control plots," dense forest and maqui thickets should be converted into much richer and less inflammable semiopen park forests by thinning, coppicing, and pruning the woody sclerophyllous vegetation and preventing their regeneration from roots and suckers by controlled browsing of domestic livestock, especially goats, or by wild ungulates. In order to reduce fuel loads around denser woody stands and to ensure the regeneration and rejuvenation of light-demanding plants, prescribed burning should be applied judiciously. Its timing and spacing should be adapted to the requirements of the major target species. But much more research is needed on these aspects (Naveh, 1990).

2. *Multipurpose afforestation and restoration* in degraded scrubland and wherever denudation has reached an advanced stage, as well as in man-created habitats on roadsides, camping grounds and recreation sites and buffer zones for urban and industrial complexes.

Such semi-natural, multilayered park forests with multiple ecological, economical and social benefits should replace the one-layered, species-poor conifer plantations, which are highly vulnerable to fire, pests and air pollution. Flow charts of management options, cybernetic models of the mutual impacts of relevant variables, including biodiversity, and balance sheets of over-all cost/benefits of their production, protection and regulation functions showed that by replacement of inferior indigenous woody plants with introduced plants, with greater values as ornamentals, soil stabilizers, fodder and/or honey producers, highest over-all benefits could be achieved.

For such multibeneficial strategies, a major problem is the quantification of "soft," intangible values and "noneconomic" richness and their relative weighing against marketable goods. Advances computer programs of knowledge engineering in expert systems opens the way for efficient integration of the complex ecological, sociological, and economical quantitative as well as qualitative information and their incorporation into the decision making process. It enables one to capture "soft," qualitative landscape values mathematically as fuzzy sets and to deal with them in algorithmic fashion as regular numbers (Negoita, 1985).

In most Mediterranean countries present ecological knowledge and experience is not sufficient for the development of such complex management strategies. There is, therefore, urgent need for (presently almost nonexistent) interdisciplinary ecosystem and landscape research on the ecological and socioeconomic implications of different landscape management options.

Ultimately, however, the success of this holistic landscape-ecology policy will depend on the creation of awareness and understanding amongst all those who care for, live from, and deal with these landscapes at all levels of decision making. Above all, it is vital to bridge the communication gap between the academicians and the professionals, the conservation-minded ecologists and the production-minded foresters, agronomists, economists, and engineers and educate a new breed of interdisciplinary landscape managers with broad ecological, economical, sociological, and technological backgrounds who can serve as "integrators." Let us hope that they will soon replace the one-track-minded scientists and professionals who are the products of a highly fragmented and rigidly discipline-oriented education.

A first step in this direction is the preparation of "Green Books of Threatened Mediterranean Landscapes." These should provide updated information on the fate of especially valuable landscapes with practical recommendations for alternative conservation management. In order to serve as an efficient tool for decision makers and for public education, they should be prepared in appealing nontechnical ways by documenting and illustrating the interdependence of the welfare of people and that of their open landscapes. For this purpose, a joint, interdisciplinary Working Group of the World Conservation Organization Commission on Environmental Strategies and Planning (IUCN-CESP) and of the International Association of Landscape Ecology (IALE) is presently preparing a blueprint of such a Green Book and a proposal for Red Lists of Threatened Landscapes. The Green Book blueprint will be based on a case study, carried out by a multinational team in western Crete, and other examples, using the ecotope mapping and integrating methods described above, as well as dynamic Geographical Information Systems.

In these case studies, special attention will be paid to the threats of global climatic changes and the predicted rise in sea levels and temperatures, which will aggravate all the detrimental human impacts on Mediterranean landscapes.

Discussion and conclusions

The description of the degradation of the seminatural and agricultural landscapes show clearly the interdependence of biodiversity and landscape heterogeneity, induced by closely interwoven ecological, demographical, socioeconomic, and cultural factors. By approaching biodiversity from a holistic view of landscapes as complex natural and cultural gestalt systems of the total human ecosystem, the strictly biological connotations of biodiversity can be expanded into a broader transdisciplinary concept of landscape ecodiversity. This concept incorporates the biological, ecological, and cultural landscape heterogeneity at different spatial and perceptional scales. For its assessment and conservation management further innovative methods will be necessary, transcending the realms of natural sciences into those of human ecology and overruling conventional mechanistic, scientific paradigms. These should, however, at the same time enable the quantification of "soft" and intangible landscape values and functions, including the intrinsic values of biodiversity and scenic beauty.

For effective conservation management of biodiversity and landscape ecodiversity a clear understanding of the ecological and cultural processes and their perturbations is essential. This is true also for natural and close-to-natural landscapes and for parks and reserves. These cannot be managed in isolation, but must become integral parts of regional and national masterplans for the conservation and restoration of all our open landscapes. For the maintenance of their integrity and diversity not only natural perturbations, like fire cycles, but also human-caused perturbations may be vital.

In conclusion, dynamic conservation management for the preservation and restoration of the ecodiversity of our open landscapes in Europe and elsewhere concerns the future of our total human ecosystem. Therefore it is our obligation not only to produce meaningful scientific information, and to show how to achieve this in the best possible way, but to transform this semantic information, communicated in lectures and in written publications, into pragmatic information, which becomes meaningful through its feedback action on the receiver, namely the public and the decision makers in future land uses.

References

Agger, P., & Brandt, J. (1988). Dynamics of small biotopes in Danish agricultural landscapes. *Landscape Ecology*, **3**, 227–240.

Baudry, J., & Merriam, G. (1988). Connectivity and connectedness: Functional versus structural patterns in landscapes. In *Connectivity in Landscale Ecology,* ed. K.-F. Schreiber, pp. 23–28. Proceed. 2nd Internat. Seminar Intern. Assoc. of Landscape Ecology, Muenster 1987. Paderborn: Ferdinand Schoeningh.

Bridgewater, P. B. (1987). Connectivity: an Australian perspective. In *Nature Conservation of Native Vegetation,* pp. 195–200. M. Hopkins Surry Beaty and Sons Pty.

Bruns, D. (1988). Planning concepts and management strategies for nature conservation in agricultural regions of South West Germany. In *Connectivity in Landscape Ecology,* ed. K.-F. Schreiber, pp. 191–195. Proceed. 2nd Internat. Seminar Intern. Assoc. of Landscape Ecology, Muenster 1987. Paderborn: Ferdinand Schoeningh.

Burrough, P. A. (1984). The use of geographical information systems for cartographic modeling in landscape ecology. In *Methodology in Landscape Ecological Research and Planning,* ed. J. Brandt & A. P. Agger, pp. 3–13. Proceed. First Intern. Seminar, IALE, Roskilde University Centre, October 15–19, 1984. Vol. 1. Roskilde, Denmark: Roskilde Universitetsforlarg.

Dale, V. H., Gardner, R. H., & Turner, M. G. (1989). Predicting across scales – comments of the guest editors of Landscape Ecology of special issue. *Landscape Ecology,* **3,** 147–152.

Davis, F. W., Stoms, D. M., Estes, J. E., & Scepan, J. (1990). An information systems approach to the preservation of biodiversity. *Int. J. Geographical Systems,* **4,** 55–78.

Egler, F. (1964). Pesticides in our ecosystem. *Amer. Scientist,* **52,** 110–136.

Gardner, R. H., & O'Neill, R. V. (1991). Pattern, process, and predictability: The use of neutral models for landscape analysis. In *Quantitative Methods in Landscape Ecology. The Analysis and Interpretation of Landscape Heterogeneity,* ed. M. G. Turner & R. H. Gardner, pp. 289–308. New York: Springer-Verlag.

Green, B. H. (1989). Agricultural impacts on the rural environment. *J. Appl. Ecology,* **26,** 793–802.

Haber, W. (1990a). Using landscape ecology in planning and management. In *Changing Landscapes: An Ecological Perspective,* ed. I. S. Zonneveld & R. T. T. Forman, pp. 217–233. New York: Springer-Verlag.

Haber, W. (1990b). Basic concepts of landscape ecology and their application in land management. In *Ecology for Tomorrow. Physiology and Ecology in Japan,* 27, ed. H. Kawanabe, pp. 131–146.

Harms, W. B., Stortelder, A. H. F., & Vos, W. (1987). Effects of intensification on nature and landscape in the Netherlands. In SCOPE 32 *Land Transformation in Agriculture,* ed. M. G. Wolman & D. C. A. Fournier, pp. 357–379. Chichester, UK: Wiley.

Hoover, S. R., & Parker, A. J. (1991). Spatial components of biotic diversity in landscapes of Georgia, USA. *Landscape Ecology,* **5,** 125–136.

Ilijanic, L. J., & Hecimovic, S. (1981). Zur Sukzession der Mediterranean Vegetation auf der Insel Lokrum bei Dubrovnik. *Vegetatio,* **46,** 75–81.

Kwakernaak, C. (1984). Information applied in ecological land classification. In *Methodology in Landscape Ecological Research and Planning,* ed. J. Brandt & A. P. Agger, pp. 5–15. Proceed. First Intern. Seminar IALE, Roskilde University Centre, October 15–19. Vol. III. Roskilde, Denmark: Roskilde Universitetsforlarg.

Merriam, G. (1984). Connectivity: a fundamental ecological characteristic of landscape patterns. In *Methodology in Landscape Ecological Research and Planning,* ed. J. Brandt & A. P. Agger, pp. 5–15. Proceed. First Intern. Seminar,

IALE, Roskilde University Centre, Oct. 15–19, 1984. Vol. I. Roskilde, Denmark: Roskilde Universitetsforlarg.

Merriam, G. (1988). Landscape ecology: The ecology of heterogeneous systems. In *Landscape Ecology and Management,* ed. M. R. Moss, pp. 17–34. Proceed. First Symposium of the Canadian Society of Landscape Ecology and Management. University of Guelph, May 1987. Montreal: Polyscience Publications, Inc.

Merriam, G., Kozakiewicz, M., Tsuchiya, E., & Hawley, K. (1989). Barriers as boundaries for metapopulations and demes of *Peromyscus leucopus* in farm landscapes. *Landscape Ecology,* **2,** 227–236.

Moss, R. (ed.). (1988). *Landscape Ecology and Management.* Proc. First Symposium of the Canadian Society of Landscape Ecology and Management. University of Guelph, May 1987. Montreal: Polyscience Publications, Inc.

Naveh, Z. (1984). The vegetation of the Carmel and Nahal Sefunim and the evolution of the cultural landscape. In *The Sefunim Prehistoric Sites Mount Carmel, Israel,* ed. A. Ronen, pp. 23–63. BAR Intern. Ser. 230 Oxford.

Naveh, Z. (1987). Biocybernetic and thermodynamic perspectives of landscape functions and land use patterns. *Landscape Ecology,* **1,** 75–85.

Naveh, Z. (1990). Fire in the Mediterranean – a landscape ecological perspective. In *Third Intern. Symposium on Fire Ecology, Freiburg University, Germany 16–20 May 1989,* ed. J. G. Goldhammer, pp. 1–20. The Hague: SPB Academic.

Naveh, Z., & Whittaker, R. (1979). Structural and floristic diversity of shrublands and woodlands in Northern Israel and other Mediterranean regions. *Vegetatio,* **41,** 171–190.

Naveh, Z., & Lieberman, A. S. (1983). *Landscape Ecology Theory and Applications, 2nd ed.* New York: Springer-Verlag.

Negoita, C. V. (1985). *Expert Systems and Fuzzy Systems.* Menlo Park, California: Benjamin/Cummings.

Noss, R. (1987). Corridors in real landscapes: A reply to Simberloff and Cox. *Conservation Biology,* **1,** 159–164.

Noss, R., & Harris, L. (1986). Nodes, networks and UMs: Preserving diversity at all scales. *Environ. Manage.,* **10,** 299–309.

Pedroli, G. M. B., Vos, W., & Dijlestra, H. (eds.). (1988). *The Farma River Barrage Effect Study.* Marrilio Editiori, Venezia: Giunta Regionale Toscana.

Phipps, M. (1984). Rural landscape dynamic: The illustration of some key concepts. In *Methodology in Landscape Ecological Research and Planning,* ed. J. Brandt & A. P. Agger, pp. 17–27. Proceed. First Intern. Seminar, IALE, Roskilde University Centre, Oct. 15–19, 1984. Vol. I. Roskilde, Denmark: Roskilde Universitetsforlarg.

Ricklefs, R. E., Naveh, Z., & Turner, R. E. (1984). *Conservation of Ecological Processes.* Gland, Switzerland: Commission on Ecology papers No. 8, IUCN.

Romme, W. H. (1982). Fire and landscape diversity in subalpine forests of Yellowstone National Park. *Ecol. Monogr.,* **52,** 199–221.

Rose, C. (1987). *Reducing Nitrogen Oxides Pollution to Conserve Nature.* Gland, Switzerland: World Wildlife Fund.

Schaller, J., & Haber, W. (1988). Ecological balancing of network structures and land use patterns for land-consolidation by using GIS-Technology. In *Connectivity in Landscape Ecology,* ed. K.-F. Schreiber. Proceed. 2nd Internat. Seminar Intern. Assoc. of Landscape Ecology, Muenster 1987. Paderborn: Ferdinand Schoeningh.

Schreiber, K.-F. (ed.). (1988). *Connectivity in Landscape Ecology.* Proc. 2nd Inter-

nat. Seminar Intern. Assoc. of Landscape Ecology, Muenster 1987. Paderborn: Ferdinand Schoeningh.
Turner, M. G. (ed.). (1987). *Landscape Heterogeneity and Disturbance.* New York: Springer-Verlag.
Turner, M. G., & Gardner, R. H. (eds.). (1990). *Quantitative Methods in Landscape Ecology.* New York: Springer-Verlag.
Vos, W., & Stortelder, A. H. F. (1988). *Vanishing Tuscan Landscapes. Landscape Ecology of a Submediterranean-montane Area (Sonalo Basin), Tuscany, Italy.* Wageningen: PUDOC.
Waddington, C. H. (1975). A catastrophe theory of evolution. In *The Evolution of an Evolutionist,* pp. 253–266. Ithaca, New York: Cornell University Press.
Zonneveld, I. S. (1990). Scope and concepts of landscape ecology as an emerging science. In *Changing Landscapes: An Ecological Perspective,* ed. I. S. Zonneveld & R. T. T. Forman, pp. 3–20. New York: Springer-Verlag.

11

Making a habit of restoration: saving the eastern deciduous forest

LESLIE SAUER

Introduction

Native habitats are deteriorating everywhere and the rate of decline is accelerating. There is no such thing as a preserved landscape. We dedicate parkland and restrict development from selected natural areas, but this does not eliminate continued impacts from surrounding development on a local scale or from environmental change on a global scale. The eastern deciduous forest, though less renowned than the tropical rainforest or coral reef, is no less imperiled. Reduced to a patchwork of fragments, the remaining forest lands in the eastern United States are afforded almost no regulatory protection at the federal, state, or municipal levels. In our tended landscapes, instead of being simply neglected, the forest is more systematically eliminated, replaced by lawns and clipped shrubs, kept back with herbicides and mowers. It is in our home landscape that we are first introduced to our cultural disdain for the patterns of nature.

The consequences are seen in the accelerating displacement of complex native communities by a few aggressive exotic invasive species, such as Norway maple and Japanese knotweed, and by a host of diseases, such as beech blight and ash yellows, which are decimating the native populations. Climate changes associated with global warming are likely to deal the last blow to many already stressed species that will be unable to move geographically as rapidly as their suitable ranges shift northward. Sustaining even a fraction of the diversity that persists today will be an uphill battle. The only hope is to begin right now to make a habit of restoration. From New England to the south along the East Coast, the fate of our forests is in our hands. Saving the forest systems east of the Mississippi will require a revolution in our attitudes toward our native landscape and fundamental changes in the management of our remaining forest lands at every scale, from the home garden to the national park. This chapter will focus on addressing the situation of the eastern deciduous forest and six major areas where it is critical to take immediate action, but the problems described are global in nature.

The ongoing loss of species diversity in even our largest preserves is evidence that the forest system is already too fragmented, yet this disruption is proceeding unabated as development continues. The narrow corridors of forestland, which should serve as vital biological connections, are being severed into ever smaller and more isolated patches of forest remains. At a time when we already face alarming species losses, the prospect of rapid global warming presents us with the likelihood of even greater levels of extinction. Despite extensive investment in horticultural research, propagation, and planting, only a fraction is directed toward preserving the biodiversity of native plant communities.

The current focus in our yards, gardens, and parks is on horticultural conventions rather than on natural diversity. There is something quite preposterous about our devotion to the landscape styles of the seventeenth, eighteenth, and nineteenth centuries at a time when so much of our thoughts must address the future. Beneath our lawnmowers and asphalt is an invisible forest that represents the richest and most complex landscape type that can be attained given our climate. After centuries of beating back the forest, we now find ourselves sadly winning this age-old battle. It will take equal effort to bring the forest back.

Throughout the region, a major shift in natural hydrology is occurring as rain water, which was previously recharged into the ground, is being intercepted by impermeable surfaces and discharged from pipes as stormwater. Current regulations address primarily flood control and, more recently, pollution reduction, but, with few exceptions, do not yet recognize the need for effective recharge to support the water table and the base flow of streams. The introduction and widespread dissemination of alien species into an environment where there are no natural controls also has been devastating. Despite general agreement on this problem among scientists and natural area managers, there is almost no public awareness or regulatory control and many invasive species are still widely planted.

The bulk of landscape management practices still comprises largely destructive actions, such as extensive mowing, reliance on herbicides and pesticides, poor erosion and sedimentation control, and a narrow horticultural perspective on restoration. Interest and experimentation in the management and restoration of native landscapes are growing daily, although efforts are confined to small-scale, often labor-intensive, projects and are greatly hampered by the lack of comprehensive information on effective techniques and adequate long-term monitoring.

Establishment of a system of forest reserves.

Our success in preserving biodiversity will depend ultimately on the quality of wilderness. A recent study, which mapped wilderness areas (defined as at least 400,000 hectares in size with few visible signs of human impact, such as roads or structures) found that wilderness has been eliminated from two thirds of the earth's landscapes, and almost none remains in the lower 48 states of this country, except for three areas in the west (McCloskey & Spalding, 1990). Here, on the East Coast, it is important to raise our expectations. The Great Smoky Mountains National Park, for example, which is the center of biodiversity for the eastern deciduous forest, is only 209,000 hectares in size and is laced with roads, and by McCloskey and Spalding's definition would not even qualify as wilderness. Our concept of a wilderness reserve must include the highest levels of trophic organization and the reintroduction of the largest predators. Natural processes, such as wildfire, should prevail to the maximum extent feasible. Primary intrusions into wildlands, such as roads and other infrastructure, should be reevaluated and removed wherever possible. Our largest reserves must be reassessed from this perspective and enlarged and consolidated. No amount of management within a reserve can make up for a site that is simply too small or too intruded upon. Though it seems difficult enough to hold our ground, we must set our sights higher. This is the most fundamental crisis facing the eastern deciduous forest and must be tackled directly, for until it is resolved true preservation will remain an illusion.

The term biodiversity is usually quite loosely applied by biologists and, in most cases, is taken to mean simply the number of species in a given area. The term species richness is also frequently used and the two terms are often used synonymously. Norton (1987) has reviewed these terms and makes the important point that "it must be remembered that other factors such as evenness of distribution and the phylogenetic distance between species" are a part of the concept of diversity. Diversity also includes the coevolutionary bonds evolved over millions of years, such as predator–prey and plant–pollinator relationships. Diversity is an aspect of pattern, including the landscape mosaics we see around us. Some qualitative aspects of diversity are illustrated in the following: there may be 30 million species of life forms on earth; Ecuador has 25,000 species of plants of which 5,000 are endemic; Malaysia has 5,000 species of orchids alone; Pennsylvania has about 3,000 species of native plants (Wherry et al., 1979); and the Brogdale Agricultural Experiment Station in Kent, England has over 2,500 varieties of apples.

The biodiversity maximum for a given area may, in fact, be the wilderness that existed there prior to human disturbance or would exist there if left undisturbed. This varies with latitude, type of climate, size of land area, local relief, and other factors. There is no substitute for wilderness as a reserve for biological diversity, and all other actions must be built on this foundation. It is the necessary premise for every other recommendation. A major hurdle to this goal is our failure to recognize the importance of forest in the East. Many of those who champion rainforests in the tropics and old-growth forests in the Pacific Northwest often dismiss the eastern forest systems as already gone, or too fragmented and modified to merit being the focus of major acquisition and preservation. Only recently tens of thousands of hectares of forestland in New York and New England were acquired by developers, despite efforts of local environmentalists. We are likely to be the last generation to be offered the opportunity to acquire tracts of such significant size and we cannot afford to forego any large parcels.

There are also large gaps in our knowledge. At present, there is no comprehensive national natural inventory, nor a regional one for the eastern deciduous forest, although a wealth of data has been gathered. At present, we do not even know the extent of the problem – how much still supports native plant communities, rich in diversity, and how much is declining?

Recent efforts to protect wetlands, both nationally and locally, have led to mapping and monitoring as well as developmental controls, and our awareness has increased commensurately. Wetland losses, though reduced, have continued and mitigation projects have revealed their limitations, leading to stricter regulatory controls in most states and a currently stated national policy of no net loss. Despite the failures of the program, the goal of wetland protection has been largely accepted and there is a real knowledge base for measuring success.

A similar effort must be directed toward forest protection. It is unlikely, however, that sweeping regulatory controls will soon be forthcoming. While we may see increased federal and state involvement and support for data gathering, forest protection will likely be very dependent on activism at the grass-roots, municipal, and regional levels.

Outside of regional-scale reserves, the protection of biodiversity will be dependent on the smaller reserves. Critical remaining wildlands should be protected and missing links should be reestablished through appropriate corridor management. In reality, however, the opposite is currently the case. Rural areas adjacent to East Coast cities are experiencing rapid suburbanization, further fragmenting and displacing relic forests. Even in areas where the amount of forestland is increasing, as farms are being abandoned, much of the land ultimately will be developed and many of these sites support large numbers of exotic species. At present, large-lot residential zoning is the major method of conservation in forested areas, despite the fact that habitat value is significantly diminished and forest remnants come to resemble planted landscapes over time due to continued landscape interventions by homeowners. Some indirect forest protection also is afforded by restrictions on development in wetlands, steep slopes, and sites supporting endangered species. At the municipal level, the clearance of selected specimen trees sometimes is restricted and tree replacement may be required on a numerical basis. There is, however, no explicit regulatory protection of the forest itself, of its critical patterns, and its capacity to sustain itself over time.

Two aspects of forest protection that must be incorporated into any effective regulations are configuration and continuity. The continuous forest cover, which greeted the European settlers, has been reduced to fragments and strips in an increasingly developed context. Most endangered is the forest interior and the species limited to that environment. At the same time, islands of habitat isolated from surrounding natural areas experience a decline in native species diversity and are less adaptable to stress over time. While a popular axiom holds that forest edge conditions are favored by wildlife (often "game species") and urges the creation of more edge, a more realistic view is that forest edges abound, while the amount of forest interior is diminishing steadily. A few centuries ago, forest edge was principally internal to forest in the form of gaps resulting from dead standing or fallen trees, fires, or other natural causes, including human settlement of what was then a native population (in that it was persistent and stable for centuries). More recently, edge has become increasingly external to forest or a boundary between forest and silviculture, extensive agriculture, suburbia, and other habitats of a rapidly expanding human population. A topological inversion occurred in the pattern of communities; gaps embedded in forest became increasingly forest embedded in gaps. What might be called everted edge has now become dominant and with it has come proliferating disturbance and instability to the forest, which has now become the gap in a sea of modernity.

The sprawl of human-dominated land use has reached and greatly exceeded its sustainable limit in the eastern deciduous forest, and in much of the world. The emphasis from now on has to be on preservation of whatever areas remain reasonably intact. Fundamental changes in our development patterns are necessary as well as redevelopment of those areas which we have already disturbed profoundly, such as the urban environment. While there is much attention paid to the environmental consequences of uncontrolled population growth, this is not actually the primary factor in our current expansion into the remaining fragments of the eastern forest. The population of the New York megalopolitan region, for example, has experienced very little net growth in 25 years but sprawl, fueled by the interstate highway system, has nearly doubled the size of the land area this population occupies. This is true in many older Northeast cities, where population has been declining and properties are being abandoned, as extensive areas are experiencing rapid suburban development and impoverishment of the remaining natural areas. This cycle must be reversed, disturbed areas reinhabited, and a comprehensive system of reserves and habitat corridors established.

Integrating the native landscape with development

The landscapes around us occur on a gradient of wildness, ranging from largely natural areas, with minimal direct intervention from people, to developed areas where every trace of nature seems to have been obliterated. Our goal should be to better manage forest resources at every point of this continuum, including the urban landscape. There should be no place where there is no wildness at all – no site where considerations of natural values are not deemed important. Our most developed landscapes are no more self-contained than natural areas. Their stormwater eventually reaches the water table, the stream, or the sea and their landscape plantings and introduced pests spread into the surrounding wildlands.

Even an industrial park or urban core should support some component of wildness, such as natural stream corridors, providing vital connections between larger forested tracts, rather than leaving behind a useless maze of leftover land. At the most developed end of the wildness gradient, where the capacity to support habitat is most restricted, at the very least, critical connections to adjacent habitats should be maintained. As older sites are rebuilt for new uses, gaps in the natural pattern can be restored. In fact, any rebuilding or maintenance, whether for highways, railroads, pipelines, or any other part of the urban infrastructure, should have a restoration component. Many critical connections will only be achieved by learning to retrofit existing urbanized areas to better support natural systems. Our care of the forest at this end of the gradient will influence the quality of wildness at the other end.

In reality, our care of the urban forest has been consistently at the expense of wildness. The look of the "metroforest" is familiar to us all. The ground is most glaringly different in appearance from that of an undisturbed forest. Litter is often ubiquitous and bare soil may be exposed, often over a large area. Where disturbance is chronic, invasive exotic vegetation may prevail, typically in monospecific stands. Even where forest persists, alien vegetation is almost always present, and is usually increasing. Large relic native canopy trees may still occur, but have long since ceased reproducing. Volunteer saplings of cherry and locust, usually typical of young field landscapes, may fill large gaps in the canopy. Exotics, such as Norway maple, frequently can be found in every layer of the landscape. Vines, both native and exotic, may occur in heaping mounds on the ground, draped over shrubs and saplings, and cloaking trees in the canopy. Guldin et al. (1990) described an old-growth patch of forest in Overton Park in Memphis, Tennessee where carpets of kudzu and honeysuckle effectively block recruitment of native tree canopy species. Ferns and woodland wildflowers may be conspicuously absent or reduced to one or two species. Species diversity has usually been declining for years, although human use, especially for recreation, may be steadily increasing. When speaking of biodiversity, the full range of species is usually the issue, but in the developed landscape sustaining even the biodiversity of the most common and abundant native habitats is endangered.

In rural parts of suburban areas, such conditions may now be confined to roadside edges and the margins of developed areas, creating a network of disturbance corridors, which will expand over time. When disturbance is uncontrolled, deterioration usually accelerates and a once rich area diminishes in diversity and natural value. The disturbed forest system is growing ever larger, while the extent of diverse native forests is shrinking.

Wildlife is equally impacted. Reptiles and amphibians are especially vulnerable to human depredation. Where pressure is severe, such as in Central Park, New York City, the total number of species is reduced to a fraction of what the habitat might otherwise support. Birds and small mammals, though better able to escape humans, fall easy prey to the artificially high density of predators, in the form of cats and dogs. Migratory birds, dependent for their survival on finding a continous supply of food in rich forest stopover sites for nourishment on their long distance journeys, have been decimated in past decades. A small woodland area, such as the 15-hectare Ramble in the center of New York City's Central Park becomes a critical link in an increasingly tenuous chain of forests for these migratory birds, which comprise nearly two-thirds of the avian species once found in our forests.

We are not only losing animal species with the loss of habitat, we are losing plant species with the loss of animals. Plants and animals have coevolved over the millennia, and plant reproduction is inextricably tied to wildlife, for pollination, seed transport, scarification, and planting, for regulating competition, etc. Where wildlife is impoverished, many plants have no means of effective reproduction, survival, and replacement. Habitat preserves, once delineated to protect existing areas of natural vegetation, must be reassessed with regard to meeting the requirements of wildlife as well. We cannot separate one from the other in our thinking anymore.

We may be surprised by what can be sustained in the developed environment, if the requirements of natural communities are respected. In the midst of the busy industrial shipping corridor of Newark Bay, over 29% of the entire colonial water-bird population along the Long Island–New York City Atlantic shoreline breeds on three dredge-spoil islands on the Arthur Kill, including snowy, great, and cattle egrets, black and yellow-crowned night herons, little blue herons, and glossy ibises (The Trust for Public Land et al., 1990). This waterway, which separates New York from New Jersey, was the site of no less than ten oil spills between January and September in 1990, including an Exxon pipeline leak of over 2.1 million liters of home heating oil. Recent monitoring to evaluate the impacts of this spill have revealed unexpected populations of organisms surviving under stressful conditions that would be lethal to individuals accustomed to less polluted environments. These areas may be where some of the most interesting speciation is occurring in modern evolution. It is important to remember that most of these species are living at the very edge, in habitats that are largely unprotected. Even a minor added degradation may eliminate an entire community, but with more effective protection, a valuable resource can be enhanced.

Reestablishment of natural hydrologic regimens

Throughout the developed landscape, a major shift in natural hydrology is occurring. Water that previously infiltrated into the soil now runs off, failing to replenish groundwater. Streams which once ran year round have become flashier, subject to periods of flooding and periods of drought. Dropping groundwater levels, which reduce the base flow of streams, also severely impact vegetation (Ashby, 1987). Many species, such as beech and white oak, are impacted by continued lowered water table conditions. If adequate levels of recharge are not sustained over time, even more dramatic changes in vegetation are likely to occur. Alternating periods of more prolonged drought and high rainfall associated with global warming will amplify existing hydrologic imbalances, further stressing the landscape, and underscore the need to redress the situation now.

Erosion from excessive runoff represents one of the most ubiquitous and costly sources of damage to wildlands. A single outfall from a storm sewer discharged onto a steep slope can cut a deep gully very rapidly and further drain away groundwater. Many of the stormwater problems in natural areas originate off-site in developed landscapes and may require complicated negotiations with adjacent landholders and public regulatory agencies to resolve. Sadly, many stormwater management regulations are applied only to new development and are nonexistent and unenforced in urban areas, where all runoff is simply shunted to the nearest stream via a storm sewer. Where regulations apply, stormwater design often focuses on flooding only and may not provide any detention of the smaller, but very frequent one- or two-year storms, which shape the stream channel and, if dramatically increased in volume, lead to severe disturbance, including bank undercutting, channel migration, and sedimentation.

Even seemingly remote forest tracts are not exempt from extreme changes in hydrologic regimen. Major highways routinely cause severe disruptions in natural hydrologic patterns, affecting large areas of rural land. Where logging is undertaken, even in supposedly "protected" national and state forestlands, stream corridors are ravaged and runoff is dramatically increased, at least temporarily, especially by clear cutting. The longer term impacts of the required access roads are more severe, altering drainage surface patterns and, where road cuts are used, creating permanent seeps which literally bleed the groundwater away, affecting the water table over a wide area.

Stormwater management is best addressed over the entire watershed. The simplest approach usually is to seek multiple solutions at different points, rather than a single cure-all at the point of discharge. We must look first for solutions that most closely mimic nature's solutions, which will likely maximize opportunities for recharge. Sometimes simply altering the management of landscapes can effect substantial reductions in runoff. Turf areas, when even only gently sloped, shed water nearly as rapidly as pavement. Within the developed fabric over 25 to 30 million acres are maintained as cultivated lawns, an area the size of Indiana (Hiss, 1990), a substantial portion of which is irrigated, and the bulk of which should be replaced with more diverse native habitats. Conversion of all or part of a lawn to tall grass and wildflower meadow can slow runoff velocities. In fact, a permit should be required to install a lawn. The design standards for engineers established by regulations must seek levels of recharge comparable to those that occurred under forested conditions as well as effective pollutant reduction of the recharge. If retaining runoff and maximizing recharge are vigorously pursued in the uplands, the problems of erosion and flooding in lowland areas will become more manageable.

Solutions to the problems of lowland areas should also seek to mimic analogous natural situations. The requirements of a drainage system should be determined not only by the anticipated volumes of runoff generated by a site, but by the capacity and requirements of the receiving stream. Computer modeling has greatly increased our capacity to tailor an engineering design to the conditions of a particular stream and watershed, rather than a more simplistic regulatory standard. In the course of restoration, appropriate actions include taking a stream out of a pipe, rather than putting a stream into one; reestablishing a natural meandering channel, rather than channelizing a stream; and planting trees along the stream corridor, rather than removing them. One of the most important roles a designer can play today is to solve site problems in a manner that does not take such a toll on our remaining natural areas. The natural hydrologic pattern of each stream can and should be respected, retained, and restored.

Native species banking and dissemination

The long-term consequences of global warming are still matters of speculation, but it is probable that we will see a dramatic decrease in biotic diversity as the range of suitable climate for a given species is displaced more rapidly northward than can be accomplished by means of natural reproduction. Current projections made at the University of Minnesota suggest that the climatic range for beech will move northward 700–900 km in the next century. After the last glacial retreat, beech dispersed itself on an average of only 20 km per century (Davis & Zabinski, 1992).

The movement of species will be further hampered by the fragmented condition of the forest today. Outlying patches from which and into which species might colonize are now separated by urban developed lands and subject to ongoing development. As difficult as migration will be for trees, it is likely to be far more difficult for woodland herbaceous species, which may require stabler conditions for reestablishment (Davis & Zabinski, 1992; Graham et al., 1990).

Even within the range of a species, stresses due to climate changes are likely to increase. There will be more severe storms, accelerating erosion. Concurrently, droughty years will occur more frequently. At greater risk because of these stresses, vegetation will be more vulnerable to insects and disease, increasing mortality. The impacts will probably mirror and magnify those we are already witnessing in the landscapes that surround us. Even without the dire predictions of future changes, what is already happening now before our eyes is alarming. A native forest community cannot maintain itself without periodically successful regeneration through successful reproduction and survival through the early life stages of each species that make up the community. This essential process is called recruitment, a term taken from both forestry and population biology. In the present highly disturbed state of the forest, recruitment of many species is blocked or impeded in many ways, so much so that we can speak of a recruitment crisis that is contributing to rapid decline of the forests as we have known them (Guldin et al., 1990). These failures are due to exotic invasive species, exotic pathogens, altered water tables, pollution, altered fire patterns, changes in populations of grazing animals, landscape fragmentation, acid rain, and direct disturbance by humans (all-terrain vehicles being an extreme example). Where wild pigs eat all the acorns or excessive deer populations consume all seedling trees, there is inadequate recruitment. The situation is the same where exotics outcompete natives and inhibit native regeneration, or where wildlife integral to pollination and seed transport have been eliminated.

The recruitment crisis, which is relatively serious now, is likely to be exacerbated by increased stresses, such as global warming, increased ultraviolet radiation and, no doubt, phenomena which we do not even know about yet. Every institution involved with landscapes and living things should begin some program to confront this situation. Where we are unable to reestablish conditions required for effective recruitment, or when climate change is too rapid for natural processes to keep pace, people will have to be the agents of recruitment. The system of biological reserves may ultimately be most important as a source of seeds and seedlings to be introduced into new potential ranges. Today, however, in the East, fewer than 10% of native plants in any given area are commercially propagated. We must make an immediate commitment to learning how to propagate the full range of species within a region. We must also learn how to sustain them in the wild. The goal must be to get plants into the ground where they belong, not merely the warehousing of genes in germplasm banks. We cannot simply assume we will be able to transplant wildlings effectively. Past plant rescue efforts have been seriously flawed. At present, transplants typically survive a few years before disappearing altogether from their new location (Fahselt, 1988).

Despite the enormity of the idea of building an ark, we are well equipped to meet this challenge. The U.S. Department of Agriculture, Soil Conservation Service, and Forest Service are all involved in plant propagation and distribution programs. Indeed, they have facilitated the spread of some of our most problematic pest species, such as multiflora rose and kudzu. With a broader, more ecological mandate, these agencies could apply the same expertise and resources to foster the replacement of lost species and ecotypes with those more suitable to changing climate conditions. There are well over 150 institutional members of the American Association of Botanical Gardens and Arboreta (AABGA, 1990) in the eastern United States involved with new species and varieties, as well as a National Arboretum in Washington, D.C. It would not be difficult to establish a central clearing house for information on the status of native species in each locality, and to document and prioritize propagation and dissemination needs. Enormous resources are already allocated to horticultural efforts; it is only a national commitment to native species that is lacking.

Once common species closer to the edge of extinction, such as the American chestnut, deserve special efforts, not unlike those directed toward rare and endangered species. The American Chestnut Foundation is developing disease resistant strains suitable for introduction into natural forests. Restoration of the once dominant canopy tree of major areas of the eastern forest and its wildlife supportive value may become possibile. Unfortunately, horticulturalists and foresters are often quick to abandon a native plant under stress, such as the current vogue to plant Chinese chestnut and Korean dogwood where native populations are diseased. We must continue to maintain populations of stressed plants in the landscape and ecosystem regardless of whether we fear some individuals may be short-lived.

We must revolutionize how we deal with the individual plant. If you read a nurseryman's catalog today, it is sometimes difficult to remember that it is describing a living thing; instead, plants have patent numbers. Grotesqueries are prized, especially abnormal color variations and dwarfism. The same attitudes are being applied to indigenous plants as horticulturists seek to collect superior native varieties suitable for mass marketing. Although some native plants are grown for landscaping purposes, many of them are grafts and tissue cultures. These methods are increasingly relied upon to produce a saleable plant more quickly; however, by replacing reproduction by seed, genetic diversity is reduced. We must turn our attention to propagating species by means which foster greater biological diversity. Similarly, botanic institutions known for their collections have the opportunity to bring ecotypes of a species together from the limits of their ranges, fostering evolutionary opportunities that would be unlikely under natural conditions but which might expand the climatic tolerances of a given species.

The possibilities for existing institutions and agencies to play a positive role in the preservation of native plant communities are almost limitless and require only a shift in program and product emphasis. Concurrently, there is a great need for public education to link the public's preferences to a broader understanding of the environmental consequences and benefits of the choices they make.

Control of exotics

One of the most visible aspects of environmental damage in natural landscapes is the spread of exotic invasive vegetation that displaces native communities. As forests throughout the larger region are increasingly disturbed and fragmented, they are ever more vulnerable to invasion, and rapidly overwhelmed if adequate control is not undertaken quickly enough. The introduction and often widespread dissemination of an alien species, such as Norway maple and Japanese honeysuckle, planted in an environment where there are no natural controls or defenses, have been devastating (Harty, 1986; Bratton, 1982; Overlease, 1978, 1987; Guldin et al., 1990). If, for instance, an invasive plant species is introduced into a plant community the diversity or species richness could initially be considered to have been increased by one species. Later, however, if that species spreads and crowds out various other species to create, as is frequently the case, a monospecific stand of the single invading species, diversity is substantially reduced. In reality, we are experiencing numerous invasions occurring simultaneously. While it is true that over time natural systems will adapt to the presence of a new entity or disturbance, it is also true that this change can replace extensive areas of native habitat and limit the capacity for recovery in a system already severely hampered by a wide range of other environmental stresses. In the Great Smoky Mountains National Park, exotic species are currently estimated to comprise 17 to 21% of the total species and 19 to 24% of species in Shenandoah National Park (Loope, 1992).

The focus here is primarily on plant invaders, but the effects of introduced animals are also severe and facilitate the spread of plant exotics. Various insect introductions have drastically changed forest assemblages, such as the ongoing invasion of the gypsy moth. Of course, the most disruptive invasive animal in the eastern forest has been the European settler, but among mammals a second is the feral pig. In the literature of biological invasions, a frequent generalization is that highly diverse natural areas are resistant to invaders, yet pigs moved quickly into the most diverse of the eastern forests, the Great Smoky Mountains region, and their control is still a problem there and throughout many forest areas today. Their rooting behavior has created ideal disturbance conditions for subsequent plant invasion (Ramakrishnan and Vitousek, 1989).

Though not all introduced exotic species become invasive, the success of a few species is more than enough to jeopardize virtually every native habitat. Unlike many animals and pathogen introductions that were accidental or can be traced to a single source, like the gypsy moth, most plant introductions have been deliberate and systematic. And most arboreta and state agencies continue to introduce new species and varieties with very little attention given to potential impacts on native communities. When kudzu was in vogue, and falsely perceived of as a quick-fix for erosion, over 4 million seedlings were distributed from a single government nursery in Georgia. Once thought to be confined to the south, kudzu has begun a slower but still effective invasion to the north. Honeysuckle, like kudzu, was once widely perceived as an excellent ground stabilizer and was introduced on a massive scale with a concerted effort coming from the railroads, which used honeysuckle to quickly cover steeply sloped embankments. In urbanized corridors, in the mid-Atlantic states, it is difficult to find a fragment of habitat that is uncontaminated. It is generally acknowledged that the spread of honeysuckle has irrevocably altered the course of forest succession in the eastern forest (Butler et al., 1981; Evans, 1984; Guldin et al., 1990).

Some of these species, such as Japanese knotweed, have not been grown commercially for decades yet are increasing in extent and range very quickly, disseminated primarily by people's activities, such as highway construction and floodplain mismanagement.

Birds have been as important to the distribution of honeysuckle and other exotics as people. Because of this, honeysuckle and other invasive species are occasionally defended as valuable to wildlife. However, this is a very short-sighted perspective. Plenty of honeysuckle is likely to always be with us; however, the continued loss of habitat diversity has been devastating to birds and other wildlife. Nonetheless, many state nurseries and conservation groups continue to propagate and advocate the widespread planting of Japanese and shrub honeysuckles, Russian and autumn olives, and multiflora rose on public lands for wildlife purposes.

Equally invasive are the Norway and sycamore maples, which are gradually overwhelming many forests in the Northeast. Because they do not have the heaping vine form, but look to most people like any other tree in the forest, the evidence of disturbance may be less apparent. In late fall, however, when the leaves from other trees have fallen, the butter-yellow foliage of the Norway maple, for example, reveals a continuous understory of saplings with no native saplings present at all. When these mature, the woodland may be entirely Norway maple, displacing hundreds of native species with one or two aliens. Because almost no other species can coexist with it (except possibly sycamore maple which is equally problematic), the soil is often bare beneath its canopy and subject to erosion. What is created has been termed an "exotic disclimax" where the natural succession of native species has been completely arrested for an indeterminant length of time (Butler et al., 1981). There are few other exotic species that appear to pose such a threat to native populations in the Northeast. Like many other disturbance species, its vigor in our landscape is part of what made the Norway maple so popular and it is still popular. Today, it is a standard on most recommended street tree lists. Typically, a nursery will have more Norway maples than all other maples combined and demand for the plant is still large enough to warrant its cultivation on a massive scale, despite its devastating impact on native forest communities. The widespread dissemination of Norway maple continues unabated.

There are, at present, no effective natural controls for these plants. The future of our native forests depends on our taking immediate action. As early as 1977, President Jimmy Carter signed the Exotic Organism Executive Order 11987, which empowers executive agencies to "restrict the introduction of exotic species on lands and waters which they own, lease, or hold for purposes of administration; and shall encourage the States, local governments, and private citizens to prevent the introduction of exotic species into native ecosystems." In Minnesota, the Purple Loosestrife Coalition, an interagency work group, guided a bill through the 1986–87 Minnesota State Legislature which established the Purple Loosestrife control program, the nation's first comprehensive state-level effort to control an invasive plant (Harper, 1988). Other states are following suit. The state of Illinois, for example, introduced legislation in 1987 for an Exotic Weed Control Act to prohibit the sale and planting of problem exotic plants throughout the state and has ceased cultivation of them in state nurseries, turning instead to native species.

It is strongly recommended that no species which has demonstrated itself to be a successful invader at the expense of native habitats in the region, or is even suspected of being a pest, be planted at all because of the threat posed to natural areas. This is a very conservative policy, but the consequences of being too optimistic have been and will be very costly to remnant habitats.

A ban may seem extreme, however, invaders are extremely difficult to eradicate once established and none is so critical to landscape character or native wildlife that it cannot be replaced by a native plant. Based on their demonstrated impact on native populations, the following species should be banned from being planted in the Northeast and control programs initiated: Norway maple (*Acer platanoides*), sycamore maple (*Acer pseudoplatanus*), Russian olive (*Eleagnus angustifolia*), autumn olive (*Eleagnus umbellatus*); barberry (*Berberis japonica*), amur honeysuckle (*Lonicera maackii*), Tatarian honeysuckle (*Lonicera tatarica*), multiflora rose (*Rosa multiflora*); porcelain berry (*Amelopsis brevipedunculata*), Oriental bittersweet (*Celastrus orbiculata*), Japanese honeysuckle (*Lonicera japonica*), kudzu (*Pueraria lobata*); purple loosestrife (*Lythrum salicaria*), and Japanese knotweed (*Polygonum cuspidatum*). This list is far from complete and the levels of naturalization vary considerably from area to area. As more monitoring and assessment are undertaken, various tiers of control may be appropriate to reflect local as well as regional-scale problems.

These species are all very difficult to eradicate once entrenched, and sites might then typically require complete replanting and a high degree of continued maintenance. Management must be initiated as soon as possible because the problem only gets worse and management must be continuous. An effective prioritization of effort in some heavily disturbed areas is to concentrate our energies on protecting sites that presently support predominantly native species and where invasion is just starting.

Our ability to ultimately control exotics depends heavily on controlling the spread of environments that favor them. In older botanical literature, the phrase "characteristic of waste places" was commonly used to describe numerous weedy and often exotic species that established themselves in the habitats ruined by various human activities. In later literature, the term "ruderal" is more often applied, which blunts the impact in a euphemistic way that what is being described is literally a waste of biotic resources. If we make a habit of restoration, then waste places will no longer be tolerated and no project or activity will be allowed to create a barren landscape. Just as abandoned buildings are subject to condemnation, so should there be a similar procedure to insure that biotic resource values and environmental quality be protected and restored.

Landscape management and restoration

Conventional landscape management generally consists of a set of typical practices that are repeated over and over again. In our region, conventional practices such as extensive mowing, specimen plants, and heavy reliance on pesticides continue the centuries-old effort at beating back the forest. Our treatment of nature in the garden mirrors treatment of nature elsewhere. In the larger context it is control that is the problem; our intolerance of wildness. Management is always purposeful. When our goal becomes to foster the diversity of our forest system, to preserve wildness, our practices can and will change dramatically. Unfortunately, past management often has tipped the scales heavily against native communities, warranting a concerted effort to mitigate the impacts. The task of restoration is immense and recovery will not happen overnight. This paper attempts to set priorities rather than trying to deal with everything, for despite the complexity of our developed landscapes, it is a few major stresses that account for the bulk of the disturbance. The object is to gradually shift the impact of our actions from being largely negative to being largely positive, to turn the tide, as it were, to reverse the trend of deterioration and to initiate recovery.

A major goal of management should be to affect as much area as possible by using our resources economically. Where natural processes regulate and sustain the habitat, the need for outside management diminishes, while the health of the whole landscape system is improved. Practices that allow more natural fires and sustain natural hydrologic cycles foster habitats more resistant to invasion by exotic plants and animals as well as to debilitating diseases and pests, both introduced and naturally occurring. As noted earlier, in many cases the configuration of parks and other natural areas may need to be altered to make them more manageable and primary infrastructure such as roads may need to be removed in order to reduce ongoing disturbance and the need for management to offset these impacts.

The management and restoration of disturbed landscapes is limited by a lack of both information and experience. Natural resource managers may be trained in the workings of natural systems, but often have no familiarity with disturbance ecology. Many park managers may be trained in recreation planning, but have no awareness of the needs of natural systems. Recommendations made by foresters and horticulturists often are very inappropriate in native habitats, such as the common directive to thin the canopy to stimulate shrub growth, which often simply opens the landscape to exotic invasion.

The art and science of managing native landscapes are being developed right now. We will be successful only if we embrace natural models as our standard of measure. An important consequence of wetlands regulation is a greater recognition of the value of a natural marsh. The first priority must be to preserve existing wetlands, in part because of the failure of the majority of "created" wetlands to provide even a fraction of the ecological value of the sites they were designed to mitigate. The first rule of good management must always be "do not disturb a site if it is healthy." Once a site has been disturbed, our standards for site restoration are currently very low, focusing almost exclusively on erosion and sedimentation control with almost no attention paid to native community replacement, or to minimizing the impact on natural vegetation in the first place. As long as gabions commonly replace natural streamside vegetation in the name of stabilization or perennial rye replaces a woodland as revegetation, we haven't really been asking the right questions or solving the larger problem. These activities are actually destabilization and devegetation when the primary intent should have been restoration, of a whole stream corridor, not simply a failing bank, of a whole site and its environs, not simply the area adjacent to construction. There are over a million miles of major gas transmission pipelines in the United States alone. If we add to this powerline rights-of-way and rail and road corridors we are looking at a network of connected landscapes that presently serve as seams of disturbance but which could be restored and managed as vital conduits for the movement of plants and sometimes animals. The focus of restoration must be the whole environment, not just those lands we perceive as "natural" areas. The ragged fabric of our landscapes must be made whole again, for only in this context can we hope to sustain wildness. Biodiversity is not expressed simply in the places most rich in species, but is reflected in the richness of all landscapes.

Discussion and conclusions

Due to their direct and indirect impacts on natural environments, people have become the principal shapers of our landscape. Wilderness in this country has nearly been eliminated in the five centuries since the European colonization began along with much of the native communities of plants and animals. Today we also must face the impacts that will be brought about by global warming. It is a certainty that we have brought about mass extinctions; this is already underway. The only question is how much we can save if we take aggressive action now. The response by government at the federal, state, and municipal levels, however, is conservative and will not provide the comprehensive regulatory support required for this effort without public and political pressure. Our first priority must be public education and a basic recognition that saving biodiversity is the preeminent value that will determine the quality of life in the future. Virtually every aspect of our use of the land must be fundamentally revised to reflect this imperative, including development patterns, energy use, modes of transportation, and lifestyle. Norton (1987) has echoed others in stating that if total diversity is valuable then "every species should be accorded some positive value." A corollary is that every piece of land must also have positive value.

References

American Association of Botanical Gardens and Arboreta (AABGA). (1990). Personal communication. P.O. Box 206, Swarthmore, PA 19081.

Ashby, W. C. (1987). Forests. In *Restoration Ecology,* ed. W. R. Jordan III, M. E. Gilpin, & J. D. Aber, pp. 89–108. Cambridge, UK: Cambridge Univ. Press.

Bratton, P. (1982). The effects of exotic plant and animal species on nature preserves. *Nat. Areas Jour.,* **2**(3), 3–13.

Butler, T., Stratton, D., & Bratton, S. P. (1981). *The distribution of exotic woody plants at Cumberland Gap National Historic Park.* U.S. National Park Service, Southeast Research/Resources Management Report No. 52. Washington, DC: U.S. National Park Service.

Davis, M. B., & Zabinski, C. (1992). Changes in geographical range resulting from greenhouse warming: effects on biodiversity. In *Consequences of Greenhouse Warming to Biodiversity,* ed. R. L. Peters & T. E. Lovejoy. New Haven: Yale Univ. Press.

Evans, E. (1984). Japanese honeysuckle (*Lonicera japonica*): A literature review of management practices. *Nat. Areas Jour.,* **4**(2), 4–9.

Fahselt, D. (1988). The dangers of transplantation as a conservation technique. *Nat. Areas Jour.,* **8**(4), 238–243.

Graham, R., Turner, G., & Dale, H. (1990). How increasing CO_2 and climate change affect forests. *BioSci.,* **38**(1), 575–587.

Guldin, J., Smith, J. R., & Thompson, L. (1990). Stand structure of an old-growth upland hardwood forest in Overton Park, Memphis, Tennessee. In *Ecosystem*

Management: Rare Species and Significant Habitats, NY State Museum Bull., **471,** 61–66.
Harper, B. L. (1988). Purple loosestrife named a noxious weed, illegal for sale, Minnesota. *Restoration & Management Notes,* **6,** 2–95.
Harty, F. M. (1986). Exotics and their ecological ramifications. *Natural Areas Journal,* **6,** 4–20.
Hiss, T. (1990). *The Experience of Place.* New York: Knopf.
Loope, L. L. (1992). An overview of problems with introduced plant species in national parks and reserves of the U.S. In *Alien Plant Invasions in Hawaii: Management and Research in Near-native Ecosystems,* ed. C. P. Stoone, C. W. Smith & J. T. Tunison. Honolulu: Cooperative National Park Resources Study Unit, Univ. of Hawaii, Honolulu.
McCloskey, J., & Spalding, M. (1990). The world's remaining wilderness. *Geographical Magazine,* August 1990: 14–18.
Norton, B. G. (1987). *Why Preserve Natural Variety?* Princeton: Princeton Univ. Press.
Overlease, W. R. (1978). A study of forest communities in Southern Chester County, Pennsylvania. *Proc. of the Pennsylvania Academy of Science,* **52,** 37–44.
Overlease, W. R. (1987). One hundred and fifty years of vegetation change in Chester County, Pennsylvania. *Bartonia,* **53,** 1–12.
Ramakrishnan, P. S., & Vitousek, P. M. (1989). Ecosystem-level processes and consequences of biological invasions. In *Biological Invasions. A Global Perspective,* ed. J. A. Drake & H. A. Mooney. New York: John Wiley & Sons.
The Trust for Public Land in conjunction with New York City Audubon Society. (1990). The Harbor Herons Report: A Strategy for Preserving a Unique Urban Wildlife Habitat and Wetland Resource in Northeastern Staten Island. New York: NY City Audubon Society.
Wherry, E. T., Fogg, J. M., Jr., & Wahl, H. A. (1979). *Atlas of the Flora of Pennsylvania.* Philadelphia: The Morris Arboretum of the University of Pennsylvania.

12

Landscapes and management for ecological integrity

JAMES R. KARR

Landscape ecology deals with the patterns and processes of biological systems in spatially and temporally heterogeneous environments (Risser et al., 1984). Landscape ecology does not fit into any of the conventional branches of ecology because it is a synthesis of many related disciplines that focus on spatial and/or temporal pattern. Four components are central to any effective exploration of landscape ecology: *scale, dynamics*, and *linkages* among *patches* (the elements of the landscape).

As initially developed in Europe, landscape ecology deals with geographic areas on the scale of 10–10,000 km^2 (Naveh & Lieberman, 1984; Forman & Godron, 1986), a spatial dimension defined by the human-oriented context of European landscape ecology. In contrast, the spatial scale of landscape ecology in North America is less restricted because it tends to be less human oriented. The study of spatial heterogeneity and its influence on many kinds of organisms and ecological systems (Risser et al., 1984; Wiens, 1989; Wiens & Milne, 1989; Merriam, 1990) provides a more diverse theoretical and applied underpinning to the discipline in North America. This distinction is important because the human scale is by no means the standard for all life forms. For example, mountain lions and jaguars range over hundreds of square kilometers while the spatial scale (landscape) of an ant colony is on the order of tens of square meters. Although stone-age humans interacted with their landscape on a scale like that of the mountain lion, the landscape of modern human society is broader.

Technology and our ability to transport food, raw materials, consumer goods, people, and garbage, and other wastes over long distances change the scale of human interaction with the environment. Because the appropriate spatial scale for a study is determined by the organism(s) and question(s) under investigation, landscape ecology as a scientific discipline must avoid a tendency to be constrained to a narrow range of landscape dimensions.

Understanding ecological pattern and process across spatially diverse landscapes requires more than knowledge of spatial scale, however. Understanding of *linkages*, the redistribution of materials, energy, and/or individuals among landscape elements, is also an essential feature of landscape ecology. This redistribution implies the *dynamic* context that is integral to landscape ecology. Until recently, the ecological sciences largely ignored the dynamics of heterogeneous environments. Those dynamics are critically important to understanding exchanges within and between the elements of the landscape (i.e., *landscape patches*). Two factors are important in determining these exchanges: 1) landscape attributes that influence permeability and 2) attributes of organisms that influence their ability to move across the landscape. To what extent does the specific landscape enhance or impede exchanges within and between landscape elements? How does intervening habitat type and between-patch distance influence transfer of propagules among patches? What abilities does an organism have to move as adults or via propagules across the landscape? How do biogeochemical processes influence the flow of energy or nutrients across landscapes?

Finally, effective resolution of many resource management issues requires understanding and application of a landscape perspective. Insightful use of landscape ecology can reduce the environmental degradation that results from human actions and can improve the success of programs to restore lost natural resource values. Coupling of knowledge of spatial scale and the dynamics of redistribution processes is central to human efforts to manage heterogeneous environments.

These general observations lead me to three points. First, the human species is no less dependent upon its landscape than are all the other species of the Earth's biota. But the scale over which we interact has changed rapidly over the past few centuries. Many current environmental problems result from our inability to recognize and act upon the consequences of that change in scale. In an evolutionary sense we have not been prepared to recognize problems at those scales. Second, management of landscapes at a human scale can be improved by studying the dynamics of human and other species in natural and disturbed landscapes. Third, studies of ecological heterogeneity in space and time are likely to provide major advances in ecological theory and application because they transcend the conventional hierarchy of ecology; that is, they require ecology to become the truly integrative science it has long claimed to be.

The purpose of this chapter is to clarify and illustrate these points through several case studies. First, definitions of ecosystem and landscape ecology are considered. Second, I discuss the biological resources of aquatic ecosystems to illustrate how human society can detect its pervasive influence on landscapes. Third, I describe recent work on birds that demonstrates the importance of a landscape perspective in the protection of biodiversity. Finally, using the Kissimmee River in Florida as an example, I demonstrate the role of a landscape perspective in planning an ecological restoration project. These examples illustrate that the concept of the landscape in ecology provides a working unit for the study of spatial heterogeneity and its cause. Further, the examples clarify how patterns across the landscape result from the dynamic interaction of biotic and abiotic factors of natural or cultural origin (Wessman, 1990).

Ecosystem management: definition of terms

To many ecologists, landscape ecology is a fuzzy concept, although most people are convinced they understand the meaning of the word "ecosystem." However, at least three usages of ecosystem are common. Textbooks (Ehrlich & Roughgarden, 1987) and dictionaries (Webster's, 1983) define an *ecosystem* as the biological community plus the physical environment with which it interacts. In another usage, an ecosystem is a relatively homogeneous physical, chemical, and biological system, such as an oak–hickory forest, a tallgrass prairie, or a coastal *Spartina* marsh. Yet a third common usage is, for example, the matrix of biological communities that occupy a watershed.

The common definition that distinguishes communities from ecosystems implies that only ecosystem ecologists deal with the interactions of the biota and the physical environment despite the fact that population or community ecologists have long recognized the importance of the physical environment (Andrewartha & Birch, 1954; Lack, 1954). I suggest that the community/ecosystem dichotomy has less to do with the way biological systems are organized than it does with the different world views used by community and ecosystem ecologists to practice their craft. *Community ecology* generally approaches the subject from an evolutionary perspectives – the process of natural selection is central to understanding biological pattern. In contrast, *ecosystem ecology* is founded on the principle of the cybernetic or thermodynamic structure of ecosystems, including recent efforts to tie this approach to evolution (Patten & Odum, 1981; Odum & Bevier, 1984). In my view, both community and ecosystem ecology depend on understanding of organisms *and* their physical environment so the conventional definition does not reflect a true hierarchical difference.

The long-held concept of an ecosystem as a relatively homogeneous area is now discredited as a general ecological truth, as is the idea of an equilibrium community (Botkin, 1990). Like the perfect vacuum of physics, theoretical constructs such as the homogeneous ecosystem, the equilibrium community, or the stable population do not exist in the real world. The concept of a homogeneous ecosystem is appealing *and* theoretically useful. However, if the theoretical construct takes on the aura of an assumed reality, advances in ecological understanding are not likely.

Finally, to use ecosystem as a designator for a geographic area of interest, such as a watershed, is appealing but it does not implicitly suggest spatial or temporal heterogeneity. Reference to such watersheds might more appropriately use the concept of a landscape. Widespread recognition of the importance of spatial and temporal heterogeneity (in the landscape) to ecological dynamics at all levels adds another dimension to the conventional population, community, and ecosystem approaches. Because biological systems are heterogeneous, valid ecological questions at the single species (population) or multispecies (community with evolutionary or ecosystem with thermodynamic conceptual foundation) levels must generally be examined with explicit recognition of landscape heterogeneity. Much mathematical theory in ecology from the past two decades illustrates the influence of assumed equilibrium and homogeneity in ecology.

Case 1: landscapes and water resources – detecting degradation

Rivers are "expressions of their terrestrial watersheds" (Sioli, 1975) in much the same way that blood samples from a mammal may provide insight about the health of the individual. Unfortunately, few efforts have been made to use the condition of the biota of rivers to assess and convey the condition of the landscapes of those rivers. For decades the primary tools to assess the quality of water resource systems and their terrestrial landscapes have been the physical and chemical attributes of the water (e.g., dissolved oxygen, temperature, nitrogen, phosphorus, heavy metals, pesticides, and suspended solids). Growing recognition that a focus on these attributes neither protects water resources nor adequately evaluates the condition of terrestrial landscapes has stimulated efforts to strengthen monitoring programs to assess water resources (Karr, 1991). Sport and commercial fisherman and environmental biologists (Richardson, 1928) have long known of the decline in water resources but others have been reluctant to recognize the biological foundations of the problem.

Laurentian Great Lakes

The decline of water quality and the biological resources of the Laurentian Great Lakes has been documented in the scientific (Smith, 1972; Regier &

Hartman, 1973; Francis et al., 1979; Colburn et al., 1990) and popular (Ashworth, 1986; Egerton, 1987) literature. Contributing factors responsible include: selective overfishing; extensive watershed modification, including drainage of lake-margin wetlands; introductions of exotics, especially sea lamprey, alewife, and salmonids; and progressive chemical modification of lake environments. Chemical modification such as nutrient enrichment and siltation of shallow areas accelerated natural processes, and the discharge of synthetic chemicals and chemical wastes added new chemical stresses. This complex of factors is clearly tied to human alteration of the landscape.

Recently, the international organizations responsible for protection of the Great Lakes adopted a landscape approach to protection of the integrity of the Great Laurentian River basin (Edwards & Regier, 1990). Although they refer to this pioneering new approach as adoption of an "ecosystem approach" to restoring integrity to the Great Lakes, their methods are more reflective of a landscape perspective as defined above (see also Allen & Bandurski, 1992).

Midwestern rivers

Changes in the fish communities of two midwestern rivers (Illinois and Maumee) during the past 140 years provide another illustration of the complexity of factors responsible for degradation. They also illustrate the complex biological dimensions of resource degradation stimulated by human actions (Karr et al., 1985).

The joining of the St. Mary's and St. Joseph's Rivers in Ft. Wayne, Indiana forms the Maumee River, the largest tributary of the Great Lakes. The Maumee flows northeasterly until it enters Lake Erie near Toledo, Ohio. The Illinois originates in northeast Illinois and flows west for nearly 100 miles before turning abruptly to the southwest, following an old channel of the Mississippi River to the modern Mississippi. Beech–maple and elm forest occupied most of the Maumee basin while the Illinois was dominated by bluestem prairie, oak–hickory forest, and floodplain wetlands before the arrival of humans. Today, both watersheds are dominated by agriculture with considerable area in urban and suburban land use. The Maumee watershed (17,000 km^2) is about one-fourth the size of the Illinois (73,500 km^2) while the flow of the Maumee (133 m^3/s) is about one fifth of the Illinois (633 m^3/s).

Ninety-eight species of fish from 21 families are known from the Maumee watershed while 140 species from 27 families are known from the Illinois. Karr et al. (1985) compiled information on the distribution and abundances of those fishes and discovered that 44% of Maumee River species and 67% of Illinois River fishes experienced major population declines or were extirpated from their respective watersheds since 1850. Fish characteristic of small streams suffered heavy losses in both watersheds while large-river fish were more affected in the Illinois than the Maumee (Fig. 12.1). Riverine fish communities worldwide are experiencing similar declines. In the rivers of California, 67% of the fish have declined or disappeared in this century (Moyle & Williams, 1990). At national levels, ten fish taxa have become extinct in the United States since 1979 and 217 others are endangered or threatened (Williams et al., 1989). During a recent 4-year survey of the freshwater fishes of Malaysian rivers, only 122 of 266 species (46%) known from Malaysia were found (Diamond, 1991).

Loss of fish species is not the only evidence of degradation of the landscape and aquatic systems. In 1908 the commercial carp catch in the Illinois River was over 6.8 million kg. One 200-mile stretch of the river produced 10% of the total U.S. catch of freshwater fish (Jeffords, 1989). By 1950 the commercial catch declined to 1.8 million kg and to 100,000 kg by 1973. In 1910 over 2,600 commercial mussel fisherman operated on the Illinois River while virtually none remain today. Further evidence of degradation due to human influence includes poor health of individual fish in the Illinois River (Sparks, 1977), perhaps because food supplies are limited, and high frequencies of tumors, eroded fins, and skeletal anomalies (J. R. Karr, personal observation), indicating prevalence of toxic materials. Massive modification of rivers and their adjacent landscapes is the obvious factor responsible for these declines. Of 5.2 million kilometers of streams in the continental United States, only 2% are

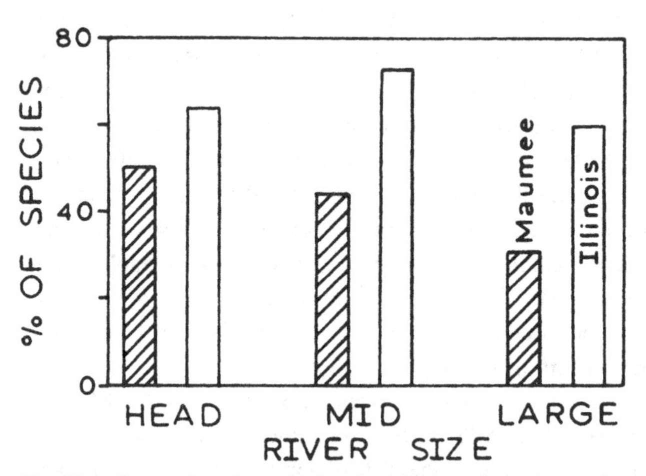

Fig. 12.1. Proportion of species that have been extirpated or with significantly declining populations as a function of river size (Head: headwater; Mid: mid-sized river; Large: large river) in the Illinois and Maumee Rivers. (From Karr et al., 1985).

healthy enough to be considered high quality and worthy of protection (Benke, 1990). Of large rivers (>1,000 km long), only the Yellowstone River is not widely altered, and only 42 medium sized rivers (>200 km long) have not been dammed.

The geographic extent of this degradation in the ecological health of the river biota and aquatic systems in general and the numerous specific ramifications of the decline clearly indicate the cumulative impact of human conversions of the landscapes of these rivers. Reports on the magnitude of this degradation date back to the early decades of this century (Kofoid, 1908; Forbes & Richardson, 1913; Richardson, 1928). While these documents went largely unheeded, recent calls for development of methods to detect degradation through routine monitoring (Karr & Dudley, 1981; Karr et al., 1986; Fausch et al., 1990) have attracted the attention of both scientists (Steedman, 1988; France, 1990) and regulators (Ohio EPA, 1988; Plafkin et al., 1989; USEPA, 1990).

Case 2: birds, forests, and landscapes – preventing degradation

On a global scale, according to the International Council for Bird Preservation, 11% of the world's bird species are endangered and 60–70% having declining populations or reduced ranges. Early recognition of the role of landscapes in protecting bird populations developed from the efforts of waterfowl biologists and hunters (e.g., Ducks Unlimited) to protect breeding, wintering, and migration habitat. The success of their management program was dependent on a continental landscape strategy. Recent studies of birds in the forest of Central America (Karr, 1982a, 1985; Stiles & Clark, 1989) and of the Pacific Northwest (Thomas et al., 1990) illustrate the importance of a landscape perspective in understanding the biology of birds and the development of management programs to protect endangered bird species.

Forest birds in Panama

The diversity and relative ease of observation of tropical forest birds makes them ideal for ecological research. In addition, they play a critical role in the pollination or dispersal of 40–80% of tropical forest plants (Gentry, 1990). Their status is, thus, informative about the health of the entire forest biota.

My research in Parque Nacional Soberania, Panama, shows that each bird species interacts with the landscape, although the scale of the interaction varies strikingly among species. Individuals of some species occur as sedentary residents in areas of less than 100 ha while others undertake annual migrations that cross two continents (Table 12.1). Understanding the diversity of those dy-

Table 12.1. *Heirarchy of geographical scales covered by birds that occur in the vicinity of Limbo Camp, Soberania National Park, Panama.*

Scale of Movement	Distance	Example
Intercontinental	4000 km	Chestnut-sided warbler Kentucky warbler
Regional		
Attitudinal	30 km	Yellow-eared toucanet
Lowland, across Isthmus	15 km	Blue ground-dove
Local	<1 km	Tawny-crowned greenlet Ochre-bellied flycatcher

namics is important to understanding and to protecting the Panama avifauna. Well over 100 species of neotropical–temperate migrants travel between North and Middle America each year (Karr, 1976, 1980, 1985). Migrants from the north begin to arrive in Panama as early as July, with abundances peaking in October (Karr, 1976). Some species, such as the Kentucky warbler, establish winter territories, while others, such as thrushes, seem more nomadic in their movements on the nonbreeding grounds. Other species (e.g., Tennessee warbler) vary in their territorial behavior as they face changing food abundances with the changing seasons (Morton, 1980).

This diversity of tropical–temperate migrant strategies is complimented by a similar range of obligatory and facultative movements by permanent resident species in the tropics. Nectarivores and frugivores are especially well known as altitudinal migrants, a pattern of seasonal movement in Costa Rica that includes nearly half of the high altitude avian community (Stiles, 1983, 1988). Aerial insectivores like swifts and soaring raptors (kites, hawks, and eagles) engage in daily altitudinal movements (Stiles & Clark, 1989), while seasonal migration seems to dominate among the fruit- and nectar-feeders. Bellbirds migrate seasonally over an altitudinal gradient of nearly 2000 m while snowcaps undertake seasonal migrations that cover only about 800 m in elevation (Fig. 12.2). Some frugivores of forest undergrowth at La Selva, Costa Rica even have permanent and migratory subpopulations (Blake & Loiselle, 1991). Most of these altitudinal migrants require forest habitat throughout their range.

Altitudinal migrations of birds in Central Panama are less marked than in Costa Rica because the altitudinal range in Central Panama is more limited (Fig. 12.3a). Further, habitat fragmentation limits the necessary landscape connections among elevations (Fig. 12.3b), especially following construction of a transisthmian highway in the 1940s (Fig. 12.3a; Karr, 1985). The lack of a high-elevation barrier in central Panama provides an opportunity for seasonal

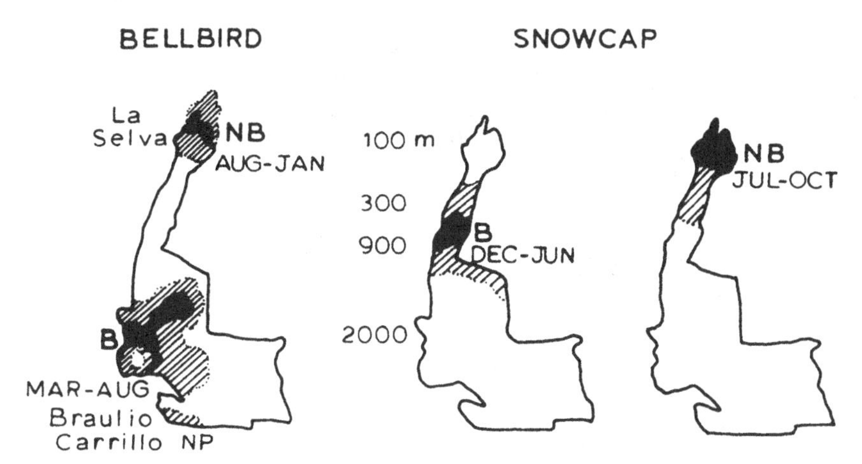

Fig. 12.2. Sample altitudinal migration patterns of Central American birds at the Braulio Carrillo–La Selva transect in Costa Rica. Bellbird: *Procnias tricarunculata*; snowcap: *Microchera albocoronata*. Breeding (B) and nonbreeding (NB) season distributions. (Modified from Stiles & Clark, 1989).

migration between dry Pacific and wet Atlantic coastal forest (Fig. 12.3a). Dry forest species such as blue ground-dove and clay-colored robin vary in abundance from year to year near Limbo Camp in Parque Soberania. While they normally breed in the dry and second growth forest of the Pacific coast, they cross the continental divide to wetter forest in drier years (Karr, 1985). Finally, forest birds "perceive" their habitat as a heterogeneous landscape on even smaller scales (Karr & Freemark, 1983). Thus, avian use of local, regional, and intercontinental landscapes is diverse and dynamic at a variety of time scales.

Population dynamics on a landscape scale account in part for the pattern of avian extinctions on Barro Colorado Island (BCI) since its isolation in Gatun Lake, and provide information important in the design of nature preserves. Extinction of birds on BCI has been attributed to loss of the largest species in guilds (ecological truncation – Wilson & Willis, 1975), differential loss of understory bird species (Wilson & Willis, 1975; Karr 1982b) and of wet and foothill forest (Karr, 1982a), loss of species with variable populations (Karr, 1982b; Karr, 1990c), high nest predation rates (Loiselle & Hoppes, 1983; Sieving, 1992), and low mainland survival rates (Karr, 1990c). Neither abundances on the mainland nor dominant food resources were good predictors of species likely to be missing from BCI (Karr, 1982a,b, 1990c). Many of these factors are associated with extinction probability because they relate to the way the species uses a complex landscape. The presence of suitable areas seasonally or during periodic dry years provides refuges during those periods. The absence

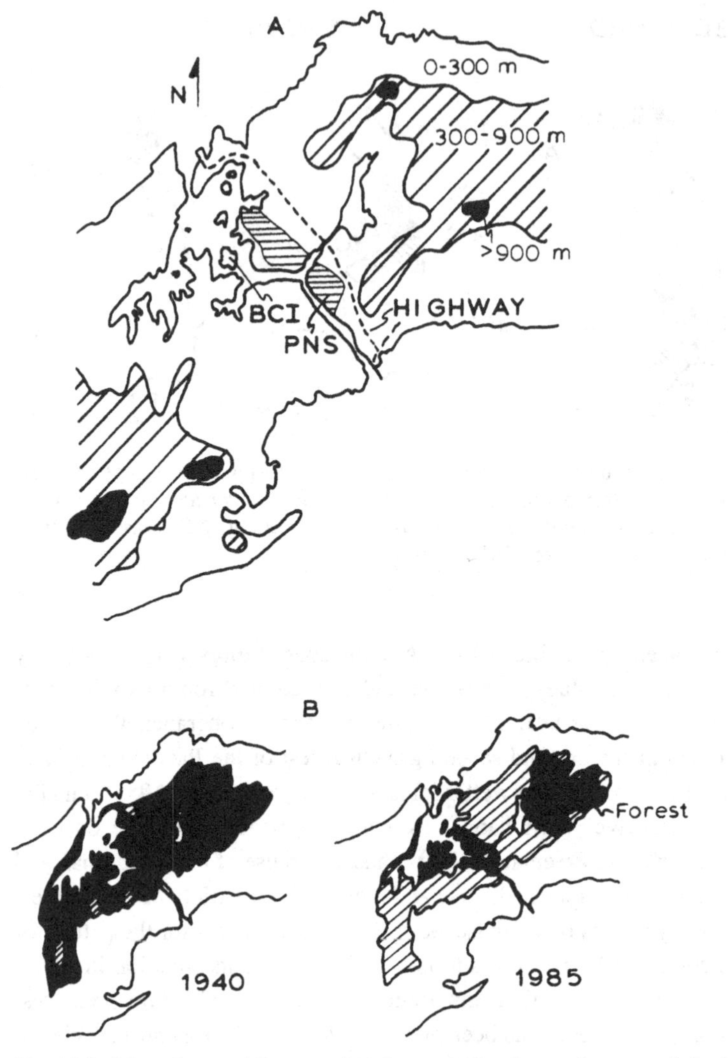

Fig. 12.3. Map of central Panama showing major landscape features of significance to permanent resident birds. A. Major geographical features. B. Approximate extent of forest in the canal watershed during 1940 and 1985.

of such refuges on BCI has resulted in the extinction of species requiring those refuges.

Although his terminology differed, Levins (1969, 1970) first called attention to the importance the complex of local populations, the "*metapopulation*," within a landscape. Some patches produce excess offspring (*sources*) while others (*sinks*) do not (Pulliam, 1988). A patch may even "wink" on (acting as a source) and off (acting as a sink) with natural environmental variation. In this way, the local landscape influences extinction rate. The selection of reserves must incorporate efforts to include the appropriate complex of habitats in the landscape. Proper mixes of habitat types and areas in a reserve system could reduce the number of extinctions (Quinn & Karr, 1992).

Four examples of this influence can be seen in central Panama. First, if an area equivalent in size to BCI had been isolated at the Limbo Camp area in Soberania National Park rather than BCI, fewer species of forest birds would probably have been lost. The Limbo area, unlike BCI, contains permanent streams and sheltered valleys that retain moist conditions during severe dry seasons (Karr, 1982a). Second, Soberania lies on the east bank of the canal with its long axis extending across the rainfall gradient of the Isthmus. As a result, its forests include areas from the moderately dry Pacific coastal forest to the wet forest of the Atlantic sector of Panama. That gradient provides the dry forest species with refuges in wetter areas during severe dry periods. Third, connections between the forest of Soberania and foothill forest to the east permit seasonal altitudinal migrations. I have recorded foothill species such as white-throated robin, yellow-eared toucanet, and sicklebill in the area around Limbo Camp. However, construction of the Transisthmian Highway during the 1940s and deforestation along that transportation corridor during the 1960s and 1970s isolated Soberania from the foothills. As a result, those foothill species appear less regular in the last decade than they were in the 1960s and 1970s. Extinctions from the east bank forest might be expected in the years ahead. Fourth, the west bank of the canal also contains forest but the patches are smaller and more isolated than those of Soberania. The loss of species in those forests has already proceeded to a level similar to that of BCI (Karr, unpublished) although those areas are not isolated by water. These examples demonstrate the merit of the program to preserve landscape connections across the full altitudinal range between La Selva and Braulio Carrillo National Park in Costa Rica (Fig. 12.2). That decision will prove to be instrumental in preserving the regional biota.

Spotted owl and old growth forest

The northern spotted owl and old growth forest controversy is another timely example of the use of landscape ecology in species management. Conservative estimates place reduction in northern spotted owl habitat since 1800 at about 60% (Thomas et al., 1990). Further, remaining habitat is badly fragmented, often occurring in patches too small or isolated to provide for owls; thus, small subpopulations are isolated. Finally, many individuals also face major fragmentation of forest within their home range, exposing them to predation and competition (e.g., with their congener the barred owl).

An initial spotted owl management plan called for designation of spotted owl habitat areas (SOHAs) to encompass areas that approximate the annual home range of a pair of spotted owls. These circles average about 2 miles (3.2 km) in radius. Prescribed habitat for owls within the circles vary from 1000 to 3000 acres (2400 to 7400 ha), depending on physiographic province. In some areas, SOHAs were clumped into groups of three. Actual owl habitat within a SOHA may be fragmented and some forested areas within the circle may be designated for cutting. Suitable habitat between SOHAs may also be cut, further fragmenting the landscape. Because of fire history and past and planned logging, SOHA habitat is seldom contiguous. Thus, apparently suitable owl habitat is often unsuitable at both the landscape scale (the SOHA network) and within a SOHA. Finally, the Forest Service and the Bureau of Land Management implemented plans for owl protection that were not consistent among agencies and regions. Overall, the northern spotted owl was not well protected by the SOHA approach at the subspecies or regional subpopulation level because no well-coordinated, biologically based management plan that covered the entire range of the species was in place. In the absence of such a plan, the continued decline of the species seemed inevitable.

A panel was formed to conduct a comprehensive review and propose a scientifically sound conservation strategy for the spotted owl (Thomas et al., 1990). They developed a two-part approach to ensure the viability of northern spotted owl populations over the next 100 years. The first stage involves protection of habitat to ensure the owl's long-term survival. The second step calls for research and monitoring to evaluate the success of the program and to make improvements where and when that seems appropriate. They proposed to replace the SOHA approach with a plan to protect larger blocks of habitat, termed Habitat Conservation Areas (HCAs). HCAs are planned to protect groups of pairs that are spaced closely enough to facilitate dispersal among blocks within an HCA as well as among HCAs. The developers of this new strategy integrated knowledge of spotted owl biology and the current distribution of spotted owl habitat, and used sophisticated approaches to population modeling and genetic theory. Because the exact landscape, its connectedness, and owl densities vary among the regions of the owl's distribution (e.g., habitat in the Olympics of Northern Washington will probably never support 20 pair SOHAs), they developed a general landscape goal but modified it where appropriate to accommodate local and regional landscape ecology. Although this strategy allows for the loss of perhaps one half of the current population, it has a high probability of protecting the species over the next 100 years (Thomas et al., 1990).

The spotted owl strategy (Thomas et al., 1990) is an important management effort that provides a comprehensive and sophisticated model for the protection of an endangered species. However, it is equally important to note that the strategy for this species is not likely to preserve all species found in old growth forests in the Pacific northwest. As Sharpe and Zhao (1990) recently showed for small woodlots in Wisconsin, protecting forests based on size and other similar attributes is not likely to protect all endangered plants. Indeed, the spotted owl strategy allows for the loss of many small isolated woodlot areas, habitat patches that are likely to contain unique plant species.

The spotted owl strategy is an excellent example of applying existing knowledge and theory, in a landscape context, to protect a natural resource. However, with its focus on a single species, it is unlikely to protect all regional biological diversity. The landscape of the owl is not coincident with the landscape scale needed to protect all threatened plants and animals.

Case 3: restoration of a wetland landscape – the Kissimmee River

For hundreds of years, the Kissimmee River of south Florida and its adjacent floodplain wetlands were broadly connected to the chain of lakes to the north (Fig. 12.4). To the south, much of the fresh water that traveled through Lake Okeechobee and the Everglades came from the Kissimmee. Beginning in 1961 a channelization project converted 165 km of natural meandering channel into a 90 km canal (Loftin et al., 1990a). That canal was designed to provide flood control for areas surrounding the Kissimmee Chain of Lakes.

Excavation of the canal and the resulting deposition of spoil destroyed 56 km (33%) of the original river channel and about 2,800 ha (14%) of the floodplain wetlands. Hydrological changes associated with channelization led to further degradation of remaining river and floodplain habitat. The ecological consequences of habitat degradation combined with increased nutrient loads transported into Lake Okeechobee generated widespread concern after the work was completed in 1971.

Since 1971, the people and politicians of Florida have called for efforts to restore lost natural resources (Loftin, et al., 1990a,b) and several legislative initiatives have mandated programs to accomplish that restoration. However, passage of informed environmental legislation or development of good regulations does not assure the implementation of good policy (see Karr, 1990a for several examples of this). Talk rather than action has characterized programs for Kissimmee restoration over much of the last 20 years. However, the past three years have shown major advances with adoption of a landscape approach to restoration.

Development of a sound landscape restoration program depends on knowledge of the disruption processes within the Kissimmee landscape that resulted

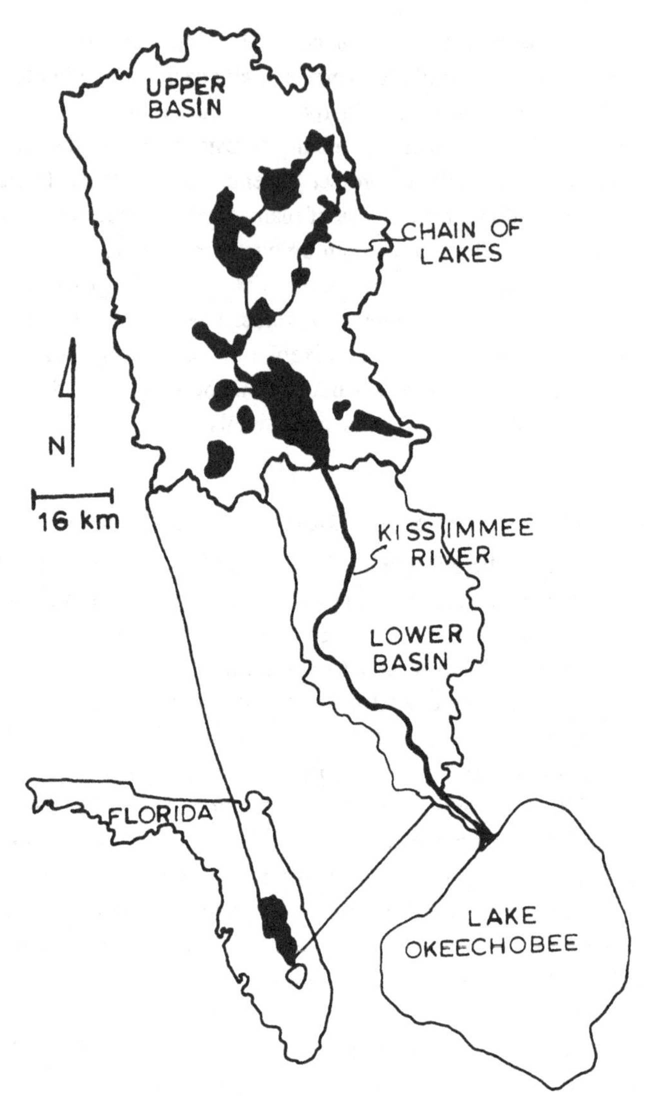

Fig. 12.4. Map of Kissimmee River System showing location of Chain of Lakes, Kissimmee River, and Lake Okeechobee, Florida.

in severe environmental damage, especially major loss of biological resources (Karr, 1990b; Loftin et al., 1990b). Channelization impacts included direct physical destruction of river and floodplain habitat, alteration of hydrological regimes within the river channel and adjacent floodplain wetlands, and a cascade of direct and indirect biological impacts. These effects included 1) transformation of the river channel to a wide, deep canal, most of which is biologically sterile; 2) lowering of the water table and consequent dewatering of most floodplain wetlands; 3) degradation of remaining wetlands due to maintenance of constant water levels; and 4) alteration of the seasonal pattern of flow. The latter includes timing of peaks and lows due to lake regulation and secondary drainage, no flow in remaining sections of river channel, and flows through the new canal so low that the system resembles a reservoir more than a river.

These changes destroyed riverine fish and wildlife habitat and the complex food webs that were once supported by the river and floodplain systems (Toth, 1990; Loftin et al., 1990b). Wintering and resident waterfowl use of the Kissimmee declined by 92–94% and wading birds made limited use of the area after channelization (Perrin et al., 1982). The largemouth bass fishery of the Kissimmee was reduced and at least six species of fish were extirpated (Perrin et al., 1982). In short, the channelization project altered the functional integrity of the river, its floodplain, and their interactions. The historical landscape of the Kissimmee watershed was destroyed, a loss in natural resources that cannot be calculated with any precision.

Planning for restoration

Soon after completion of the channelization project, public demands called for reestablishment of the "environmental values" of the Kissimmee River system. Passage of the Kissimmee Restoration Act in 1976 set the stage for the restoration effort. This Act called for restoration measures to be constrained by three primary goals: 1) using natural and free energies of the river system, 2) restoring natural seasonal water level fluctuations, and 3) restoring conditions favorable to increases in abundances of the native biota (Loftin et al., 1990a). Upon appointment of a Coordinating Council in 1983, the specific objective was to "reestablish the natural ecological functions of these natural systems in areas where these functions have been damaged" and successfully "restore and preserve these unique areas."

A symposium held in 1988 reiterated these points and presented detailed studies based on a demonstration project to examine the feasibility of restoring old river channels and wetlands to their former condition (Loftin et al., 1990b). That symposium clearly emphasized that the Kissimmee River restoration effort should focus holistically on the Kissimmee landscape, rather than individual biological components (e.g., waterfowl, wading birds, or sport fish). However, an additional constraint had to be met; that is, all restoration plans must maintain flood protection to private property as provided by the existing flood control project. Following that symposium a hierarchical framework was developed for continuing studies of restoration alternatives. This four-tiered hierarchy included restoration *goals, guidelines, objectives*, and *criteria*. All restoration alternatives were to be evaluated uniformly against these standards.

The *goal* was establishment of ecosystem integrity, defined as the capability to support and maintain biological communities with a species composition, diversity, and functional organization comparable to that of the natural habitat of the region (Frey, 1975; Karr & Dudley, 1981). The general *guideline* for attaining this goal involved returning control of the system to natural hydrologic processes. That is, given a chance, natural hydrological processes will restore complex ecosystem attributes and thus the system's environmental values. Critical components include both the channel and floodplain flow regimes and their interactions. This, however, is no simple task. It requires restoring the form and function of the landscape with minimal management efforts by human society. The *objective* is to do that over as much of the river/floodplain landscape as possible. Finally, detailed restoration *criteria* were established by studying the prechannelization hydrology of the Kissimmee River. Five key hydrologic criteria were defined (Loftin et al., 1990a):

1. *Continuous flow with duration and variability characteristics comparable to prechannelization records.* This criterion includes both regular seasonal cycles of flow and stochastic discharge variability.
2. *Average flow velocities between 25 and 55 cm/sec when flows are contained within channel banks.* The goal here is to keep flows within the preferred range of resident fish and insure maximum habitat availability.
3. *A stage–discharge relationship that results in overbank flow along most of the floodplain when discharges exceed 1,400–2,000 cf/s.* This criterion is selected to ensure important physical, chemical, and biological interactions between the river and floodplain.
4. *Stage recession rates on the floodplain that typically do not exceed 30 cm/month.* This criterion prevents rapid dewatering and thus provides habitat stability for both wetland and river plants and animals. In addition, it protects important aspects of river water quality.
5. *Stage hydrographs that result in floodplain inundation frequencies comparable to prechannelization hydroperiods, including seasonal and long-term variability characteristics.* This criterion ensures that wet–dry cycles along the periphery of the floodplain are preserved.

Alternative restoration plans

Four different plans were evaluated under this set of criteria: wier plan, earthplug plan, level I backfilling, and level II backfilling (Loftin et al., 1990a). Only the level II backfilling, replacing for most of the canal length the material dredged to form the canal, met all the criteria. The other three failed because of excessive river channel velocities, rapid stage recession rates, or inadequate floodplain inundation. In short, they would not restore the ecological integrity of the Kissimmee landscape. The level II backfilling plan met restoration goals and criteria by reestablishing prechannelization hydrologic characteristics along 84 km of contiguous river channel and 9,700 ha of floodplain. Overall, the plan would restore ecological integrity to about 9,000 ha of river ecosystem. Costs of the level II backfilling plan are estimated at just over $400 million, four times the cost of the wier plan. However, the natural resource benefit would be much greater and the long-term operation and maintenance costs of the backfilling plan are about one fourth of the wier plan. Finally, for comparison the cost of restoration is about the same as the cost of construction of 4 miles of interstate highway around Miami.

The Kissimmee River Restoration Project is an excellent example of an effort to use a landscape perspective in a well-reasoned effort to manage natural resources. The South Florida Water Management District (including its predecessor) has been involved in that effort since 1972. Their pioneering decision, that environmental values could be restored by returning the control of the system to hydrological processes, established an approach that was scientifically sound while it avoided the controversy associated with selection of specific natural resource goals (e.g., maximizing wintering waterfowl populations or sport fishery harvest).

Management of the human landscape

The examples selected for discussion in this paper deal with landscape ecology and protection of biological diversity or other natural resource values. The human species is dependent on the landscape, as are all species. A test of our environmental wisdom will be our success in using that reality to balance the four different kinds of environments (Fig. 12.5) necessary for a high-quality human landscape (Odum, 1969). This classification identifies the four basic functional environments required by human society: *productive, protective, urban–industrial,* and *compromise.* Although I take issue with characterizing one environment as productive because that implicitly suggests that the other three types are nonproductive, I preserve here the original terminology of Odum. *Productive* environments yield agricultural commodities such as food and fiber and include aquaculture and monoculture forestry. *Protective* environments are natural areas that protect biological integrity, the elements and processes implicit in the protection of biodiversity (Karr, 1990a). *Urban–industrial* environments provide land for factories, housing, transportation

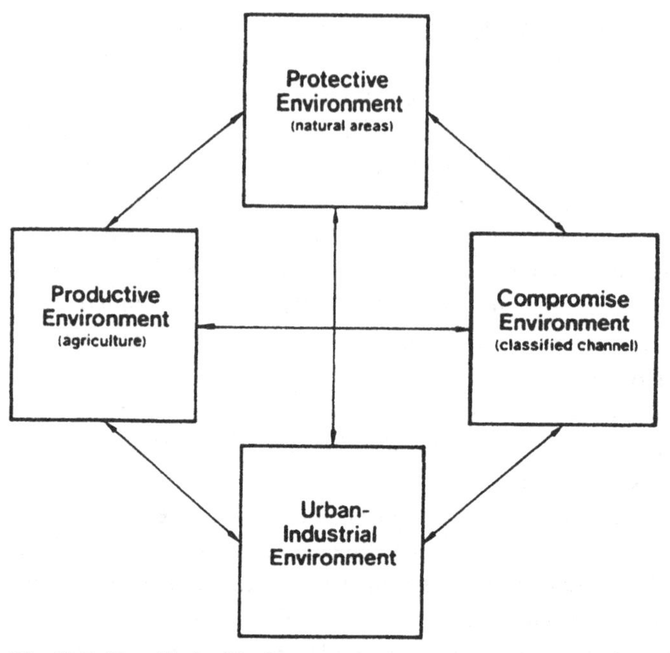

Fig. 12.5. Four kinds of landscape units essential to the maintenance of a quality human landscape. (From Karr & Dudley, 1981, after Odum, 1969).

corridors, and urban centers. The most important constraint in selecting these areas is that environmental considerations must play a major role. Historical destruction of wetlands, rivers, and other unique habitats has been possible because of emphasis on economic arguments to the exclusion of environmental costs. The fourth environment type is the *compromise* environment. The concept of the extractive reserve in Brazil is one type of compromise environment as is low-density housing that preserves regional habitat for many species. Extractive reserves permit selective sustained harvest of natural products (e.g., hunting, harvest of fruits or exudates such as rubber, and selective logging).

Past development plans often assumed that technology can provide substitutes for the benefits derived from the protective and compromise environments. The depth and extent of human environmental problems clearly demonstrates that we cannot continue under that assumption, spending ever-increasing amounts to reverse the environmental disasters that inevitably result. We are slowly recognizing that we can neither afford to technologically mimic the benefits of a healthy environment nor to technologically resolve environmental problems associated with many technological approaches.

Conclusion

Degradation of global life-support systems is obvious and expanding rapidly due to growing human populations and a proliferation of technology. Success at reversing this trend has been limited for a number of reasons, not the least of which is the lack of a landscape perspective in science and in resource management. Although we do not know enough now (and may never know enough) about ecological dynamics in natural and human-altered landscapes, we do know enough to make more informed decisions.

Discussions of landscape ecology often point to the fragmentation of landscapes as a driving force in reduction in biodiversity. Other important forces are the fragmentation of responsibilities among agencies and the lack of a coherent legal framework to protect ecological integrity. Even when laws are sound, they are often poorly implemented (Karr, 1990b).

No simple blanket solutions exist to protect ecological or landscape health, perhaps most importantly because of variation in the idiosyncratic natural history of species and assemblages of species. Ecologists often wring their hands because no single, small set of general principles applies everywhere. In fact, landscapes are too complicated to have six, or eight, or ten general principles. For this reason, environmental problems often require unique solutions for the same reason that engineers must create unique designs for each building, bridge, or dam to accommodate different construction materials (e.g., wood, concrete, plastic, or glass). Ecologists must consider the local landscape with all its components in efforts to protect natural resources.

The increased demand for ecological understanding, at landscape and other scales, requires increased funding from a society that demands more informed judgments from ecologists. Establishment of long-term monitoring approaches that clearly and concisely communicate the condition of landscapes must become a high priority. Our continued success as a society requires incorporation of all relevant scales in time and space into environmental planning. The landscape perspective provides a new and more integrative dimension to learning about the world around us and to plan our interactions with that world. Progressive biotic impoverishment, the systematic reduction in the capacity of the Earth to support living systems (Woodwell, 1990), is arguably the most important and ominous signal that the human landscape, and thus the landscape for all other species, is in jeopardy.

Acknowledgments

Support from numerous agencies and organizations (National Science Foundation, U.S. Environmental Protection Agency, U.S. Forest Service, Earthwatch/Center for Field Research, Smithsonian Tropical Research Institute) has been instrumental to the development of the ideas expressed in this chapter. J. Brawn, M. Dionne, C. Kellner, and B. Kerans commented constructively on an early draft of this paper.

References

Andrewartha, H. G., & Birch, L. C. (1954). *The Distribution and Abundance of Animals.* Chicago: University of Chicago Press.

Allen, T. F. H., & Bandurski, B. (1992). The Ecosystem Approach. Report of the Ecological Committee, Great Lakes Science Advisory Board, International Joint Commission, Ann Arbor, Michigan.

Ashworth, W. (1986). *The Late Great Lakes: An Environmental History.* New York: Knopf.

Benke, A. C. (1990). A perspective on America's vanishing streams. *J. N. Am. Benthol. Soc.*, **9**, 77–88.

Blake, J. G., & Loiselle, B. A. (1991). Variation in resource abundance affects capture rates of birds in three lowland habitats in Costa Rica. *Auk*, **108**, 114–130.

Botkin, D. (1990). *Discordant Harmonies: A New Ecology for the Twenty-first Century.* New York: Oxford University Press.

Colburn, T. E., Davidson, A., Green, S. N., Hodge, R. A., Jackson, C. I., & Liroff, R. A. (1990). *Great Lakes, Great Legacy?* Washington, D.C.: The Conservation Foundation.

Diamond, J. M. (1991). World of the living dead. *Natural History*, **9/91**, 30–37.

Edwards, C. J., & Regier H. A. (eds.). (1990). *An ecosystem approach to the integrity of the Great Lakes in turbulent times.* Spec. Publ. 90–94. Great Lakes Fishery Commission, Ann Arbor, Michigan.

Egerton, F. N. (1987). Pollution and aquatic life in Lake Erie: early scientific studies. *Envir. Rev.*, **11**, 189–205.

Ehrlich, P. R., & Roughgarden, J. 1987. *The Science of Ecology.* New York: MacMillan.

Fausch, K. S., Lyons, J., Karr, J. R., & Angermeier, P. L. (1990). Fish communities as indicators of environmental degradation. *Am. Fish. Soc. Symp.* **8**, 123–144.

Forbes, S. A., & Richardson R. E. (1913). Studies on the biology of the upper Illinois River. *Bull. Ill. Nat. Hist. Surv.*, **9**, 1–48.

Forman, R. T. T., & Godron, M. (1986). *Landscape Ecology.* New York: Wiley.

France, R. L. (1990). Theoretical framework for developing and operationalizing an index of Zoobenthos Community: Application to biomonitoring with zoobenthos communities in the Great Lakes. In *An Ecosystem Approach to the Integrity of the Great Lakes in Turbulent Times*, ed. C. J. Edwards & H. A. Regier, pp. 169–193. Special Publication 90–4. Great Lakes Fishery Commission, Ann Arbor, MI.

Francis, G. R., Magnuson, J. J., Regier, H. A., & Talhelm, D. R. (eds.) (1979). *Rehabilitating Great Lakes Ecosystems.* Technical Report 37, Great Lakes Fisery Commission, Ann Arbor, Michigan.

Frey, D. (1975). Biological integrity of water: An historical approach. In *The Integrity of Water*, ed. R. K. Ballentine & L. J. Guarraia, pp. 127–139. Washington, D.C.: USEPA.

Gentry, A. (1990). Tropical forests. In *Biogeography and Ecology of Forest Bird Communities*, ed. A. Keast, pp. 35–43. The Hague, Netherlands: *SPB Academic Publishing.*

Jeffords, M. R. (1989). *The Illinois River: A lesson to be learned.* Champaign, Illinois: Illinois Natural History Survey.

Karr, J. R. (1976). On the relative abundances of migrants from the north temperate zone in tropical habitats. *Wilson Bull.* **88**, 433–458.

Karr, J. R. (1980). Patterns in the migration systems between the north temperate zone and the tropics. *Migrant Birds in the Neotropics: Ecology, Behaviour, Distribution and Conservation*, ed. A. Keast & E. S. Morton, pp. 519–543. Washington, D.C.: Smithsonian Institution Press.

Karr, J. R. (1982a). Population variability and extinction in a tropical land-bridge island. *Ecology*, **63**, 1975–1978.

Karr, J. R. (1982b). Avian extinction on Barro Colorado Island, Panama: a reassessment. *Am. Natur.*, **119**, 220–239.

Karr, J. R. (1985). Birds of Panama: biogeography and ecological dynamics. In *The Botany and Natural History of Panama: La Botanica e Historia Natural de Panama*, ed. W. G. D'Arcy & M. D. Correa *Monographs in Systematic Botany*, **10**, 77–93. St. Louis, Missouri: Missouri Botanical Garden.

Karr, J. R. (1990a). Biological integrity and the goal of environmental legislation: lessons for conservation biology. *Cons. Biol.*, **4**(3), 244–250.

Karr, J. R. (1990b). Kissimmee River: Restoration of degraded resources. In *Proceedings of the Kissimmee River Restoration Symposium.* October 1988, Orlando, Florida, ed. M. K. Loftin, L. A. Toth, & J. T. B. Obeysekara, pp. 303–320. West Palm Beach, FL: South Florida Water Management District.

Karr, J. R. (1990c). Avian survival rates and the extinction process on Barro Colorado Island, Panama. *Cons. Biol.*, **4**, 391–397.

Karr, J. R. (1991). Biological integrity: a long-neglected aspect of water resource management. *Ecol. Appl.*, **1**, 66–84.

Karr, J. R., & Dudley, D. R. (1981). Ecological perspective on water quality goals. *Envir. Management*, **5**, 55–68.

Karr, J. R., & Freemark, K. E. (1983). Habitat selection and environmental gradients: dynamics in the "stable" tropics. *Ecology*, **64**, 1481–1494.

Karr, J. R., Toth, L. A., & Dudley, D. R. (1985). Fish communities of midwestern rivers: A history of degradation. *BioScience*, **35**, 90–95.

Karr, J. R., Fausch, K. D., Angermeier, P. L., Yant, P. R., & Schlosser, I. J. (1986). Assessment of biological integrity in running water: A method and its rationale. Illinois Natural History Survey Special Publication No. 5, Champaign, IL.

Kofoid, C. A. (1908). The plankton of the Illinois River, 1894–1899, with introductory notes upon the hydrography of the Illinois River and its basin. Part II. Constituent organisms and their seasonal distribution. *Bull. Ill. Nat. Hist. Surv.*, **8**, 1–360.

Lack, D. (1954). *The Natural Regulation of Animal Numbers.* London: Oxford University Press.

Levins, R. (1969). Some demographic and genetic consequences of environmental heterogeneity for biological control. *Bull. Entomol. Soc. Am.*, **15**, 237–240.

Levins, R. (1970). Extinction. *Lectures in Mathematical and Life Sciences*, **2**, 75–107.

Loftin, M. K., Toth, L. A., & Obeysekera, J. T. B. (1990a). *Kissimmee River Restoration: Alternative Plan Evaluation and Preliminary Design Report.* West Palm Beach, Florida: South Florida Water Management District.

Loftin, M. K., Toth, L. A., & Obeysekara, J. T. B. (eds.). (1990b). *Proceedings of the Kissimmee River Restoration Symposium.* October 1988, Orlando, Florida. West Palm Beach, FL: South Florida Water Management District.

Loiselle, B. A., & Hoppes, W. G. (1983). Nest predation in insular and mainland lowland forest in Panama. *Condor*, **85**, 93–95.

Merriam, G. (1990). Ecological processes in the time and space of farmland mosaics. In *Changing Landscapes: An Ecological Perspective*, ed. I. S. Zonnevald & R. T. T. Forman, pp. 121–133. New York: Springer-Verlag.

Morton, E. S. (1980). Adaptations to seasonal changes by migrant land birds in the Panama Canal Zone. In *Migrant Birds in the Neotropics: Ecology, Behavior, Distribution, and Conservation*, ed. A. K. Keast & E. S. Morton, pp. 437–453. Washington, D.C.: Smithsonian Institution Press.

Moyle, P. B., & Williams, J. E. (1990). Biodiversity loss in the temperate zone: decline of the native fish fauna of California. *Cons. Biol.*, **4**, 275–284.

Naveh, Z., & Lieberman, A. S. (1984). *Landscape Ecology: Theory and Application.* New York: Springer-Verlag.

Odum, E. P. (1969). The strategy of ecosystem development. *Science*, **164**, 262–270.

Odum, E. P., & Bevier, L. J. (1984). Resource quality, mutualism and energy partitioning in food chains. *Am. Natur.* , **122**, 45–52.

Ohio Environmental Protection Agency (1988). *Users manual for biological field assessment of Ohio Surface Waters.* 3 Volumes. Surface Water Section, Division of Water Quality, Monitoring and Assessment. Columbus, Ohio: Ohio Environmental Protection Agency.

Patten, B. C., & Odum, E. P. (1981). The cybernetic nature of ecosystems. *Am. Natur.*, **118**, 886–895.

Perrin, L. S., Allen, M. J., Rowse, L. A., Montalbano, III, F.,K Foote, . J., & Olinde, M. W. (1982). *A report on fish and wildlife studies in the Kissimmee River basin and recommendations for restoration.* Okeechobee, Florida: Florida Game and Freshwater Fish Commission, Office of Environmental Services.

Plafkin, J. L., Barbour, M. T., Porter, K. D., Gross, S. M., & Hughes, R. M. (1989). *Rapid bioassessment protocols for use in streams and rivers: Benthic macroinvertebrates and fish.* EPA/444/4-89-001. Washington, D.C.: Office of Water, U.S. Environmental Protection Agency.

Pulliam, H. R. (1988). Sources, sinks and population regulation. *Am. Natur.*, **132**, 652–661.

Quinn, J. F., & Karr, J. R. (1993). Habitat fragmentation and global change. In *Biotic Interactions and Global Change*, ed. P. M. Kareiva, J. G. Kingsolver, & R. B. Huey, pp. 451–463. Sunderland, Massachusetts: Sinauer Associates.

Regier, H. A., & Hartman, W. L. (1973). Lake Erie's fish community: 150 years of cultural stresses. *Science*, **180**, 1248–1255.

Richardson, R. E. (1928). The bottom fauna of the Middle Illinois River, 1913–1925: its distribution, abundance, valuation, and index value in the study of stream pollution. *Bull. Ill. Nat. Hist. Surv.*, **17**, 387–472.

Risser, P. G., Karr, J. R., & Forman, R. T. T. (1984). *Landscape ecology: directions and approaches.* Special Publication #2. Champaign, Illinois: Illinois Natural History Survey.

Sharpe, D. M., & Zhao, Y. (1990). Fragmentation, disturbance, and species loss from remnant forest patches in the Upper Midwest, U.S.A. Abstract. Vth International Congress of Ecology, Yokohama, Japan. August 23–30, 1990.

Sieving, K. (1992). Nest predation and differential insular extinction among selected forest birds of central Panama. *Ecology, 73*, 2310–2328.

Sioli, H. (1975). Tropical rivers as expressions of their terrestrial environments. In

Tropical Ecological Systems: Trends in Terrestrial and Aquatic Research, ed. F. B. Golley & E. Medina, pp. 275–288. New York: Springer-Verlag.
Smith, S. H. (1972). Factors of ecological succession in oligotrophic fish communities of the Laurentian Great Lakes. *J. Fish. Res. Board Canada*, **29**, 717–730.
Sparks, R. E. (1977). *Environmental inventory and assessment of navigation pools 24, 25 and 26, upper Mississippi and lower Illinois rivers: an electrofishing survey of the Illinois River.* Special Report 5, Water Resources Center, University of Illinois, Urbana, Illinois.
Steedman, R. J. (1988). Modification and assessment of an index of biotic integrity to quantify stream quality in southern Ontario. *Can. J. Fish. Aq. Sci*, **45**, 492–501.
Stiles, F. G. 1983. Birds: Introduction. In *Costa Rican Natural History*, ed. D. H. Janzen, pp. 501–530. Chicago, Illinois: University of Chicago Press.
Stiles, F. G. (1988). Altitudinal movements of birds in the Caribbean slope of Costa Rica: implications for conservation. In *Tropical rain forests: diversity and conservation*, ed. F. Almeda and C. Pringle, pp. 243–358. *Mem. Cal. Acad. Sci.* #12, San Francisco, California.
Stiles, F. G., & Clark, D. A. (1989). Conservation of tropical rain forest birds: a case study from Costa Rica. *Am. Birds*, **43**, 420–428.
Thomas, J. W., Forsman, E. D., Lint, J. B., Meslow, E. C., Noon, B. R., & Verner, J. (1990). *A conservation strategy for the Northern Spotted Owl.* Portland, Oregon: Interagency Scientific Committee to Address the Conservation of the Northern Spotted Owl.
Toth, L. A. (1990). Impacts of channelization on the Kissimmee River Ecosystem. In *Proceedings of the Kissimmee River Restoration Symposium*, pp. 47–56. West Palm Beach, Florida: South Florida Water Management District.
United States Environmental Protection Agency. (1990). *Biological criteria: National program guidance for surface waters.* EPA 440/5-90-004. Washington, D.C.: Office of Water, USEPA.
Webster's New Universal Unabridged Dictionary. (1983). New York: Dorset and Baber.
Wessman, C. A. (1990). Landscape ecology: analytical approaches to pattern and process. In *An ecosystem approach to the integrity of the Great Lakes in turbulent times.* ed. C. J. Edwards & H. A. Regier, pp. 285–299. Great Lakes Fish. Comm., Ann Arbor, MI. Special Publication 90–94.
Wiens, J. A. (1989). Spatial Scaling in ecology. *Funct. Ecol.*, **3**, 385–397.
Wiens, J. A., & Milne, B. T. (1989). Scaling of 'landscapes' in landscape ecology, or, landscape ecology from a beetle's perspective. *Lands. Ecol.*, **3**, 87–96.
Williams, J. E., Johnson, J. E., Hendrickson, D. A., Contreras-Balderas, S., Williams, J. D., Navarro-Mendoza, M., McAllister, D. E., & Deacon, J. E. (1989). Fishes of North America endangered, threatened, or of special concern: 1989. *Fisheries* (Bethesda), **14**(6), 2–20.
Wilson, E. O., & Willis, E. O. (1975). Applied biogeography. In *Ecology and evolution of communities.* ed. M. L. Cody & J. M. Diamond, pp. 522–534. Cambridge, Massachusetts: Harvard University Press.
Woodwell, G. M. (ed.). (1990). *The earth in transition: patterns and process of biotic impoverishment.* Cambridge: Cambridge University Press.

Part Five

Socioeconomics of Biodiversity

13

Economic valuation of biodiversity

ROBERT D. WEAVER

Introduction

The economic valuation of goods and services is often viewed as a heartless, quantitative process that is inappropriate as an aid to many of the great decisions faced by individuals or society as a whole. Within the context of biodiversity, many argue that it is absurd to attempt to place an economic value on a species or ecosystem. Nonetheless, society is increasingly faced with decisions that will directly affect the survival of species and ecosystems. Such decisions are no longer abstract or hypothetical, but have boiled down to the meanest terms of spotted owls versus lumber and jobs, or snail darters versus water management and power generation projects. Faced with these imperatives for decisions, society is increasingly forced to recognize that biodiversity must be explicitly "provided," that is, society must make explicit decisions to dedicate the use of economic goods and resources to ensure the existence of or influence the extent of biodiversity. In this sense, society is forced to recognize the cost of "provision" of biodiversity. Faced with such imperatives, a clear social interest in the economic value of biodiversity emerges naturally and understandably.

In this chapter the role of economic valuation of biodiversity will be considered. In the next section of this chapter the paradox of the value of public goods will be reviewed, establishing the universal problem of economic valuation of public goods. The following section will establish that characteristics of biodiversity may be viewed as public goods. Within this context, the final section of the chapter will more specifically consider the role of economic valuation for biodiversity.

The paradox of value for public goods

To build a basis for understanding the paradoxical value of biodiversity, this section will consider the more general problem of valuation of what economists call public goods. Based on the distinctions between public and private goods made in this section, the unique and complex economic nature of biodiversity will be considered in the following section. As will become clear, biodiversity, biological resources, and landscapes provide many public good characteristics for human consumption. To begin, the economic nature of goods and services will be established and "private" goods will be defined and distinguished from "public" goods. For both cases, a characteristic that is key to the distinction will be the extent to which the good or service is limited in supply. Goods and services that are limited in supply are usefully thought of as "resources," which may be used in production or consumption.

The logic of economic value: private goods[1]

Value as an economic concept is central to the management of limited resources. The distinction of "limited" is loaded with connotations, the understanding of which are crucial to an appreciation of the nature and role of economic value. Consider the motivation that might warrant use of the term "limited." Clearly, if a resource could be consumed without depletion, the distinction of limited might be irrelevant. In such a case, every consumer could benefit from the resource without impinging on the opportunity of other consumers to benefit from the resource. In contrast to this situation, consumption of most resources or economic goods is both *exclusive* and *exhaustive*. That is, for most resources, consumption does involve depletion, implying that one consumer's use must occur at the expense of another's. This predicament provides motivation for what might be thought of as *an exclusive right to consume.* When consumption is exclusive and exhaustive, and the resource is finite or limited in supply, some means of allocating it among potential consumers must be established. One means of achieving this would be to establish exclusive rights to consumption. Institution of such rights renders the allocation problem one of allocating rights to consume. Equivalently, such rights establish ownership over finite amounts of the resource and the benefits of consumption are exclusively *appropriable* by the consumer. The role of rights to consume in allocation of limited goods and resources is equally crucial even in nonmarket economic systems. However, in market economies it is implicit that acquisition of the right to consume can only be accomplished through direct economic competition in the marketplace among those who wish to consume the resource. In nonmarket econo-

[1]The concepts presented in this and the following section are commonly used in the economic literature. The noneconomist reader may find Nicholson (1990) an interesting source for futher reading.

mies, competition may be resolved through other social, political, or cultural means. Resources and economic goods of this type are called *private goods*. Examples of private goods include what comes to mind most often when economic goods are considered: Food products, nondurable and durable goods, housing, etc.

Markets represent one institutional means of resolving consumer competition for consumption rights. When limited resources are consumed in the process of making other economic goods, producers also must compete for limited resources. The process of competition for the right to consume such resources forces both potential consumers and producers of limited resources to establish a *personal valuation* of the resource. Within a competitive market, the limited resource is allocated across competing uses through a process of arbitration of differing personal valuations. The result is a *market valuation* or price of the resource. The conclusion can be drawn that it is the necessity of managing limited resources that constitutes the imperative for economic valuation. The function of markets is to arbitrate differing personal valuations to render a common market value that allows private goods and limited resources to be managed by markets and individual allocation decisions. Importantly, when the market is competitive, economists have shown that the price which emerges indicates the value of the goods to society, i.e. the *social value*. This social value represents the full costs to society of utilizing its limited resources to produce the good. Market competition also ensures that the social value reflects how much consumers are willing to pay for, or how much they personally value, the good. That is, the price reflects both consumer valuation of consumption, and producer costs of production. Further, the price represents a signal, or bit of information, with which competitive markets allocate use rights for the private goods among competing participants in a way that maximizes their joint welfare. Stated differently, the limited resources are managed in such a way as to create maximum social welfare.

The logic of value: public goods

Because of this role of economic value in the management of private goods, the question of determining economic value would seem universally relevant to the management of all economic goods. Despite this apparent importance of valuation for the management of limited resources or private goods, a vast number of economic goods and resources are not characterized by either exclusive or exhaustive consumption. As an extreme case, consider a resource or economic good that is not depleted by consumption. An example that is often used is national defense; others might include liberty or freedom. If such a good or resource is available, all consumers can benefit and any individual's benefit does not occur at the expense of other consumers' access to the resource. Such resources or economic goods are labeled by economists as *public goods.* Further examples include safe food, clean air, tranquility, or air traffic management. For such goods, the necessity of management and, therefore, of valuation is unclear. By definition, these goods cannot be thought of as limited, and there is no need for their allocation since their consumption is nonexclusive and nonexhaustive.

By definition, consumers do not need to compete for access to the consumption of public goods. In the absence of such an imperative for competition, consumers can *not* be expected to be willing to purchase rights to consume. Consumption rights are superfluous. Because the scope of availability of public goods is not limited, no imperative exists for the management of their consumption or preservation. Although consumers might highly value the consumption of such goods due to the personal benefits generated, no individual consumer would be willing to pay for a right to consume them. This *paradox of value* for public goods has often been interpreted as implying that concepts of economic value have little to offer in the management of such goods.

The paradox of value for public goods is resolved by recognizing that the provision of any public good requires direct or indirect utilization of limited private goods. Given this dependence, no private producer would be willing to produce the public good since revenue could not be generated by sale of consumption rights to consumers, who could access consumption without any rights, given the nonexclusive character of public good benefits.

Despite the absence of incentives for private provision of public goods, the collective benefits to consumers provides a rationale for collective action to provide the public good. It is within the context of this decision that a role for economic valuation emerges. Where the joint benefits of consumption of a public good exceed the cost or value of private goods and limited resources employed in production of the public good, the set of the consumers that would be benefited by the public good could chose to provide the public good by jointly paying for its production. The nonexclusive nature of consumption of public goods implies that production of a public good is synonymous with its provision. In practice, for most public goods, some potential for exclusion from consumption exists within some spatial or temporal sense. Economic theory suggests that the extent of the collective of consumers who join together in some governance jurisdiction to provide a public good can be limited by this potential for exclusion.

It follows that where provision of a public good involves utilization of limited resources or private goods, the management of that provision is required at the level of the collective of consumers who would benefit. Specifically, they are faced with the question of allocating private goods and limited resources to the production of the public good. This need for management immediately raises the need for valuation of the costs of provision of the public good and, importantly, valuation of the collective's benefit from provision of the public good. The costs of the private goods and limited resources can be valued at their social value by using market prices, if those goods and resources are managed at a social level by competitive markets. The social value of the benefits of consumption of the public good are, however, more elusive. Nonetheless, the need for and role of economic valuation for public goods is clear.

While private goods lend themselves to economic valuation through market processes, this is not the case for public goods. However, the imperative for valuation of public goods follows directly from the dependence of the provision of public goods on utilization of limited resources and private goods. It is this linkage with private goods that provides the imperative for economic valuation of public goods. At a conceptual level the economic value of a public good is simply the sum of the individual consumption benefits generated through the provision of the public good. Ultimately, while the economic value of a private good is limited to the benefit generated for the sole consumer of the good, the nonexclusive and nonexhaustive nature of the consumption of public goods implies that a wide range of individuals may benefit from provision of the good. Intuitively, the economic value generated by the provision of a public good must be measured as the sum of benefits realized by the consumers of the good.

At the individual consumer level, the economic value of consumption of a public good is determined by the objectives and constraints that influence consumer demand for the good. Objectives define overarching goals of the individuals and indicate how their consumption of resources and public goods contribute to the objectives. For private and public goods, the objectives and constraints that guide individual decisions provide the basis for each individual's economic valuation of the goods. For private goods, individuals are forced to reveal their individual valuations through their competition for rights to consume the goods. As already noted, competitive markets result in relative prices, which indicate the levels of these relative valuations. In any economic system, some means of allocation must be achieved to resolve the natural competition for consumption rights to private goods. However, for public goods, individuals are not forced to compete for rights of consumption.

By definition, the benefits of public good consumption are nonexclusive and nonexhaustive. Access to consumption for all individuals follows immediately when a public good is produced or "provided." It is for this reason that competitive markets fail to result in a price for public goods which indicates a market-wide economic valuation of the good. Further, it is for this reason that individuals would not be expected to accurately measure the extent of the benefits they realize from consumption of a public good. Given this nonexclusive character of public goods, no individual would be willing to pay for a consumption right, despite any positive economic value the individual may attribute to consumption of the good. The conclusion can be drawn that the problem of economic valuation of a public good is in essence one of measuring the individual economic valuations that are otherwise not revealed through market processes. In nonmarket economic systems, a similar necessity of valuing public goods emerges whenever those goods are produced by use of exhaustible, limited resources and private goods. However, in such systems, social, political, or cultural processes may be employed to establish such values.

This conclusion holds regardless of whether or not the contribution of consumption of the public good to individual objectives varies with the level of consumption of public good. Importantly, the conclusion also holds when objectives reflect *absolute goals* such that consumption of the public good contributes to individual objectives only if consumption is at a certain level or where the contribution of other consumption of private or other public goods to individual objectives is contingent on a specific level of consumption of the public good. Most commonly, objectives involve more general goals that can be achieved to a variable extent through the use of a possibly infinite variety of combinations of resources.

The hedonistic goal of achieving maximum utility as used in microeconomics is a good example. According to this theory, an infinite variety of resource combinations allow for the achievement of each level of utility. This results in the possibility of substitution of or "trading off" one resource for another while maintaining the same level of utility. The quantitative measure of this trade-off indicates each consumer's relative valuation of the resources. Hedonism and the associated effort to attain maximum utility given limited income leads consumers to align their relative valuations with that of the market as expressed by relative prices. Given competition, market forces also result in production decisions where use of limited resources establishes an equality between their prices and the value of their contribution to profits. It follows that prices determined in competitive markets are regarded as indicators of both consumer relative valuation and the relative profitability of the resource to producers.

Bases for valuing public goods

Several complications are associated with the valuation of public goods. Among them is the possibility that individuals hold nonutilitarian values for the good. Such nonutilitarian values may enter individual objectives and goals either as absolutes or they may vary with the level of consumption of the public good. Absolutes might emerge from individual acceptance of imperatives that make all other achievement of objectives conditional on the provision of a particular public good. These imperatives could have origins ranging from sacred or moral to whimsical. Classification of origins can be expected to depend on social norms and individual beliefs and values that establish the legitimacy of the imperative based on current social values. Clearly, higher laws or moral imperatives can be expected to differ over individuals, groups, and social history.

Nonetheless, the implications of such imperatives for individual objectives is that they imply objectives that can be thought of as discontinuous with respect to the consumption of the good. If the good is provided, the objective could be thought of as taking on the highest possible value, while in the absence of the provision, the objective would take on a value of zero. In either case, it is both necessary and conceptually feasible to consider this type of non-utilitarian benefit in the economic valuation of the public good. While economic valuation can accommodate either utilitarian or nonutilitarian value, it does require that the valuation be adopted or recognized by human decision makers. Thus, "intrinsic values" that somehow exist independently of human cognition necessarily fall outside of the realm of values that can be integrated into economic valuation. In contrast, if such intrinsic values are recognized by individuals, i.e., integrated into their objectives, then such values could play a role in economic valuation.

Both at a personal and at a market or social level, market based economic value reflects current *endowments* including the composition and distribution of limited resources available to the society as well as the current preferences, values, and perceptions held by individuals. These apparent dependencies present bases for questioning the usefulness of economic valuation in the management of public goods. For example, where the current distribution of limited resources is rejected on equity grounds, the usefulness of market prices as indicators of social value is also compromised. Similarly, where the provision of public goods results in benefits to future generations and the ability of the current population to represent those generations is perceived to be inadequate, the usefulness of measures of current economic value would appear to be compromised.

A final concern of substantial relevance to economic valuation for public goods is the possibility that particular groups of individuals could hold the view that the preferences, values, and beliefs of other groups in society must be rejected as bases for valuation of limited resources or public goods. This possibility for conflicts in personal valuation of the benefits of public goods can be seen for the case of economic goods where exhaustive consumption is possible, yet exclusion is infeasible.

Within this context, each individual's consumption affects the nature and extent of consumption opportunity available to others. It follows that each individual has an interest in the consumption activities of others and conflicts in individual valuations can be expected. The example of a "commons" pasture area illustrates this possibility. However, the possibility of conflicting individual valuations are also clearly possible in general for public goods. This possibility arises when the provision of a public good is dependent upon the use of private goods and limited resources. In that case, it is natural that individual preferences may conflict. Consider a specific biodiversity resource such as the spotted owl habitat in the northwest of the U.S. Clearly, at least two conflicting individual valuations can be imagined for the habitat: 1) a high value for preservation of the habitat to allow preservation of the spotted owl and 2) a high value for the use of the habitat for lumbering or other utilitarian activities.

Where individual economic valuations of a public good conflict, the evaluation of the social value of the public good is critically dependent upon how individual valuations are aggregated into a measure of economic value at a broader social level. In an egalitarian society where all individuals are acknowledged by society as having equal import, the social value of public good would simply be the sum of individual values. Where consensus is based on political rivalry won by majority voting, individual valuations are reflected by individual votes, allowing the sum of votes to indicate the sum of individual valuations. Clearly, voting offers an effective mechanism for determining perceived social value so long as the levels of individual valuations do not vary substantially across interest groups. Where substantial variation in individual valuations exist, other political processes may allow identification of the decision outcome that leads to maximum social value.

For any of these reasons, the complexity of the problem of measuring the economic value of public goods may be extended considerably past the relatively simple problem of measuring market prices and individual benefits.

Nonetheless, the essence of the challenge of use of economic valuation in decisions to provide public goods involves 1) the measurement of individual valuations of consumption benefits and 2) determination of society's relative valuation of the individual valuations.

Economic valuation and provision of biodiversity

Biodiversity as a public good

From the perspective of classifying biodiversity as a private or public good, it is essential to consider the exclusiveness and exhaustiveness of its "consumption." In the process of resolving this issue, the definition of biodiversity and what is meant by its "consumption" will be clarified. Classification of biodiversity as an economic good is complicated by the nature of and wide variety of its economic characteristics. From a strictly utilitarian perspective, biodiversity is best thought of having both private and public good features. Clearly, individual biotic elements of biodiversity, such as particular species, can be consumed as private goods to generate a variety of benefits that are exclusive to the consumer. It is also clear that such consumption often involves exhaustive consumption of the individual. Timbering, hunting, land clearing for agricultural or urban uses, pest management, and literally any human use of the landscape involves both exhaustive and exclusive use of biodiversity.

Also from a utilitarian perspective, biodiversity clearly has many consumption characteristics that are nonexhaustive and nonexclusive in consumption, rendering biodiversity a public good in these regards. For example, biodiversity can be viewed as providing a laboratory for learning about biological functions. In this case, the characteristic of biodiversity consumed could be thought of as knowledge. Other aspects of biodiversity such as provision of opportunities for *in situ* maintenance of particular species of utilitarian value could be viewed as having public good characteristics. From a utilitarian perspective, the recreactional and esthetic value of biodiversity clearly involves public good characteristics. Numerous other utilitarian benefits could be cited. For example, the water quality and soil erosion benefits of certain grassland ecosystems or knowledge of ecosystem characteristics that determine drought tolerance of the system's biodiversity.

From a nonutilitarian perspective, the continued quality and existence of human life has repeatedly been argued to be dependent on the existence of biodiversity (e.g., see Myers, 1994). Similarly, the existence of biodiversity has been argued to result in nonutilitarian benefits based on moral or ethical values and imperatives (Norton, 1994). In each case, the consumption of biodiversity is of a public good nature and the value of that consumption follows from an absolute objective. Importantly, from these perspectives, the value of biodiversity is viewed as paramount.

The need for economic valuation of biodiversity

The relevance and role of economic valuation for biodiversity can now be established. This chapter has argued that, in general, the existence of a role for economic value emanates from an imperative and capacity to manage limited resources and private goods. Economic value provides a basis for determining how limited private good resources are to be utilized in direct consumption, creation of other private goods, or provision of public goods. It was argued that the relevance of economic value follows from the necessity to determine where, when, and who will produce and consume limited resources. This conclusion was drawn for both private and public goods.

It follows that the existence of a role for economic value of biodiversity must be founded on recognizing the existence of public good consumption characteristics of biodiversity, as already discussed, as well as that biodiversity is a limited resource and/or that provision of its public good characteristics depends on the use of limited resources and private goods. Second, a role for economic value must also be founded upon the feasibility of and acceptance of responsibility for management of biodiversity. In combination, the presence of public good characteristics, dependence on limited resources and private goods, and the feasibility of "provision" of biodiversity as a public good provide an imperative for economic valuation in the provision decision. Specifically, if limited resources or private goods of economic value are to be utilized to "provide" biodiversity, their use must be rationalized by considering the economic value that biodiversity creates by its provision.

The provision of biodiversity is clearly linked, both directly and indirectly, to both private good consumption and production activities. While the natural state of biodiversity might be thought of as independent of any human intervention, given interaction between human activity and biodiversity, the continued existence of biodiversity requires direct consumption or utilization of private goods and limited resources. A prime example is the dedication of human resources to enhancement of habitats degraded by human activity.

With respect to indirect interaction with private goods, the existence and nature of biodiversity is often argued to be affected by a variety of private good consumption and production activities. These external impacts or "externalities" of human activity provide the basis for another important dependency between the "provision" or existence of biodiversity and the utilization of private goods and limited resources. Over the past decades, a near universal recognition has emerged of externalities such as industrial pollution and habitat alteration through human activity. In particular, this recognition has evolved from direct impacts such as pollution of water and associated reduction in quality of drinking water to more elusive impacts such as those of CO_2 emissions and the ozone layer or global warming. The presence of any of these externalities impacting biodiversity generated by private good production or consumption activities or limited resource utilization activities implies a direct linkage between the provision of biodiversity and private goods and limited resources. As has become increasingly apparent over the past decades, provision of biodiversity requires management of such externalities through alteration of the private good and limited resource activities generating them.

Biodiversity as a limited resource

While biodiversity can be seen as dependent upon the utilization of private goods and limited resources, to what extent can biodiversity itself be viewed as a limited resource? First, it must be acknowledged that the economic nature of biodiversity is as highly dimensioned as the physical character of biodiversity. On first consideration, biodiversity might be considered simply another label for biological resources. Like mineral resources, in this sense, biological diversity could be considered as involving a wide variety of specific limited resources. From this perspective, biodiversity could be managed as any other private good resource by cataloguing the locations of resources, evaluating their type and extent, and establishing some management methods. The fact that the resources are spatially diffuse would only complicate implementation of management.

However, this analogy between biodiversity and mineral resources is flawed by its failure to recognize three fundamental characteristics of biodiversity not found in mineral resources: 1) dynamics, 2) interrelatedness among biotic resources, and 3) dependence of existence of biodiversity on the use of private goods and limited resources. Clearly, the nature of biological resources at any point in time is merely a snapshot of highly dynamic resources in which inter- and intra-species variation over time results in a high degree of variation in the nature of the resources. It follows that biodiversity must be thought of not simply as a static resource involving a finite catalogue of all species of plants, animals, and microorganisms, but as a dynamic "resource" encompassing these elements as well as their capacity to dynamically adapt, evolve, and interrelate as ecological systems. From this perspective, the extent and nature of variation in both static and dynamic characteristics are crucial characteristics of the resource. While this definition is consistent with those offered by others (e.g.,

OTA, 1981), it goes beyond them by emphasizing the dynamic and interaction characteristics of biodiversity in comparison to biological resources.

Both the interrelatedness among biotic resources as well as the dependence of biodiversity on private goods and limited resources provides the imperative for an explicit decision to "provide" biodiversity. Importantly, this dependence influences the dynamics of biodiversity and in many cases determines the existence of species. Most notable is the interrelatedness of biodiversity with habitats that must be viewed as limited resources involving both appropriable biotic and nonbiotic resources.

The conclusion can be drawn that biodiversity represents a limited resource, despite its dynamics. Further, from an economic perspective, the provision of biodiversity requires both direct private and public good consumption as well as indirect utilization of such goods as externalities associated with their production and consumption are managed.

Determining the value of biodiversity

Having established the imperative for explicit "provision" of biodiversity and the associated rationalization of a role for economic valuation in such provision decisions, it remains necessary to consider the feasibility of and the extent of economic valuation necessary within the context of biodiversity. In the second section of this chapter, four complications that arise in the economic valuation of a public good were reviewed: 1) existence of nonutilitarian values, 2) existence of absolute goals, 3) dependence of current economic value on current endowments of resources and current preferences, and 4) the existence of conflicts in individual valuation. Within the context of economic valuation of biodiversity each of these complications could be claimed to preclude the usefulness of economic valuation in the decision to provide biodiversity.

The existence of nonutilitarian values with respect to biodiversity offers no special complications for biodiversity. As already argued, nonutilitarian values can be incorporated as an extension of standard economic analysis of individual valuation of private and public good consumption. In contrast, one of the key bases for rejecting the feasibility of economic valuation of biodiversity has been the citation of absolute goals. For example, on moral, ethical, or theological bases individuals may perceive an absolute imperative for provision of (i.e., preservation of) biodiversity. Such individuals might value biodiversity so highly that they might be willing to forego a substantial portion of their current standard of living to ensure the provision of biodiversity. Clearly, if all or a majority of individuals held such values, the role of economic analysis in resolving whether biodiversity should be provided would be quite limited. The value of biodiversity would be so high as to preclude the need for any tedious evaluation of whether its benefits exceeded costs of private good and limited resource utilization. Nonetheless, in such a situation the role of economic value would be retained, despite the absence of any need to conduct detailed analyses. In the more realistic case, where such absolute values are not universally held, the decision to provide biodiversity would rely on measurement of individual values and an aggregation of those values at a social level based on a social welfare function as discussed above. Through the use of such a social-welfare function conflicts in individual valuations would also be resolved. The concept of a social-welfare function that establishes the relative social value of individual values might seem elusive, however, as noted already, political processes are well known as means for elicitation of such welfare functions.

A final concern is one which has been argued to limit the usefulness of economic valuation for management of biodiversity: the dependence of economic values on current economic endowments and individual values. As was noted above, utilitarian values of private goods are reflected by prices determined by competitive economic markets. Economists have long appreciated the dependence of these prices on existing distributions of limited resources across individuals and on individual beliefs. With minor extension of conventional economic theory, it is also apparent that such prices can reflect expected future endowments and individual preferences and values. Where nonutilitarian values, social norms, ethical standards, and moral imperatives establish absolute goals, such absolutes must also be viewed as myopically dependent on the current human state. The conclusion follows from definition that all valuation based on or responsive to human cognition must be viewed as conditioned by the current human state. Applying this to the usefulness of economic valuation for management of the provision of biodiversity, a role for economic valuation must be acknowledged to the extent that individual valuation is based on human cognition.

Conclusions

A central emphasis of this chapter has been to clarify that the relevance of economic valuation follows first of all from the existence of limited resources or private goods that must be allocated among competing demands for exhaustive use. Economic valuation emerges from this imperative for management as a key element of the information necessary for management decisions that determine allocations or use of those resources based on maximizing whatever objectives are adopted and constraints on decisions. When use of private goods or limited resources is required for the provision of public goods, economic valuation of the public good provides the basis for rationalizing the use of those limited resources and private goods in its provision. Where the benefits generated by public goods are directly resultant from a specific use of limited resources or private goods, valuation might be relatively straightforward. However, more typically, public goods may generate benefits from a variety of uses. In such cases, the benefit of provision of the public good must be viewed as the sum of all use-specific streams of benefits. Given the non-exclusive nature of public good consumption, each use-specific stream of benefits must be viewed as the sum of benefits across all those individual consumers and producers who benefit from consumption of the public good.

I have argued that biodiversity must be viewed as a limited resource, despite its complex nature. Further, the "provision" of biodiversity was argued to be dependent on both direct and indirect utilization of private goods and limited resources. Finally, biodiversity was argued to provide consumption benefits that can be viewed as public goods. From these perspectives, the role for economic valuation of biodiversity emerges as a potentially useful aid to managing the resource. The nature of economic valuation necessary for the management and "provision" of biodiversity was argued to be dependent upon the nature of the individual values and the role in those values of utilitarian versus nonutilitarian values that can be viewed as *absolutes.* In particular, it was argued that when nonutilitarian absolutes are held by many consumers, it is possible that the social benefit of provision of biodiversity outweighs the costs of utilization of private goods and limited resources by such an obvious extent that actual measurement might be unnecessary. Clearly, this is the argument presumed by many interest groups that support preservation of biodiversity at any cost. Where such a consensus of value does not exist, a variety of means exist for implementation of economic valuation at both an individual and social level.

To conclude, this chapter has attempted to untangle a variety of issues concerning the usefulness and role of economic valuation of biodiversity resources. In doing so, it has argued that a clear role exists, that economic valuation can accommodate both nonutilitarian values and what were labeled as *absolutes.* The chapter has not directly addressed the question of implementation of economic valuation for biodiversity except to note that market processes cannot be expected to play a direct role, that elicitation of individual values is complicated by the infeasibility of exclusion of individuals from consumption, and that aggregation of individual values to establish a social-level measure of value must involve use of some form of a social-welfare function elicited through political processes or selected by political elites.

References

Myers, N. (1994). We do not want to become extinct: the question of human survival. In *Biodiversity and Landscapes*, ed. K. C. Kim & R. D. Weaver, pp. 133–150. New York: Cambridge University Press.
Nicholson, W. (1990). *Intermediate Microeconomics and Its Application*, 5th ed. Chicago: Dryden Press.
Norton, B. (1994). Thoreau and Leopold on science and values. In *Biodiversity and Landscapes*, ed. K. C. Kim & R. D. Weaver, pp. 31–46. New York: Cambridge University Press.
Office of Technology Assessment. (1981). *Technologies to Maintain Biological Diversity*. Washington, D.C.: U.S. Government Printing Office.

14

Thinking about the value of biodiversity

ALAN RANDALL

The evidence seems strong that a mass extinction of species is underway and, this time, humankind bears most of the blame. The problem seems to be that the scale of human activity on the planet has grown so large that it is imposing unprecedented stress on the biota. The planet is supporting many more people than ever before, and the lifestyle that the affluent minority enjoys and most everyone else covets is increasingly intrusive on natural systems.

The interests of humanity and the rest of the biota are not in pure opposition. In some cases, the immediate threat to biodiversity comes more from simple human carelessness than from pressing human needs. Let me offer two kinds of examples. First, Bishop (1980) documents several cases in which the future of threatened species could be secured at quite small expense. In such cases, it does not require generous assumptions about preservation benefits to justify making the effort. Second, economists (e.g., Southgate, 1988) have argued persuasively that many deforestation and land-settlement movements around the Third World – major threats to biodiversity – are motivated by inefficient subsidies and incentives. Respect for traditional concepts of efficiency would be sufficient to rein in these threats; enlightened respect for biodiversity would be laudable but not strictly necessary. In both these examples, humans could accommodate the needs of biodiversity at very little real cost, if they would simply be a little more careful.

At the other end of the spectrum, humanity and the biota have a common interest in long-term survival. At some point, the survival prospects of people and other living things will become so intertwined that, given sufficient understanding of the urgency of the situation, human self-interest would demand action to preserve the biosystems that support human life.

Humans would do well to find ways to live in greater harmony with nature, whether by avoiding needless assaults on the environment for trivial gain or by recognizing the ultimate interdependence of life forms when survival is truly in the balance. In these cases, there are few difficult choices to make because, in the one instance, the gains from the probiodiversity option are not counterbalanced by significant costs and, in the other, because the absence of acceptable options precludes meaningful choice.

But these are not the situations I want to address here. At this point in human history, there are important choices to be made. On the one hand, there are plenty of cases in which the probiodiversity option carries real, if perhaps not insurmountable, costs. On the other, the kind of ecosystemic collapse that would threaten human survival is probably not yet at hand.

There is still scope for choice, and choice has real consequences. For one important example, many of the world's richest ecosystems and many of the world's poorest people can be found in the tropics. So, it is not surprising that the legitimate human aspiration for improved living conditions frequently clashes with biodiversity objectives in the tropics. We need to develop a rationale for choosing, when conflicts between biodiversity and other human goals cannot be circumvented.

Benefits of biodiversity

There is little doubt that we are presently embarking on a mass extinction that may well be worse than any since the last major ice age. This time, humans bear most of the blame and have it within their collective power to sharply reduce its severity if decisive action can be taken. Nevertheless, the fact that biodiversity, worldwide, is under threat does not establish sufficient reason for concern. It is necessary to show also that biodiversity is worth caring about.

To many people, the most immediate reasons for caring about biodiversity are instrumental and utilitarian. A diversity of species in a variety of viable ecosystems serves as an instrument for people seeking to satisfy their needs and preferences. Many of the instrumental services that nature provides for people are obvious: food and fiber from domesticated plants and animals that were bred and selected from wild ancestors, and chemicals and pharmaceutical products with biotic origins.

Some attempts to develop a rationale for preserving biodiversity focus on these kinds of services. The argument proceeds as follows: there are many instances in which biotic resources have proven valuable to people and, together, the total value of these services is enormous. However, the majority of species on this earth are yet to be catalogued and systematically evaluated for their commercial potential. But, since many of those species we know have proven useful, it is reasonable to expect that many of the presently unknown or poorly understood species will turn out to be useful, too (Bishop, 1978). In addition, it is reasonable to expect continued technological progress, although we cannot predict its direction (Fisher & Hanemann, 1986). So, new uses may be discovered for known species not currently thought useful. By standard statistical notions, the usefulness of many species under present technologies suggests a positive probability that literally any species, known or unknown, will eventually prove useful. Thus, we should approach the potential loss of any species with the presumption that its expected value to humans is positive; that is, its preservation is worth something to humans. Nevertheless, this appeal to current and expected future usefulness as commercial raw materials is only the beginning.

The knowledge arguments – that species represent a store of genetic information for future use – have been extended to include ethnobiological knowledge. Ecosystems and indigenous human cultures coevolved in considerable harmony, it is conjectured. This implies that disruption of indigenous cultures by colonists from elsewhere threatens destruction not only of the ecosystem but also the folk knowledge that enhances its value (Southgate, 1988; Norgaard, 1984, 1988).

In the 1970's, the rapidly growing field of environmental economics established that people have demand for natural systems not only as sources of raw materials, but also as amenities. Amenity values include use values and existence values. Use values include aesthetic and recreation values and, in this latter connection, it is suggestive to observe that travel is now the fastest growing industry worldwide and adventure travel is the fastest growing segment of the travel industry. Existence value arises from human satisfaction from simply knowing that some desirable thing or state of affairs exists.

The instrumental and utilitarian arguments for preserving biodiversity recognize not only raw material and amenity values, but also ecosystem support services. Natural ecosystems serve as effective assimilators of wastes. In this way, wetlands help purify water, and forests assimilate greenhouse gases and help restore the oxygen balance. Natural ecosystems contribute to water and air quality objectives. More generally, ecosystems are complex and fragile, and it is a tenet of ecology that everything has its place in the broader scheme of things. Thus, there is a presumption that species are not only useful directly as suppliers of raw materials and amenities but also indirectly for their contribution to ecosystem support. Thus, species that have no conceivable value in providing raw materials and amenities (if there are any) could still be valued for the ecosystem support services they provide to species that are more directly valued.

The ultimate instrumentalist argument – that all species must be preserved because the loss of any would initiate processes leading inexorably to the collapse of the whole ecosystem – seems clearly false. In its place, the Ehrlichs (1981) have introduced, by analogy, a statistical argument. Their much-cited rivet popper justifies his continued removal, one-by-one, of rivets from airplane wings by reasoning that the practice must be safe since no planes have been lost yet. There is an obvious logical fallacy in the rivet popper's claim: each successive inconsequential loss of a rivet does not serve to confirm the low probability of the practice causing a crash. Rather, since the initial numbers of rivets is finite and the number needed for safer operation is smaller yet, each inconsequential rivet removal increases the probability that the next one will be disastrous.

The point of the Ehrlich's analogy is that one may concede the redundancy that is built into the ecosystem without condoning a cavalier attitude to the piecemeal sacrifice of species. Each inconsequential loss of a species increases the probability that the next one will cause serious problems.

Taken together, these various considerations amount to a convincing argument, in the instrumentalist utilitarian tradition, that preserving biodiversity should be a serious consideration for a society of rational human beings pursuing what are essentially homocentric goals. At the very least, biotic resources are resources to be allocated carefully, rather than squandered or merely wasted. They are to be valued for their amenity services as well as their usefulness as raw materials; use and existence values count; the concept of value encompasses but goes beyond commercial values; ecosystem and environmental support services are recognized; and, where ignorance is rampant, statistical arguments are used to infer positive expected values.

This account of the reasons for valuing biodiversity is consistent with standard economic thinking about the benefits of biodiversity. This is scarcely surprising; after all, mainstream economics is a homocentric and utilitarian system of thought. The concept of benefits, however, is counterbalanced by the concept of costs, including opportunity costs. To recognize that biodiversity is beneficial does not, by itself, clinch the economic case for protecting biodiversity. What if the costs outweigh the benefits? Such an outcome is always possible, under benefit cost thinking, and would do little to promote the case for biodiversity.

In search of a failsafe case for biodiversity

I suspect Ehrenfeld (1988) speaks for many preservationists who "would like to see [conservation] find a sound footing outside the slick terrain of the economists and their philosophical allies" (p. 215). For these preservationists, the homocentric, instrumentalist and utilitarian rationale for conservation is slick terrain, in that it provides a rationale for valuing biodiversity but its value is always relative to the values attached to other things. Some of the preservationists and their philosophical allies have provided clues as to what is needed, to develop a rationale for biodiversity that avoids the slick terrain. Such a rationale should not depend on:

- The human utility function. Many conservationists suspect that any rationale for preservation that depends on human preferences provides less than iron-clad guarantees. Preferences for biodiversity may wane, or simply be overwhelmed by competing claims on behalf of other things that also give us pleasure.
- Instrumentalist arguments. Arguments that biodiversity is valued not so much for its own sake but because it serves as an instrument for various human purposes – e.g., as a storehouse of genetic information for agriculture and medicine – could be undermined by technological change. Ehrenfeld (1988, p. 213) claims that pharmaceutical researchers do not in fact tramp through the jungles searching for exotic species with medicinal prospects; a strategy of computer modeling, organic synthesis of promising molecular structures, and screening of the resultant synthetic compounds is more efficient. It is not yet clear whether the emerging technologies of bioengineering will enhance the value of naturally occurring genetic material or merely substitute for it.
- "Divide and conquer strategies." Norton (1988) argues that we should refuse even to try to answer questions about the value of the biodiversity losses associated with particular policy proposals considered one by one. The piecewise, or "divide and conquer," valuation strategy would inevitably trivialize the aggregate losses from human actions. The whole of the losses far exceeds the sum of its estimated pieces. Lovejoy (1986) makes a similar argument. Norton's solution is to deny the validity of any quantified values for piecemeal losses of biodiversity, while insisting that the value of biodiversity in toto "is the value of everything there is" (p. 205).
- Trade-offs. Norton (1988, p. 204) complains that the act of asking about the value of biodiversity is itself a measure of the unique arrogance of humankind. Ehrenfeld (1988) while citing the Old Testament, offers an argument that perhaps owes more to the concept of hubris in Greek tragedy: "Assigning value to that which we do not own and whose purpose we cannot understand except in the most superficial ways is the ultimate in presumptuous folly" (p. 216). I interpret their arguments as not being opposed to the idea that biodiversity is valued in any absolute sense. Rather, what is being opposed is the idea that biodiversity can be valued in a relative sense; that one can

compare its value to that of other good things and adjust society's production and consumption for the better by making marginal trade-offs.

The kinds of rationales for biodiversity that would be developed by mainstream economists and homocentric utilitarian philosophers would surely fail the "slick terrain" test in every way: human preferences would count, instrumentalist arguments would have some force, choices would be conceptualized as typically piecemeal rather than all-or-none, and the interaction of technical possibilities and human preferences would often lead to trade-offs (although by no means always in favor of material goods at the expense of biodiversity). It is interesting to consider whether alternative philosophical approaches can avoid the slick terrain.

Consequentialist approaches

First, the homocentric, utilitarian approaches are a subset of a broader group of consequentialist theories that claim, loosely speaking, that the rightness of an action should be judged by the goodness of its consequences. Consequentialist theories do not have to be homocentric. Bentham was in principle open to the ideas that the utility of the animals could count; he just could not visualize any obvious way of incorporating it into the utilitarian calculus (1823, p. 311). More generally, utilitarianism evaluates all consequences in terms of their contribution toward preference satisfaction, whereas consequentialism is open to other ways of evaluating consequences.

Clearly, a consequentialist could accord preservation of biodiversity the status of the preeminent value. Then all proposed actions would be evaluated in terms of their effects on biodiversity, and those with the most favorable consequences for biodiversity would be chosen. Other objectives would be subordinated to biodiversity. Such a consequentialist scheme would avoid the slick terrain, but only at the cost of subordinating other worthy objectives such as enhancing the life prospects of the very worst-off people. Of course, according these other objectives the status of preeminent values too would return us to the slick terrain. The primacy of biodiversity cannot be assured in a clash of preeminent values.

Appeals to moral duty

Duty-based moral theories attempt to identify the moral obligations that bind humans, and the morally correct actions these obligations entail. Ehrenfeld (1988) offers a solution to the "slick terrain" problem: "If conservation is to succeed, the public must come to understand the inherent wrongness of the destruction of biological diversity" (p. 215). Clearly, Ehrenfeld's is a duty-based approach: right action is that which respects the moral obligation of human beings to preserve biodiversity.

When several considerations have moral forces (cannibalism is morally evil, while self-preservation is morally worthy), clashes among them (under what conditions, if any, would self-preservation justify cannibalism?) can be resolved only via deduction from higher moral principles. The slick terrain can be avoided only by asserting that preservation of biodiversity is a first principle, a trump among moral principles, that defeats all others. Without such an assertion, Ehrenfeld's "inherent wrongness" does not solve his problem. Surely, many would argue that enhancing the life prospects of the worst-off people has moral force at least powerful as that of protecting biodiversity. Again, biodiversity is on slick terrain.

Contractarian approaches

Contractarians argue that arrangements are justified if they respect the rights of all the affected parties. In duty-based reasoning, "rights" is often used to mean moral claims; one respects rights (in this sense) by observing moral obligations. In contractarian theories, rights are enforceable claims. Change occurs when all affected parties, endowed with enforceable rights, consent to it; without consent, the status quo prevails. While consent justifies change, the lack of consent for change is insufficient to justify the status quo. The starting point (or constitution) must itself be justified directly, typically by arguing that it was (or might have been) chosen by voluntary agreement among all concerned.

Contractarian approaches encounter great difficulties when taken literally. The burden of demonstrating that any existing starting point was actually chosen by consent, or that any proposed starting point might be endorsed unanimously by real choosers with diverse interests, seems insurmountable. Contractarians typical retreat to thought experiments, trying to deduce the characteristics of constitutions that might plausibly emerge from voluntary agreement under ideal conditions such as the "veil of ignorance" posited by Rawls (1971).

In serious discussion of the applied ethics of biodiversity, contractarian approaches have additional problems. Should all life-forms count equally or, at the other extreme, should the interests of the biota be represented by humans whose veil includes ignorance as to which generation they will be born into? Considering that rational choice is fundamental to contractarian approaches, should it be presumed (as it usually is, in western philosophy) that only humans are rational? Norton (1989) conducts a Rawlsian thought experiment where all potentially living things are represented behind the veil of ignorance, although it is recognized that all but those born humans will lose rationality at birth. The substantial probability of being born nonhuman would, in Norton's thought experiment, lead risk-averse participants to view with disfavor the extinction of any species. Thus, one would expect agreement on a constitution in which preservation of biodiversity is taken very seriously.

Nevertheless, it is unlikely that a contractarian thought experiment would yield iron-clad constitutional guarantees for biodiversity. It is individuals, not species, that are represented in the constitutional procedure, and there are many circumstances, other than extinction of one's designated species, that pose a real risk of an individual never being born (or being born into unrelievedly miserable circumstances). Thus, a contractarian thought experiment is likely to produce less than iron-clad commitments to biodiversity.

The consequentialist case for biodiversity is iron-clad if biodiversity is assigned preeminent value status. Similarly, the duty-based case is failsafe if conservation of biodiversity is the highest of all moral principles, and the contractarian case is secure if all participants in the constitutional process place the survival of all species above all other concerns. Recognition of other coequal or superior values, moral principles, or individual concerns returns us to the slick terrain. The case for biodiversity is always circumstantial, i.e., relative to the possibilities that are available and the strength of completing claims. Ehrenfeld (1988) claimed the high ground by resisting all the circumstantial approaches as mere manifestations of the moral repugnancy of homocentrism. But his victory is empty, since it depends on first-principle or preeminent value status for biodiversity, and such status is unlikely to survive scrutiny given the powerful appeal of many other candidates.

A strong, but circumstantial, case for biodiversity

We can do more than deny the viability of a failsafe case for biodiversity. It is possible to work, more affirmatively, toward constructing a strong but defensible circumstantial case.

First consider a duty-based approach. Assume that preserving biodiversity and enhancing the life prospects of the worst-off people are both moral goods. However, the claims of humans trump those of nonhumans. From these moral principles, it can be deduced that humans should make some, but not unlimited, sacrifices for biodiversity. This result endorses the basic idea of a safe minimum standard (SMS) rule: a sufficient area of habitat should be preserved to ensure the survival of each unique species, subspecies, or ecosystem, unless the costs of doing so are intolerably high (Ciriacy-Wantrup, 1968; Bishop, 1978). One could add an additional moral premise – marginal increments in the welfare of people count for less in the case of the already well-off than for those who are currently immiserated – and the implication is that one could identify situations in which the well-off have an obligation to subsidize the SMS in impoverished places.

The SMS rule places biodiversity beyond the reach of routine trade-offs, where to give up ninety cents worth of biodiversity to gain a dollar's worth of ground beef is to make a net gain. It also avoids claiming trump status for biodiversity, permitting some sacrifice of biodiversity in the face of intolerable costs. But it takes intolerable costs to justify relaxation of the SMS. The idea of intolerable costs, in a pluralistic society, invokes an extraordinary decision process that takes biodiversity seriously by trying to distinguish costs that are intolerable from those that are merely substantial.

At this point, consider a utilitarian approach. The benefit cost (BC) approach applies classical utilitarian principles with just one nod toward the value system of mainstream economics. The basic value data in BC analyses are those of economics – willingness to pay (WTP) for desired changes and willingness to accept (WTA) compensation for changes that are not desired – and as such reflect not only the preferences but also the endowments of the valuer. The rule that benefits and costs be aggregated anonymously – i.e., without regard to the identity and welfare status of the gainers and losers – is merely an economic operationalization of Bentham's classic rule of the greatest good for the greatest number. Thus, the BC approach is (at least) one rather direct and plausible approach of implementing a value system based on preference satisfaction. To the extent that humans value the services that biodiversity provides more than the services that would be foregone in pursuit of biodiversity, the BC approach would support biodiversity.

One of the more persistent arguments against the BC approach, and many alternative expressions of utilitarianism, is that preferences may be myopic and human understanding of the technical possibilities – in the Ehrlich analogy, the consequences of popping each and every possible combination of rivets on the airplane – may be incomplete or mistaken. Not that the BC approach does worse, in these respects, than other approaches that take citizen opinion seriously. As humans come to comprehend the technology of natural systems and how it limits the performance of anthropogenic technology, this understanding is reflected in a valid BC analysis. As human preferences extend to the amenity and existence services provided by diverse ecosystems, the valuations that emerge are fully reflected in a valid BC analysis.

Nevertheless, one must concede that human myopia is a valid concern. How can we be assured that the lure of immediate gratification will not induce us to make decisions that will surely have very unpleasant consequences later on? Elster (1979) has shown that "binding" behavior – Ulysses bound himself to the mast in advance to prevent himself from doing what he was quite sure he would do in the heat of the moment, i.e., steer his ship into the rocky waters separating it from the sirens – is consistent with both rational behavior and utilitarianism. Thus, one logically coherent utilitarian strategy would be to make policy choices on the basis of benefits and costs, but subject always to the constraint that actions we are reasonably sure we (or future generations of people we care about) will regret are forbidden. Biodiversity issues may be decided by consulting a BC analysis but subject to a safe minimum standard or similar constraint. Net benefits are maximized because benefits are good consequences, and the constraints are imposed because the consequences of not satisfying them are terrible. Again, the SMS constraint would not accord trump status to biodiversity, but would trigger a serious and searching decision process before it could be relaxed.

Let us return momentarily to the duty-based approach. We saw how the SMS could be derived from moral reasoning, but the decision rule was left incomplete. Upon what basis should people decide those many issues that do not threaten the SMS? It is hard to conceive of a plausible moral theory that does not, in the absence of overriding concerns, give a good deal of weight to the satisfaction of human preferences. Thus, we should take seriously a rule that policy issues be decided on the basis of benefits and costs, but always subject to constraints identified by moral reasoning. Net benefits are maximized because human preference satisfaction is morally worthy, and the constraints are imposed because they ensure that higher moral goods can trump preference satisfaction in the event of conflict.

Norton's (1989) contractarian thought experiment also identified the SMS constraint as a likely component of a just constitution. Preference satisfaction counts, also, in contractarian thought. However, many contractarians, naturally enough, pursue preference satisfaction via the individualistic routes of free exchange, voluntary taxation, and public decision by consent. The BC approach emerges from contractarian thinking as a kind of second-best result. If, as may well be the case, the pattern of compensating transfers to achieve voluntary agreement on policy is too complex to be feasible and the transactions costs too high, maximizing net benefits of policy becomes an attractive approach. At least it assures that the game is positive-sum. In the problem at hand, a plausible contractarian solution is to maximize net benefits (to satisfy preferences) subject to an SMS constraint (because participants in the "veil of ignorance" process would insist on it).

Interestingly, it seems that the same general kind of decision rule – maximize net benefits subject to an SMS constraint – is admissible under consequentialist, duty-based, and contractarian reasoning. If we accept that satisfaction of human preferences is at least a consideration, and that BC analysis can provide a method of systematically accounting for preference satisfaction, then estimated benefits and costs are at least a consideration when making choices about biodiversity. If environmental economists could be satisfied with claiming no more than this, and if preservationists could discipline themselves to eschew the grandstanding strategy of denouncing any and all moral theories that do not accord first-principle or preeminent value status to biodiversity, a harmonious collaboration ought to be possible.

The prospects for benefit evaluation of biodiversity

A sound argument for considering empirical estimates of benefits and costs when deciding policy with respect to biodiversity requires that these empirical estimates provide information, as opposed to misinformation or noise. In other words, the argument for considering the output of BC analysis depends on the possibility of producing estimates that reflect the underlying WTP and/or WTA with sufficient accuracy and precision that the results are informative rather than misleading or merely confusing. Those whose main line of attack is on the moral underpinnings of BC analysis in the context of biodiversity (Ehrenfeld, 1988; Norton, 1988; Sagoff, 1988) tend also to embellish their case with assertions about the inherent nonquantifiability of the benefits and costs of preserving biodiversity or the arbitrariness of the estimates obtained using the available methods. In contrast, Randall (1988) takes a relatively optimistic view of the feasibility of generating informative BC estimates.

There has been substantial progress in all aspects of environmental benefit estimation in the last quarter-century. The theory has been developed consistent with the economic theory of welfare-change measurement. Explicit concerns of environmental economics have been addressed: the benefits of complex policies and the relationship between total value and the value of policy components (Hoehn & Randall, 1989); the distinction between use values and existence values (Randall & Stoll, 1983); use value under uncertainty (Graham, 1981; Smith, 1987); a total value framework for benefit estimation (Randall, 1991); the concept of quasioption value to account for situations in which current decisions might forestall the opportunity to benefit from knowledge emerging in the future (Fisher & Hanemann, 1986); and development of the theory and methods for estimating values of nonrival, nonexclusive, and non-marketed environmental services and amenities, including contingent valuation

(Randall et al., 1974; Hoehn & Randall, 1987, Mitchell & Carson, 1989) and methods that use market observations for weak complements (Maler, 1974; Bradford & Hildebrandt, 1977). A very substantial body of empirical estimates of environmental benefits has been assembled. The practice of estimating environmental benefits has become routine in the United States policy process, encouraged by legislation, Presidential Executive Orders, and the Office of Management and Budget. There is no denying that challenging problems remain in environmental benefit estimation. However, there seems little doubt that environmental benefit estimation is a progressing research program (Lakatos, 1970) that has already achieved important successes.

Biodiversity is an especially challenging topic for environmental benefit estimation. In order to evaluate policies pertaining to biodiversity, it is necessary to develop scenarios that describe baseline conditions and project them into the future, and do the same for with-policy conditions. Yet scientific data bases and the understanding of the basic natural science relationships are often very incomplete. BC analysis will reflect this scientific uncertainty and ignorance (but, to be fair, so will any other process by which people try to come to terms with the policy issues surrounding biodiversity).

The basic economic data for BC analysis, WTP and WTA, are likely to be volatile, as new information is added to the very sparse information base that most people have about biodiversity issues. However, this volatility reflects the basic reality of decision processes that start with limited information and revise expectations as new information becomes available (Viscusi, 1985) and surely applies not only to WTP and WTA but also to any other measure of public preferences about biodiversity policy.

For several reasons – the relative importance of existence values, the importance of uncertainty and therefore of *ex ante* value (Smith, 1987; Randall, 1991), and the relative difficulty of identifying serviceable weak complements – contingent valuation seems destined to play a major role in benefit estimation for biodiversity. While contingent valuation is gaining respectability, it remains a controversial method (Mitchell & Carson, 1989). Nevertheless, a number of studies have successfully estimated existence or preservation values with contingent valuation (Bennett, 1987; Bowker & Stoll, 1988, Brookshire et al., 1983; Majid et al., 1983; Stoll & Johnson, 1984). The recent development of referendum methods of contingent valuation (Hoehn & Randall, 1987; Mitchell & Carson, 1989) will be especially important for evaluating biodiversity. These methods appeal to the public perception of public goods, such as biodiversity, and they rely on scenarios built around political rather than market institutions and should, therefore, be more readily adaptable to international contexts.

While contingent valuation will likely be the method most widely used in evaluating biodiversity benefits, there may be important opportunities to use weak complementary methods. The growing importance of adventure travel to, and vacation and retirement homes near, exotic environments suggests a limited but nontrivial role for travel cost and hedonic price methods.

In summary, biodiversity presents some of the most difficult challenges for the rapidly developing theory and methods of environmental BC analysis. Nevertheless, there are promising prospects for successful empirical applications.

Conclusions

Let me attempt to summarize the argument with a few succinct statements.

- Some preservationists and philosophers are contemptuous of any theory of value that treats the value of biodiversity as relative to the values of other concerns. Yet, it seems that any coherent duty-based or consequentialist theory will do that, unless the theory asserts that preservation of biodiversity is a first principle or a preeminent value that trumps all others. And there seems no overpowering reason to accord trump status to biodiversity. Finally, contractarian thought experiments seem unlikely to produce iron-clad guarantees for biodiversity.
- Without asserting first-principle or preeminent value status for preserving biodiversity, consequentialist, duty-based, and contractarian theories can be developed in which biodiversity counts.
- A sound argument can be made that human preference satisfaction counts morally. Thus, benefits and costs (as environmental economists conceptualize them) count in duty-based theories as well as consequentialist theories of value. This suggests that benefits and costs have a place in a more complete theory of the value of biodiversity.
- The idea that benefits and costs count in a more complete theory of biodiversity does not exclude other moral or consequential considerations. One admissible theory would be that policy should implement the strategy with the greatest net benefits, subject to the safe minimum standard (SMS) constraint. The SMS constraint could be derived from consequentialist, duty-based, or contractarian principles.
- Real and considerable difficulties exist in obtaining reliable empirical estimates of the benefits and costs of preserving biodiversity. Such estimates are likely to be volatile, reflecting the reality that the current information base is

small and new information may change perceptions dramatically. Nevertheless, and in full awareness of the challenges, I believe the effort could be made and the results should be taken seriously.

Acknowledgments

The author wishes to thank Mike Farmer, Don Hubin, and Bryan Norton for helpful suggestions and comments, and the National Science Foundation for research support (grant number BBS 8710153).

References

Bennett, J. W. (1984). Using direct questioning to value existence benefits of preserved natural areas, *Australian Journal of Agricultural Economics*, **28**, 136–152.

Bentham, J. (1970). *An Introduction to the Principles and Morals and Legislation.* Darien, CT: Hafner. Originally published 1823.

Bishop, R. C. (1980). Endangered species: an economic perspective. *Transactions of the 45th North American Wildlife and Natural Resources Conference*, pp. 208–218. Washington, D.C.: Wildlife Management Institute.

Bishop, R. C. (1978). Economics of endangered species. *American Journal of Agricultural Economics*, 60, 10–18.

Bowker, J. M., & Stoll, J. R. (1988). Use of dichotomous choice nonmarket methods to value the whooping crane resource. *American Journal of Agricultural Economics*, **10**, 372–381.

Bradford, D. F., & Hildebrandt, G. G. (1977). Observable preferences for public goods. *Journal of Public Economics*, **8**, 111–131.

Brookshire, D. S., Eubanks, L. S., & Randall, A. (1983). Estimating option prices and existence values for wildlife resources. *Land Economics*, **59**, 1–15.

Ciriacy-Wantrup, S. von (1968). *Resource Conservation: Economics and Policies* (3rd ed.). Berkeley: University of California Division of Agricultural Sciences.

Ehrenfeld, D. (1988). Why put a value on biodiversity? In *Biodiversity*, ed. E. O. Wilson, pp. 212–216. Washington, D.C.: National Academy Press.

Ehrlich, P. R., & Ehrlich, A. (1981). *Extinction.* New York: Random House.

Elster, J. (1979). *Ulysses and the Sirens.* Cambridge, UK: Cambridge University Press.

Fisher, A. C., & W. M. Hanemann. (1986). Option value and the extinction of species. In *Advances in Applied Microeconomics*, vol. 4, ed. V. K. Smith, pp. 169–190. Greenwich, CT: JAI Press.

Graham, D. A. (1981). Cost–benefit analysis under uncertainty. *American Economic Review*, **71**, 715–725.

Hoehn, J. P., & Randall, A. (1987). A satisfactory benefit cost indicator from contingent valuation. *Journal of Environmental Economics and Management*, **14**, 226–247.

Hoehn, J. P., & Randall, A. (1989). Too many proposals pass the benefit cost test. *American Economic Review*, **79**, 544–551.

Lakatos, I. (1970). Falsification and the methodology of scientific research programs. In *Criticism and the Growth of Knowledge*, ed. I. Lakatos & A. Musgrave. London: Cambridge University Press.

Lovejoy, T. (1986). Species leave the ark one by one. In *The Preservation of Species*, ed. B. Norton, pp. 13–27. Princeton, NJ: Princeton University Press.

Majid, I., J. A. Sinden, & A. Randall. (1983). Benefit evaluation of increments to existing systems of public facilities. *Land Economics*, **59**, 377–392.

Maler, K-G. (1974). *Environmental Economics: A Theoretical Inquiry*. Baltimore: Johns Hopkins University Press.

Mitchell, R. C., & Carson, R. T. (1989). *Using Surveys to Value Public Goods: The Contingent Valuation Method.* Washington, D.C.: Resources for the Future.

Norgaard, R. B. (1984). Coevolutionary development potential. *Land Economics*, **60**, 160–173.

Norgaard, R. B. (1988). The rise of the global exchange economy and the loss of biological diversity. In *Biodiversity*, ed. E. O. Wilson, pp. 206–211. Washington, D.C.: National Academy Press.

Norton, B. (1988). Commodity, amenity, and morality: the limits of quantification in valuing biodiversity. In *Biodiversity*, ed. E. O. Wilson, pp. 200–205. Washington, D.C.: National Academy Press.

Norton, B. G. (1989). Intergenerational equity and environmental decisions: a model using Rawls' veil of ignorance. *Ecological Economics*, **1**, 137–159.

Randall, A. (1988). What mainstream economists have to say about the value of biodiversity. In *Biodiversity*, ed. E. O. Wilson, pp. 217–223. Washington, D.C.: National Academy Press.

Randall, A. (1991). Total and nonuse values. In *Measuring the Demand for Environmental Improvement*, ed. J. B. Braden & C. K. Kolsted, pp. 303–321. Amsterdam: North-Holland.

Randall, A., Ives, B. C., & Eastman, C. (1974). Bidding games for valuation of aesthetic environmental improvements. *Journal of Environmental Economics and Management*, **1**, 132–149.

Randall, A., & Stoll, J. R. (1983). Existence value in a total valuation framework. *Managing Air Quality and Scenic Resources at National Parks and Wilderness Areas*, ed. R. D. Rowe & L. G. Chestnut, pp. 265–274. Boulder: Westview Press.

Rawls, J. (1971). *A Theory of Justice*. Cambridge, MA: Harvard University Press.

Sagoff, M. (1988). *The Economy of the Earth.* New York: Cambridge University Press.

Smith, V. K. (1987). Nonuse values in benefit cost analysis. *Southern Economic Journal*, **54**, 19–26.

Southgate, D. (1988). *Efficient Management of Biologically Diverse Tropical Forests.* London: IIED/UCL, London Environmental Economics Centre.

Stoll, J. R., & Johnson, L. A. (1984). Concepts of value, nonmarket valuation, and the case of the whooping crane. *Transactions of the Forty-Ninth North American Wildlife and Natural Resources Conference*, **49**, 382–393.

Viscusi, W. K. (1985). Are individuals Bayesian decision makers? *American Economic Review*, **75**, 381–385.

15

Lessons from the aging Amazon frontier: opportunities for genuine development

CHRISTOPHER UHL, ADALBERTO VERISSIMO, PAULO BARRETO, MARLI MARIA MATTOS, and RICARDO TARIFA

Introduction

"Amazonia" conjures up visions of mystery, richness, and grandness, but today these visions are intermingled with thoughts of rampant deforestation, armed land disputes, and forests aflame. There are few places on the planet where the human affront to biodiversity is more direct and damaging than in Amazonia.

Most of the deforestation in Amazonia has been concentrated in an arc extending from the State of Pará in the east through the states of Mato Grosso and Rondonia in the south. Large-scale forest clearing began in this region in the 1960s. The development process has been disorderly. A parade of actors, including farmers, ranchers, miners, and loggers, have worked to wrestle riches from the landscape. Ecological and economic evaluations of the development process conducted in the 1970s held out little hope for sustainable development (Goodland & Irwin, 1975; Hecht, 1983). But the Amazon settlement experiment is in constant evolution. Hence, previous pronouncements concerning development prospects need to be continually updated in light of new findings.

One area undergoing particularly intensive deforestation is the municipality of Paragominas, located in eastern Pará along the Belém–Brasília Highway (Fig. 15.1). In this chapter we review the history of natural resource use at Paragominas and evaluate the prospects for sustainable development in this region. The municipality of Paragominas (22,000 km^2) is an ideal location for research aimed at understanding landscape change and assessing the sustainability of Amazon development because in many ways it presents a microcosm of Amazonia, containing within its boundaries significant areas devoted to ranching, logging, slash-and-burn agriculture, and mining. Moreover, because of its more mature status, Paragominas serves as a bellwether municipality: developmental trends, innovations, and failures appear in this municipality before they are seen in most other younger development centers. Hence, lessons learned at Paragominas could have broad import in influencing regional development and natural resource conservation.

History of land use in Paragominas from the 1960s through the 1980s

The municipality of Paragominas lies three degrees south of the equator. Annual rainfall averages 1700 mm with a pronounced dry season from July to November. The terrain is rolling and cloaked with 35-m tall evergreen rain forest growing on red-yellow latosols.

Early settlers practice agriculture

When the Belém – Brasília Highway was laid down across what is today the municipality of Paragominas in the early 1960s, the region was very sparsely inhabited. Indian settlements were present to the east in Maranhão (Tembe Indians) and along the upper reaches of the Rio Capim to the west, but there is no evidence of anything but extremely sparse past indigenous settlements in the Paragominas region. The first settlers along the Belém – Brasília Highway came from São Miguel do Guama, an old riverine settlement to the north (Fig. 15.1). These colonists established farms, but generally, within a few years, most sold off their plots to ranchers and land speculators arriving from Goias and Minas Gerais to the south.

Ranching in the rainforest

Cattle ranching began in earnest in Paragominas in the late 1960s with the paving of the Belém–Brasília Highway. Land prices were never more than a few dollars per hectare, and government provisions allowing investors from southern Brazil to use their tax payments to establish cattle ranches in the North (Amazonia) were common, as were low interest loans with long payback periods. In some cases, the incentives were so generous that it was impossible to lose money. In addition, the law encouraged large-scale clearing for pastures (Hecht, 1985; Mahar, 1989). Land titling procedures specified that the only way to acquire land title in Amazonia was through clearing. And the National Institute for Agrarian Reform (INCRA) initially established that settlers could gain title to 6 hectares for every hectare they cleared, thereby providing an incentive to clear ever-larger tracts. Under this combination of conditions, hundreds of ranches were established in Paragominas. Because half of each land holding was required by law to be left in forest, a mosaic of forest and pasture was created.

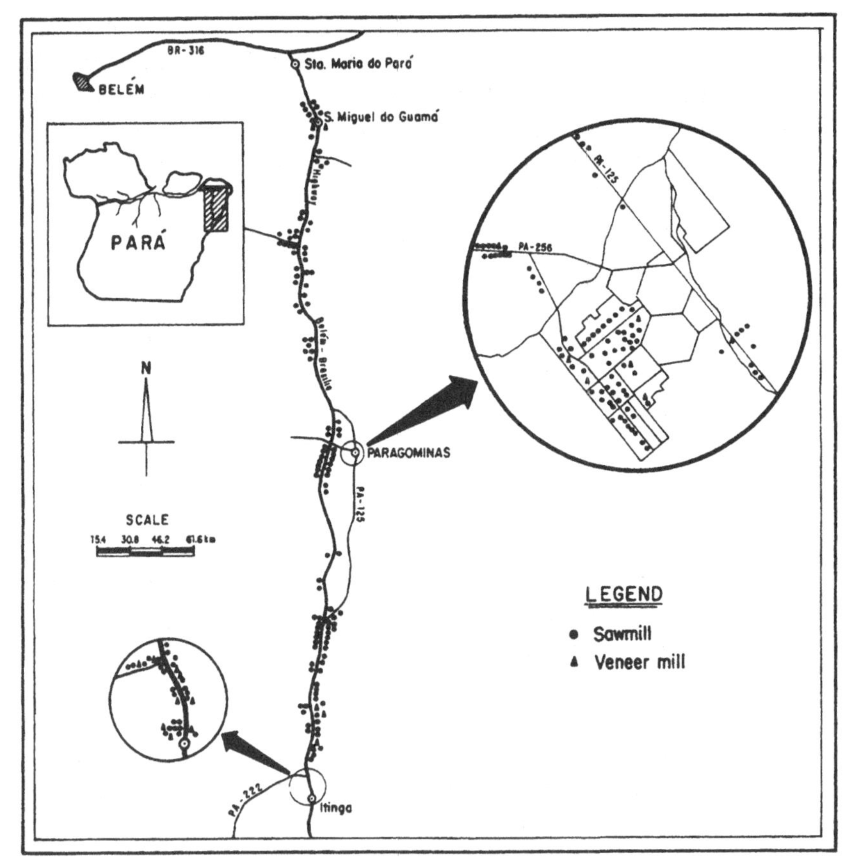

Fig. 15.1. The location of sawmills (n = 234) along a 340 km stretch of the Belém–Brasília Highway between Santa Maria and Itinga in eastern Amazonia.

Within four to six years of pasture establishment, pasture grasses (*Panicum maximum* and/or *Brachiaria* spp.) began to lose vigor because of soil infertility [phosphorus is a potentially limiting element in Amazonian pastures (Serrão et al., 1978)], insect attack (*Brachiaria* spp., in particular, are susceptible to spittle bugs (Serrão & Simão Neto, 1975)], and competition with aggressive weeds. By the late 1970s, cattle raising in the north of Pará, in general, and in Paragominas, in particular, faced severe economic and ecological constraints. A national monetary crisis prompted the government and banks to significantly reduce support to the ranching sector. Ranchers responded by reducing the rate of new forest clearing. It was at this time that sawmills first began to be installed in the region.

The arrival of the sawmill industry in the eastern Amazon provoked a profound change in the way that the forest was viewed. Previously, forest land had been accorded a value inferior to that of pasture land, even inferior to degraded pasture lands. Indeed, the forest was an impediment to land clearing. It was only in the 1980s that the value of the forest for its wood came to be appreciated. Two developments catalyzed this awakening. First, wood supplies in the south of Brazil declined steadily during the 1980s. Second, it took some 20 years to establish a reliable transport and communication system in the eastern Amazon making the establishment and maintenance of sawmills less risk-prone.

The eastern Amazon was a natural place for this burst of logging activity because it is adjacent to Brazil's Northeast with its burgeoning population and paucity of wood resources. Moreover, logging provides a rich array of work opportunities, and in Pará a whole cast of middlemen had arisen to cut the wood, haul it to the mill, transport sawn products to urban markets, use wood "waste" to make charcoal, etc. This meant that sawmills didn't require large sums of capital to get started.

The extent of growth of the wood industry in Amazonia and the predominance of Pará in this process is noteworthy. In the twelve-year period, 1976–1988, total roundwood production in Brazil's southern states (Paraná, Santa Catarina, and Rio Grande do Sul) decreased from 15 million m^3 (47% of Brazil's total roundwood production) to 7.9 million m^3 (17% of total) (FIBGE, 1988). During this same period, roundwood production in the North region (Amazonia) increased from 6.7 million m^3 (21% of Brazil's total) to 24.6 million m^3 (54% of Brazil's total).

Land use in Paragominas at the outset of the 1990s

The dominance of logging

At the outset of the 1990s logging was the dominant economic activity in Paragominas. Indeed, for every truck that left the municipality with cattle, dozens more left with sawn wood products. Most of the mills in this region were established in the 1980s. For example, of 234 mills located along the stretch of the Belém–Brasília Highway between Santa Maria and Itinga (Fig. 15.1), 79% had been established during the 1980s. Mills tend to concentrate in frontier towns where energy, communication networks, and services are available. The town of Paragominas with 77 mills is the largest wood processing center in the eastern Amazon.

Most mills are similar in design and function. In our survey of 234 mills, we found that 78% operated a single bandsaw with an average annual production of sawn timber of 4,300 m^3 (s.d. = 858). Twelve percent of the mills had more than one bandsaw with the largest mill, with seven bandsaws, producing 24,000 m^3 of lumber per year. The remaining 10% of the mills had machinery to spin logs, thereby peeling off veneer. Six of these industries had a plywood factory associated with the veneer mill.

Most (97%) of the Paragominas sawmill owners were from outside the region, with 40% coming from the southern Brazil state of Espirito Santo, a major wood producing region in the 1960s and 70s. Eighty percent of all owners had been involved in wood harvesting and processing activities previously and half of these had actually owned mills. Only 25% of the sawmills in the Paragominas region were transferred from other regions to Pará. While most mills (73%) still only produced sawn boards in various dimensions, 27% were producing more elaborate products (flooring, paneling, doors, etc.), as well as standard-dimension boards.

Sawmill owners were quick to invest profits in the acquisition of logging equipment and land. Nearly two-thirds (63%) of the mills had forest extraction teams and log transport crews in addition to their sawmill operations. Extraction generally occurs on the lands of large property holders (ranchers), although 61% of the mill owners had also invested in property and 39% of the mills were getting at least some of their wood supply from their own holdings.

The logging–ranching link

By the time loggers arrived in force to Paragominas in the 1980s, most land, cleared and forested alike, was in large holdings controlled by ranchers and land speculators who had moved into the region in the previous two decades.

In January 1990, we interviewed 14 large ranchers (average area of holding = 14,100 ha) in the environs of Paragominas, inquiring about the size of holdings, percentage of holdings in active pasture, abandoned pasture, virgin forest, logged forest, herd size, management techniques, productivity, and operating costs. On average, 35% of the total land area was still in virgin forest with 26% in logged forest. The remaining 39% of the land was in either active or degraded pasture.

Ranching is profitable provided that holdings are large. The operating expenses (in U.S. dollars) for a typical large ranch (3,500 ha of pasture clearing) are $75,000/yr divided among labor (39%), animal care (36%), infrastructure maintenance (15%), transport (6%), and taxes (4%). Profits are approximately $70,000/yr or $20/ha/yr.

As pastures age, they degrade and slowly turn into weed-infested old fields unless revitalized. This pasture revitalization process consists of: 1) clearing the land of debris with a bulldozer – the residual logs and stumps are pushed into strips and burned; 2) tilling and fertilizing; and 3) planting of *Brachiaria bryzantha*, a promising new forage species. The cost of pasture reformation is $260/ha, with more than half of this cost (54%) devoted to mechanical clearing and ploughing and the remaining investment absorbed in planting and fertilizing (37%) and infrastructure (9%) (Mattos & Uhl, in press).

The capital for pasture restoration frequently comes from the forest in the form of wood sales. Ranchers either sell the rights to extract wood from their properties or extract the wood themselves. When ranchers sell forest logging rights, they receive $70/ha (Verissimo et al., 1992), whereas the net profile is $200/ha when ranchers log their forests themselves. Hence, approximately four hectares must be logged to restore one hectare of degraded pasture when logging rights are sold ($260/$70). However, when ranchers conduct their own logging, almost one hectare of degraded pasture can be restored for each hectare of forest that is logged.

Once pastures are reformed their productivity improves markedly and the costs of maintenance and labor decline. Recent field studies by Mattos and Uhl (in press) reveal that the profits from reformed pastures are approximately $50/ha/yr, even when fertilized at five-year intervals. Overall, our analysis shows that forests now play a critical role in sustaining ranching activities and that, in effect, the sudden valuation of Amazon forest timber resources extends the ranching trial period.

Large ranchers, with appreciable areas in forest, who are willing to participate in the timber extraction process, now have a new subsidy. This is the third time that ranching has been subsidized in eastern Amazonia. The first subsidy came with the initial forest felling when the nutrients embodied in forest biomass were liquidated to help the pasture grasses grow. The second subsidy was from the government in the form of capital for ranch infrastructure establishment. The third subsidy, to restore degraded pastures, like the first subsidy, comes from nature and, although it is assumed to be free, there are potential costs (environmental degradation) involved.

The importance of forest timber resources in present-day Amazonia frontier economies cannot be overemphasized. Now, for the first time, local resources are providing the capital to drive regional development. Hence, the development model of Amazonia is shifting from the artificial government-subsidized version to one that relies on the exploration of local resource wealth. On the one hand, this can be viewed as a welcome change: resources that were previously burned are now being used. However, this shift in orientation in no way ensures that development will now be sustainable. In the classic development model, riches gained in the initial resource extraction phase are employed to build a diversified economy based on processing, refining, and services. In this process, the region's inhabitants realize gains in health care, schooling, and earning power. Of course, this does not always occur. Frequently, the riches are merely mined leaving behind a resource-depleted landscape.

Logging and the fate of Amazonia

Because of the central position of logging in the occupation of the eastern Amazon, we devote the rest of this chapter to a consideration of the social, ecological, and economic impacts of logging in the Paragominas region.

Social impacts

The 112 mills in the vicinity of Paragominas (i.e., those in the town itself and those located along the Belém–Brasília Highway on the outskirts of the town; (see Fig. 15.1) generate approximately 5,750 jobs distributed among sawmill employees, truckers, forest timber extractors, and odd-job laborers. Overall, we estimate that 56% of the urban population of Paragominas depends directly on the wood industry for sustenance. Considering that these 112 mills log approximately 32,000 ha of forest each year, each employee in the industry depends on approximately 5 ha of forest per year for sustenance.

In household interviews in the mill neighborhood, we found that most sawmill workers had come to Paragominas from other municipalities within Pará or from other states (e.g., 41% were from the poverty-stricken State of Maranhão). Fifty-five percent of the interviewees had arrived within the last five years. The great majority (90%) of these mill workers had migrated from the rural zone, having left farming for urban wage labor in hopes of a better life. But mill salaries are low: $112.00/month (n = 87, s.d. = 43). Three-quarters of the families interviewed used more than 66% of their salaries in the purchase of food and 11% used more than 90% in food purchase. Furthermore, owing to the nonspecialized nature of the work, there was no apparent relationship between years of work in sawmill service and earning power.

Given the importance of wood in the regional economy, the timber industry could generate significant tax revenues that could improve the quality of life of the regional inhabitants. For example, the 238 mills in the Paragominas region should have jointly produced approximately $28,000,000 in tax revenues in 1990. If half this money were retained in the region, there would be $200/person/year available for social services or approximately $1,000/family of five.

Environmental impacts

We studied three logging operations in the environs of Paragominas. Logging, in all cases, was intensive and highly mechanized. Bulldozers opened the primary network of extraction roads and then moved through the forest in search of prefelled trees. These trees were dragged out to 0.1–0.3 ha landings that serve as staging areas for the loading and transport of logs.

The number of trees harvested per hectare in the three study areas ranged from 2.9 to 9.3 (mean = 6.4, s.d. = 2.6), while extraction yields ranged from 18 to 62 m^3/ha (mean = 38, s.d. = 18) (Table 15.1). The number of species extracted per site ranged from 43 to 57; and the total number of species extracted, considering all sites together, was 83.

A considerable amount of damage occurred in the opening up of logging roads and the felling and extraction of trees. Logging decreased canopy cover by an average of 35% in the three study sites. While an average of 6.4 trees were extracted/ha, almost 150 trees >10 cm diameter at breast height (dbh)

Table 15.1. Damage caused in wood extraction in three logging study areas in the municipality of Paragominas, Pará, Brazil (from Verissimo et al., 1992).

	Area 1	Area 2	Area 3	Average
Characteristics of Logging:				
Study area (ha)	115	37	16	56
Number of species extracted	57	55	43	52
Trees extracted (Number/ha)	2.9	6.9	9.3	6.4
Trees cut but not extracted (Number/ha)[a]	0.2	0.7	0.4	0.4
Volume extracted (m^3/ha)	18	35	62	38
Volume cut but not extracted m^3/ha)	2.2	3.2	3.6	3.0
Damages Caused in Logging:				
Trees ≥ 10 cm dbh damaged (Number/ha)	121	130	193	148
Basal area ≥ 10 cm dbh damaged (m^2/ha)	5.0	6.6	7.6	6.4
Volume ≥ 10 cm dbh damaged (m^3/ha)	47	63	77	62
Canopy opening (m^2/ha)[b]	2500	4500	4400	3800
Damage Indices:				
Trees damaged per tree extracted	41	19	20	27
m^3 damaged per m^3 extracted	2.6	1.8	1.2	1.9
m logging road per tree extracted	37	38	43	39
m^2 road and patio per tree extracted	186	219	249	218
canopy opening per tree extracted	862	652	473	663

[a]Not extracted because of defects.

[b]Considering only canopy openings caused by logging activities.

were severely damaged/ha (Table 1). Almost half (mean = 48%, s.d. = 5) of the damaged trees were uprooted, 41% had broken stems, and the remaining 11% suffered severe bark loss. The average size of damaged trees was 20 cm dbh (range = 10–93). Many of these trees were of potential economic interest. For example, 32% were being sawn in the Paragominas mills with an additional 44% used in rural construction, although not frequently sawn at present. The remaining 24% of the damaged trees apparently have no present or potential wood-related importance.

Secondary impacts are also associated with intensive logging. Of particular concern is fire. Forest cutting for any purpose leaves slash on the ground, increasing potential fuel loads; and opening of the forest canopy, by increasing the amount of radiation reaching the forest floor, dries this slash. During the six-month dry season, a period of five or six rainless days is sufficient to dry fuels below a threshold of combustion in these disturbed forest stands. As fire is commonly used for pasture establishment and weed control, an ignition source is widespread during the latter part of the dry season. Hence, these logged ecosystems represent an entirely new and unique fire environment in Amazonia: the presence of logging slash residues increases fuel loads, the microclimate is decidedly drier, and anthropogenic ignition sources are common.

Overall, this analysis reveals that the forest resource is being mined with little thought for the future. Indeed, there appear to be very few serious attempts at forest management in the eastern Amazon. If present practices continue, it is likely that the forest will gradually be degraded and thereby lose its wealth-generating potential. But need this be so?

Forest management as a strategy to reconcile development and conservationists aims

Could the eastern Amazon forest be managed to sustainably produce timber? Sawmill owners claim that it makes no sense for them to manage the forest. Indeed, forest management does imply costs and when resources are abundant and cheap, as is the case for wood in the eastern Amazon, it makes more sense to just buy logging rights, log the stand as quickly as possible, and move on. Forest management, by contrast, would involve the diversion of time, effort, and money away from immediate money-making activities into future-oriented forest improvement. Here, we combine our data on logging and wood processing yields and profits to consider the economic feasibility of forest management.

Economic feasibility of forest management at Paragominas

The most fundamental timber management measures that could be taken at Paragominas to increase timber production are: 1) a preharvest stand inventory to determine the location of desirable trees and to plan the felling directions and skidder pathways; 2) vine cutting one year prior to harvest to reduce felling damage and reduce competition for light; and 3) general stand refinement and liberation thinning together with vine cutting to open up growing space for desirable trees at one, ten, and twenty years following logging. The costs of these management measures are estimated, in very general terms, at about $180/ha and divided as: 1) preharvest stand inventory ($20/ha); 2) preharvest vine cutting ($25/ha); and three postharvest thinnings ($45.00/ha each) (Barreto, unpubl.; see also: DeGraaf, 1986; Jonkers, 1988; Hendrison, 1990).

A typical mill with one bandsaw that also engages in logging would need to manage 242 ha/yr to supply its raw material needs at an estimated cost of $43,560 (242 ha × $180/ha) (Table 15.2). The actual cost over the first two years of management (the year prior to logging and the first year following logging) would be $90, with an additional investment made after 10 and 20 years. However, for mills truly engaged in managing forest estates, investments would be required in recently logged, 10-year-old and 20-year-old parcels each year. Hence, although not applied to the same parcel, an average of $180/ha would be spent on 242 ha of forest land each year. Given an annual mill profit of approximately $216,000, these management costs would consume 20% of total profits or 7% of total annual receipts (Table 15.2). Even if no benefits accrued from management, profit margins would only be shaved from 32% to 26% given a management investment.

However, simple forest management techniques do result in increases in volume accumulation. For example, when vine cutting and girdling of non-economic species are done following unplanned logging operations, commercial trees ≥ 30 cm dbh, achieve long-term annual diameter increments of 0.6–1.0 cm compared to 0.1–0.4 cm typical of untreated stands (DeGraaf, 1986; Jonkers, 1988). Our projections, based on the postharvest stand characteristics of our three forest study sites, and given 2% annual mortality and annual growth increments of 0.8 (managed stands) and 0.3 (unmanaged stands), reveal that the difference in accumulated bole volume between managed and non-managed stands, considering just commercial species ≥ 30 cm dbh, will be 22 m^3 after 35 years (Barreto, unpubl.). Moreover, these simulations reveal that there would be an adequate timber stocking for future harvests.

If a preextraction inventory and vine cutting are added to the management program, logging damage might be reduced by as much as 50% (Marn & Jonkers, 1981; Appanah & Putz, 1984; Hendrison, 1990). Hence, we might

Table 15.2. Preliminary summary of annual costs and profits in the extraction and processing of wood in the municipality of Paragominas, Pará, considering a typical sawmill with one bandsaw that engages in forest extraction activities.

Production:	
Volume extracted (m^3/ha)	38
Total volume extracted (m^3/year)[a]	9,200
Hectares logged/year[b]	242
Costs (US$):	
Cost of logging[c]	$122,200
Cost of transport[d]	$83,800
Costs of processing[e]	$247,800
Total costs of extraction, transport, and processing	$453,800
Value of Production and Profits:	
Total value of sawn wood produced[f]	$670,000
Total profits	$216,000
Profits/ha	$900
Profits/m^3 extracted	$24
Profit margin	32%
Viability of Management:	
Total cost of management[g]	$44,000
% of total mill profits necessary for management	20%
Cost of management/ha	$180
Projected return on management investment/ha	0–5%

[a]A logging team extracts 1310 m^3/month (mean of 11 logging teams). Logging teams are composed of ten people and work with bulldozers and chainsaws for seven months each year (dry season).
[b]There are, on average, 38 m^3 of wood extracted per hectare. Therefore, a typical sawmill will log approximately 242 ha/yr.
[c]Logging expenses include the cost to buy forest logging rights (14% of total costs), salary, and food for the logging team (24%), fuel and maintenance of tractor and chainsaws (24%), equipment depreciation (20%), taxes (12%), and capital costs (9%).
[d]The costs of transport includes salaries (13%), fuel for trucks and log loaders (22%), maintenance of machinery (18%), depreciation (37%), and capital costs (9%).
[e]We consider that 770 m^3 of wood are processed/month producing 360 m^3 of sawn wood (i.e., processing efficiency – 47%). The cost to process 1 m^3 of roundwood is $27.
[f]This estimate assumes a 47% wood-use efficiency and a weighted sawn wood market price of $155.00 (9,200 m^3 × 0.47 × $155.00 = $671,000).
[g]Given a total annual production of 9,200 m^3 and a per hectare yield of 38 m^3, it is necessary to manage 242 ha/yr. The costs of management are $180.00/ha.

expect that, in the case of our three study areas, some 24 individuals/ha ≥ 10 cm dbh presently valued for their wood would be spared damage (Barreto, unpubl.). Given our assumptions for growth (0.8 cm/yr) and annual mortality (2%) for managed stands, this additional stock would increase the volume accumulated by commercial trees ≥ 30 cm dbh by an average of 10 m^3 after 35 years.

If we add the additional 10 m^3 gained with planned extraction to the 22 m^3 resulting from vine cutting and thinning treatments, the total difference in volume accumulated between managed and nonmanaged stands is projected at 32 m^3. Meanwhile, nonmanaged stands in our simulations have about the same volume of commercial wood after 35 years as that present just after harvest. This is due to high mortality and slow growth rates. More than a decade of research in tropical forests in Suriname (DeGraaf, 1986; Jonkers, 1988; Hendrison, 1990), reveals that volume accumulation of commercial species in logged stands that are not managed is extremely low over the first several decades following logging (0–0.25 m^3/ha/yr). Hence, timber harvest rotation times for nonmanaged stands are projected to be well beyond 50 years.

There is no straightforward way to calculate the return on forest management investments. At one extreme, the return might be regarded as the value of the extra wood volume generated by management prior to cutting. At present, loggers pay \$1–3/m^3 for the right to extract standing timber. But the investment necessary to produce 1 m^3 in the management example is approximately \$30 (using a 6% interest rate). At the other extreme, one might consider the return on the management investment to be the value of the wood volume resulting from management after it is sawn. Even in this case, given a net profit of \$23 for each m^3 of roundwood that is processed, the final value of the extra wood volume generated by management would be \$740 (32 m^3 × \$23). Considering the time pattern of the management investments, the return on investment would be 4.9%. While neither of these accounting approaches is entirely satisfactory, our analysis, at the very least, reveals that: 1) simple forest management approaches could lead to substantial increases in commercial wood volume accumulation; and 2) mill revenues are generally adequate to cover management costs.

Strategies to promote forest management

Although the return on investment is low for forest management given current prices, the large profit margins of mills permit management. Even if management had no benefits and were construed as an ecological tax, the application of management would reduce mill profits by only 6% – from 34 to 28% (Fig. 15.2).

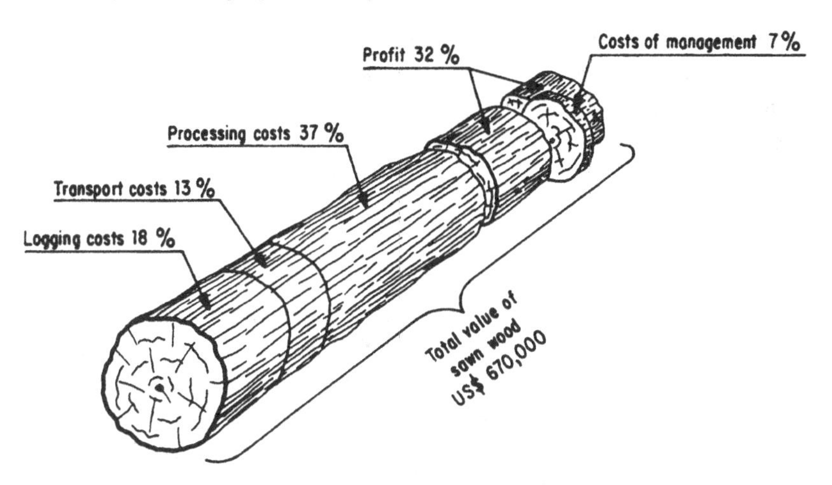

Fig. 15.2. Annual costs involved in logging, wood processing, and forest management in comparison to profits for a typical sawmill in the Paragominas region of eastern Amazonia.

At present, forest timber resources are abundant and cheap in Amazonia and, hence, there is no incentive to manage, only to harvest the biggest and best trees as quickly as possible. Under these conditions, management will not occur naturally – it will have to be encouraged through government policies. These policies might be designed to make timber resources artificially scarce. This could be done by allowing logging only in designated areas of State Forests and prohibiting sawmill owners who operate in these areas from relocating their operations. In return, each operator could be given a license to log a specified area of forest, adequate to supply his mill indefinitely, if properly managed.

In addition, there are at least three economic tools that could be employed to promote management. First, management could simply be mandated by law.[1] In this case, mill profit margins would decline but still remain robust (above 25%). Second, management costs could be passed on to the consumer as a 7% increase in product price (ecology tax). Third, commercial tax revenues (ICM) could be reduced from 12% of total mill receipts to 5%. The tax savings for a typical mill (about $45,000) would be enough to cover management costs.

Logging is likely to be a key economic activity in Amazonia throughout much of the next century. If practiced carelessly, regional biodiversity will be

[1]Given the large annual fluctuations in mill profit margins in Brazil's unstable economy, and given our conclusion that the return on forest management investments will probably be very low in the foreseeable future, it would only be appropriate to mandate that mill establishments shoulder management costs if steps are taken to assure stable wood product prices.

put at risk and profits will be ephemeral and only accrue to a select few. If concrete forest management measures are taken, as we recommend here, the integrity of the region's ecosystems will be maintained and biodiversity preserved.

Conclusion

In closing, we wish to stress four points. First, Amazonia presents wealth in various forms and for this reason many different actors have maneuvered to gain control of the region's resources over the past thirty years. First came peasants in search of land (Fig. 15.3, Period 1960). They were followed by ranchers and land speculators in search of tax shelters and good investments (Period 1975). Now that an infrastructure of highways is well established, the lumberman has appeared (Period 1990).

Second, logging and wood processing came into prominence just as agriculture and ranching were floundering. Because of a national financial crisis, government financing was largely unavailable to Amazonian ranchers and farmers in the 1980s. But now ranchers are able to use profits gained from wood extraction in their forest reserves to reform badly degraded pastures.

Third, forest timber resources are abundant and cheap in the eastern Amazon. Hence, there is little incentive on the part of the wood industry to engage management. Management will only begin to make sense if and when forest timber resources begin to become scarce. If this were to happen, our economic data reveal that sawmill industries could manage the forest for sustainable timber yields and still realize profits.

And, finally, moving into the 1990s, it appears that the Paragominas landscape could evolve in two distinct directions. In the first case (Fig. 15.3, Period 2000, bottom), the land will continue to be mined with little thought for sustainable management. Pastures will degrade and eventually be abandoned. These lands will slowly revert to forest but frequent fires may lead to the development of shrub–oldfield ecosystems. Meanwhile, the remaining forest tracts will be intensively logged leaving behind forest fragments that are fire-prone and increasingly dominated by vines. As the value accorded to this landscape declines, land invasions and spontaneous settlements for subsistence farming will increase and the productive capacity of the land will steadily decline.

This grim scenario need not take place. Indeed, there are pieces of the Paragominas landscape where humans are caring for the land (Fig. 15.3, Period 2000, top). Degraded pasture lands can be reformed with well-adapted germplasm, and these improved pastures are economically viable. A variety of

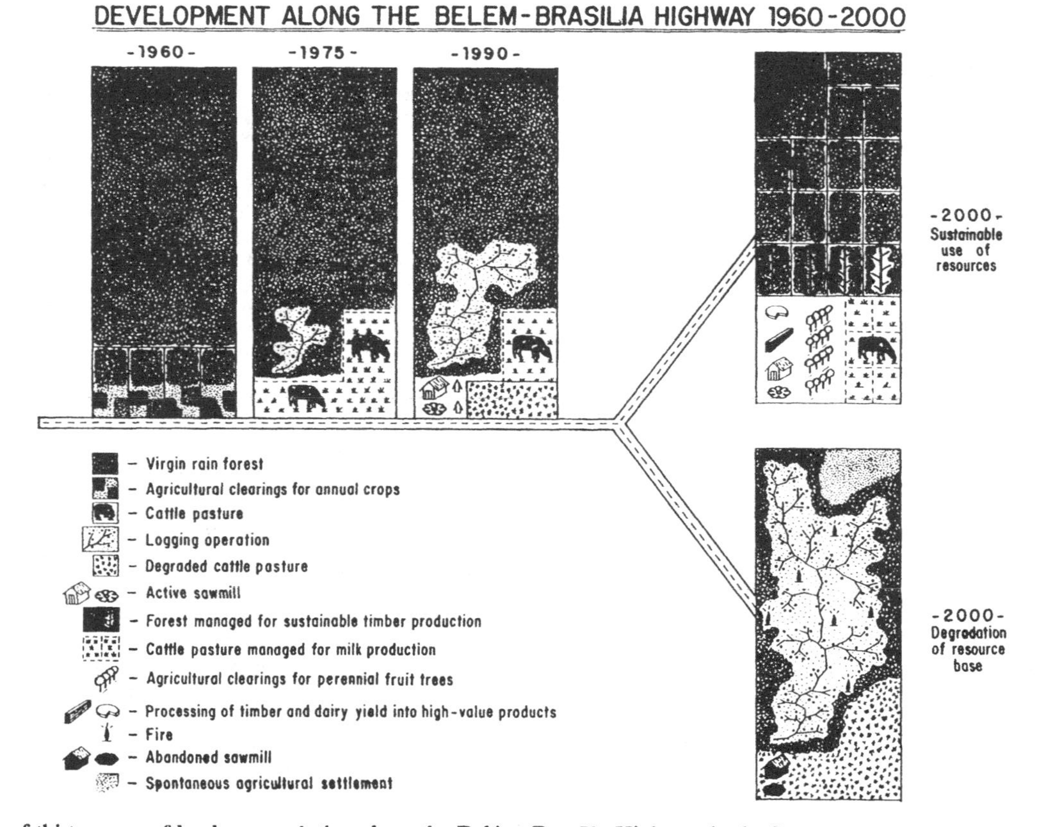

Fig. 15.3. A summary of thirty years of land-use evolution along the Belém–Brasília Highway in the Paragominas region of eastern Amazonia.

agricultural tree crops such as citrus, capuaçu, caju, and mango, grow well in the region, even when introduced into abandoned farm fields and pastures. Logging, if done carefully, can also be sustainable. Indeed, the quality of life at Paragominas could be improved substantially just by putting the 6000 km^2 that have already been cleared into a vital productive form. Meanwhile, the remaining 75% of the municipality that is still forested could be managed for timber production. With respect accorded to both the land and its human inhabitants, this landscape could be managed while maintaining its biotic diversity intact.

Acknowledgments

The results summarized in this chapter are based on field research supported by the W. Alton Jones Foundation and The World Wildlife Fund for Nature. The Fulbright Commission of Brazil supported C. Uhl during the field research.

References

Appanah, S., & Putz, F. E. (1984). Climber abundance in virgin dipterocarp forest and the effect of pre-felling climber cutting on logging damage. *The Malaysian Forester*, **47**.

Barreto, P., Uhl, C., & Yared, J. (In press). O potencial de produção sustentavel de madeira na Amazonia Oriental na região de Paragominas, Pará: considereações economicas e ecologicas. *Pará Desenvolvimento*. Instituto do Desenvolvimento Economico-social do Pará, Belém, Pará, Brasil.

De Graaf, N. R. (1986). *A silvicultural system for natural regeneration of tropical rain forest in Suriname*. Wageningen, The Netherlands: Agricultural University.

FIBGE. (Fundaçõ Instituto Brasileiro de Geografia e Estatística). (1988). *Produção e rendimento total do Estado do Pará*. Delegacia do Pará.

Goodland, R. J. A., & Irwin, H. S. (1975). *Amazon jungle: green hell to red desert?* New York: Elsevier.

Hecht, S. B. (1983). Cattle ranching in the eastern Amazon: environmental and social implications. In *The Dilemma of Amazonian Development*, ed. E. F. Moran, pp. 155–188. Boulder, CO: Westview Press.

Hecht, S. B. (1985). Environment, development and politics: capital accumulation and the livestock sector in eastern Amazonia. *World Development*, **13**, 663–684.

Hendrison, J. (1990). *Damage-controlled logging in managed tropical rain forest in Suriname*. Wageningen, The Netherlands: Agricultural University.

Jonkers, W. B. J. (1988). *Vegetation structure, logging damage and silviculture in a tropical rain forest in Suriname*. Wageningen, The Netherlands: Agricultural University.

Mahar, D. J. (1989). *Government Policies and deforestation in Brazil's Amazon region*. Washington, D.C.: World Wildlife Fund and The Conservation Foundation. Published by arrangement with The World Bank.

Marn, H. M., & Jonkers, W. B. (1982). Logging damage in tropical high forest. In

Tropical Forests – Source of Energy through Optimization and Diversification, ed. P. B. L. Srivastara, et al., pp. 27–38. Proceedings of an international conference held 11–15 November 1980 at Penerbit University Pertanian, Serdang, Selangor, Malaysia.

Mattos, de M. M., & Uhl, C. (in press). Perspectivas economicas e ecologicas sobre pecuária na Amazônia Oriental na decada 90. *Pará Desenvolvimento.* Instituto do Desenvolvimento Economico-social do Pará, Belém, Pará, Brasil.

Serrão, E. A. S., & Simão Neto, M. (1975). The adaptation of tropical forages in the Amazon region. In *Tropical Forages in Livestock Production Systems*, pp. 281–310. Special Publication No. 24. Madison, Wisconsin: American Society of Agronomy.

Serrão, E. A. S., Falesi, I. C., da Veiga, J. B., & Teixeira Neto, J. F. (1978). Productivity of cultivated pastures on low fertility soils of the Amazon of Brazil. In *Pasture Production in Acid Soils of the Tropics*, ed. P. R. Sanchez & L. E. Tergas, pp. 195–226. Proceedings of a symposium held at Centro International de Agricultura Tropical (CIAT), Cali, Colombia.

Verissimo, A., Barreto, P., Mattos, M., Tarifa, R., & Uhl, C. (1992). Logging impacts and prospects for sustainable forest management in an old Amazonian frontier: the case of Paragominas. *Forest Ecology and Management*, **55**, 169–199.

Part Six

Strategies for biodiversity conservation

16

Market-based economic development and biodiversity: an assessment of conflict

ROBERT D. WEAVER

Introduction

The conflict between market driven economic activity and the environment would seem to be a broadly accepted fact. The conflict is so generally perceived that the feature film "The Medicine Man" symbolizes the conflict graphically with scenes of bulldozers devastating an Amazonian rainforest defended by a field biochemist seeking a cure for cancer in the treetops of the forest. When I saw this film, the audience quietly watched, apparently accepting the validity of this dramatic illustration of the impacts of economic activity on biodiversity and the environment. The same logic is also apparently accepted for urban development, which in the developed world is often viewed as posing a threat to remaining grasslands, forests, and wetlands.

These images provide a view of the paradoxical conflict facing humanity that has been raised throughout this volume. The conflict can be summarized as follows. Extinction of species is occurring at a substantially greater rate today relative to past periods (Myers, 1994). This high extinction rate is due in part to human economic activity and development and the associated pressures, e.g., population growth (Holdgate, 1989), resource extraction (Soule & Wilcox, 1980), and waste disposal (World Bank, 1992), placed on limited resources and ecologies. The extinction process is of such magnitude that many conclude it may threaten human life itself (Myers, 1988: Raven, 1988). Thus, the paradox is that humans depend for survival on biodiversity, nature, and ecological resources that are subsumed in landscapes, yet, at the same time, human survival apparently threatens the survival or existence of those very resources.

Conservation strategies are cited as providing one recourse (McNeely et al., 1990), though this strategy can generally be viewed as likely to be ineffective unless the basic conflict that lies at the root of the paradox is resolved (Kim & Weaver, 1994). In this sense, the past relationship between human activity and biodiversity has been viewed by many as incompatible with sustained increases in human consumption (Myers, 1988). This leads some observers to the conclusion that a fundamentally different course for economic activity must be taken to allow and ensure a sustainable relationship between human activity and biodiversity (Katz, 1994: Cairns, 1994).

Other chapters in this volume have examined the extent and underlying bases of the conflict that lies at the root of the paradox. These chapters consider the extent of the conflict, the current manifestations of the conflict, and the options available for resolution of the conflict. In this chapter, I will consider these issues within the context of sub-Saharan Africa as a case for study. To start, I would like to review the nature and extent of the impacts of development on biodiversity and the environment in sub-Saharan Africa. Given this overview of the extent of the problem, I will review current approaches to managing those impacts. As a bottom line for the chapter, I would like to summarize, by illustration, the complexity of the conflict between human activity and biodiversity, as well as the extent of innovative human effort that is being put behind solving that conflict.

Linkages between economic activity and biodiversity

First, the nature and extent of economic development experienced in sub-Saharan Africa during the past decades is illustrated by consideration of the growth of gross domestic product (GDP). GDP measures the economic value created within a nation's borders through the use of available resources and imported inputs. During the past decade, GDP in most sub-Saharan countries has expanded substantially (World Bank, 1992). Despite this economic growth, expanding population has exceeded economic growth in many countries, resulting in an ever more meager standard of living for their growing populations (Figs. 16.1 and 16.2). Even when Gross National Product (GNP) (includes returns earned from economic activity in foreign countries) is used to measure economic growth, economic growth on a per-capita basis has been generally meager (Fig. 16.3).

Over the past three decades, population growth rates have been reduced in some countries (Zaire, Kenya, Zimbabwe, Cote d'Ivoire, Botswana, and Gabon): however, in most countries population growth rates have actually increased. The conclusion must be drawn from these numbers that when we discuss economic development in sub-Saharan Africa, we are, for now, discussing expansion of economic activity necessary to keep pace with expanding population to slow or prevent further degradation of already extremely low standards of living (World Bank, 1992). In this sense, the consideration of the case of sub-Saharan Africa allows a view of the paradoxical conflict between economic activity necessary for survival and biodiversity. This case is clearly

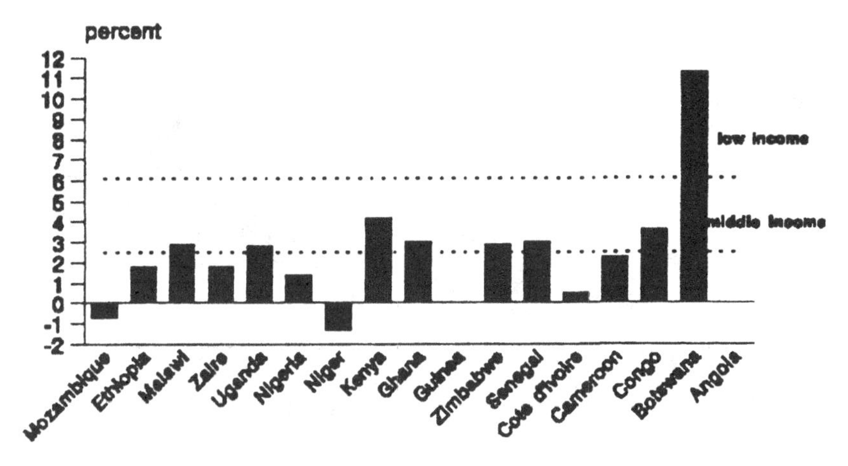

Fig. 16.1. Average annual GDP growth rate, 1980–1990 (percent). Source: World Bank, 1992.

differentiable from the situation in many developed countries where the conflict follows from demands for increases in standards of living beyond subsistence levels. It is an imperative for survival that motivates economic expansion in most sub-Saharan countries.

A third trend that has accompanied expanding populations in sub-Saharan countries has been substantial increases in urban migration (Fig. 16.4). This

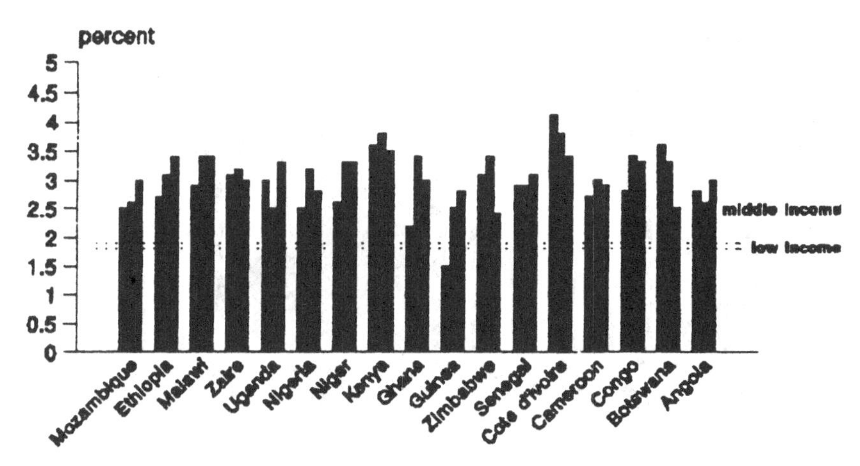

Fig. 16.2. Average annual growth of population (percent). Left column = 1965–1980; middle column = 1980–1990; right column = 1990–2000 (projected). Source: World Bank, 1992.

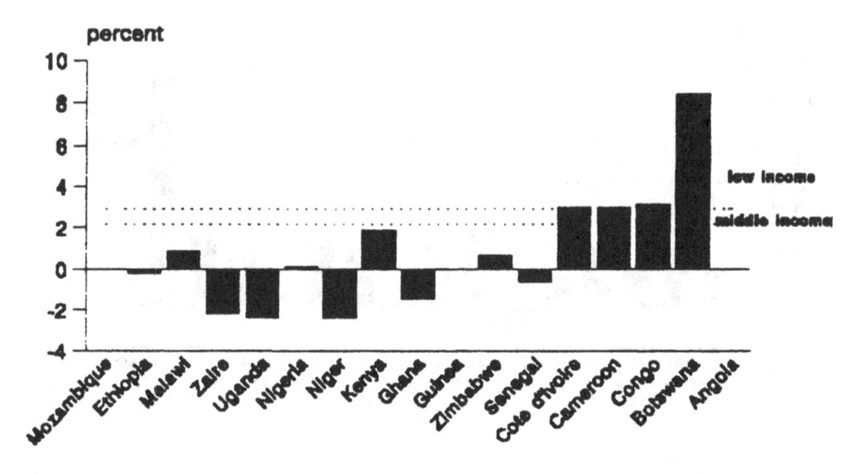

Fig. 16.3. Average annual growth of GNP per capita. Source: World Bank, 1992.

process has been driven by demands for consumer goods, safe drinking water, health services, education services, and employment opportunities not available in rural settings. Accompanying urbanization has been growth in rural towns and villages, often outstripping the capacity of water supplies and available infrastructure. In both settings, urban and rural water and air pollution accompanies expansion of economic activity.

Increased population has led to increased demand for subsistence food crops.

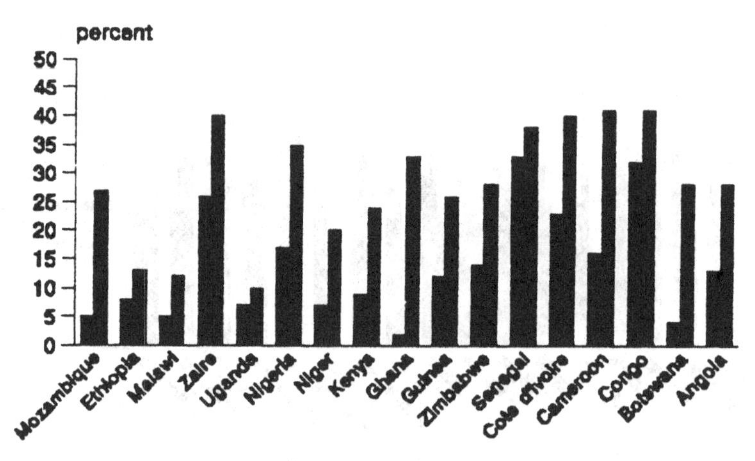

Fig. 16.4. Urban population as percentage of total population. Left column = 1985; right column = 1990. Source: World Bank, 1992.

At the same time, global demand for agricultural and forest products has expanded. Combined, these forces have placed increased demands on limited soil, water, and forest resources in sub-Saharan countries (Goodson, 1988). These forces have been manifested in increased clearing of savanna and forests, reduced fallowing periods in slash and burn farming systems, intensification of use of fragile lands subject to high risk of erosion and loss of nitrogen, and increasing pressures on remaining wildlife populations (McNeely et al., 1990).

Expansion of agricultural production in most of the sub-Saharan countries has progressed at 1–2% per year over the last decade (Fig. 16.5). However, it is well known that these growth rates have not been accomplished by expansion of yields, (e.g., World Bank, 1992). In most countries yields of primary food crops have fallen or remained flat. Instead, area cultivated has expanded to allow increased production.

The accompanying changes in land use must be put in perspective (Fig. 16.6). The percentage of closed forests deforested each year has been extremely high. Nonetheless, on a global basis, sub-Saharan Africa does not suffer the highest annual rate of deforestation (Myers, 1988). Three important demands by growing populations contribute to deforestation: 1) firewood demand, 2) demand for timber, and 3) demand for cropland. Demand for cropland leads to deforestation as well as degradation of soil resources as fallowing periods are reduced and savanna converted to cropland.

The net effect of these processes in sub-Saharan Africa is loss of habitat (Figs. 16.7 and 16.8). This presents a picture of the extent of loss of habitat,

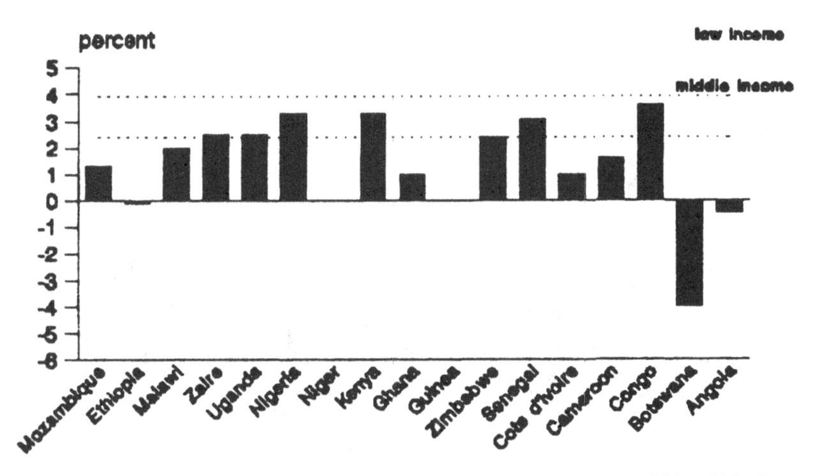

Fig. 16.5. Average annual agricultural production growth rate, 1980–1990. Source: World Bank, 1992.

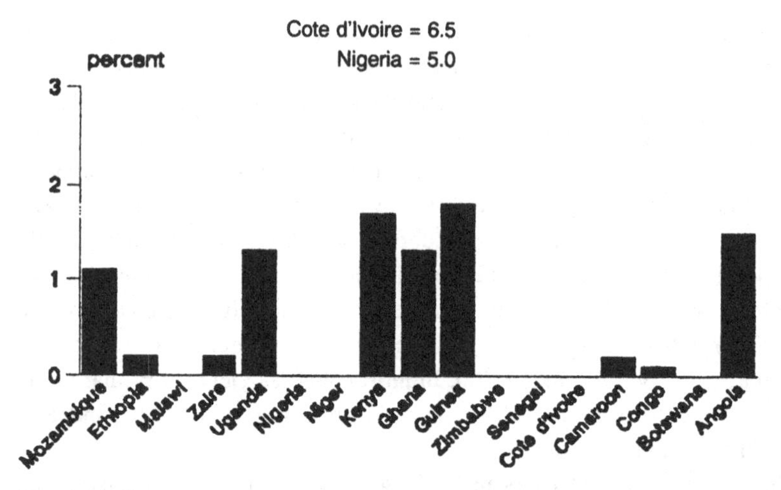

Fig. 16.6. Percentage of closed forest deforested per year. Source: FAO, 1981.

based on estimates of original land occupation, that has been extreme (IUCN/UNEP, 1986a,b). Clearly, questions of changes in grasslands, wetlands, or fragmentation of habitats are difficult to assess. Expectedly, the data on such visible blocks of habitat, such as tropical forests, and their evolution over the past decades are also very weak. Nonetheless, the biological richness of tropical forests make them of special interest. Estimates indicate that somewhere between 800 and 1,200 million hectares remain worldwide. However, for sub-

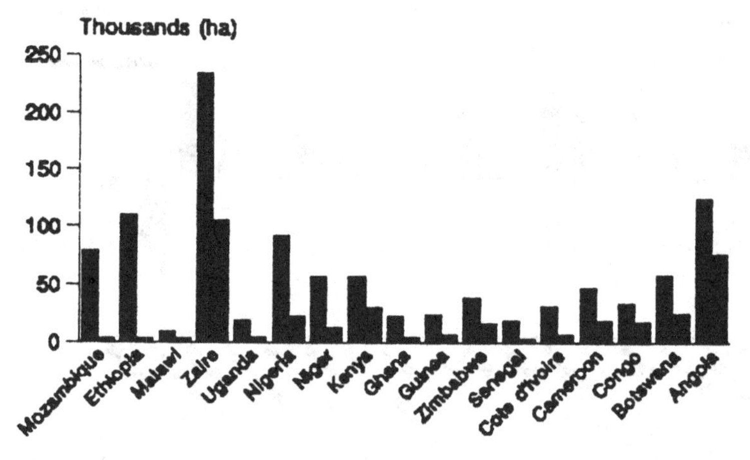

Fig. 16.7. Wildlife habitat loss in sub-Saharan Africa. Left column = original wildlife habitat; right column = amount remaining. Source: IUCN/UNEP, 1986.

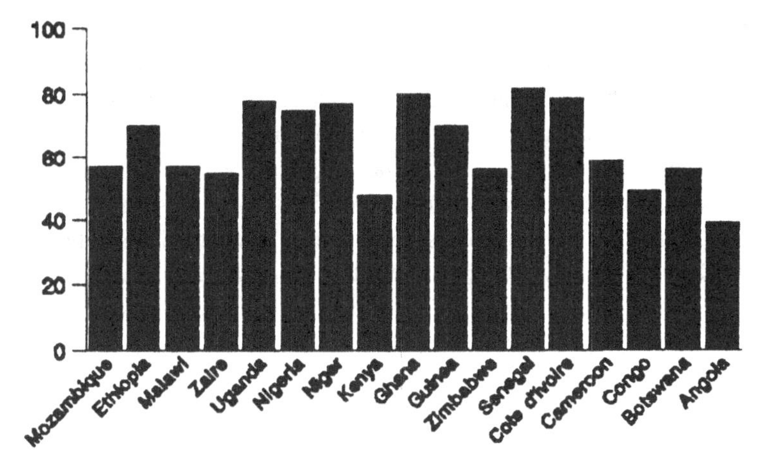

Fig. 16.8. Habitat loss (percent). Source: IUCN/UNEP, 1986.

Saharan Africa, the rate of deforestation was estimated as very high, ranging from an extreme of 6.5% per year for Cote d'Ivoire to 0.1% for the Congo (FAO, 1981).

Viewed together, these facts illustrate a dramatic decline in habitat throughout much of sub-Saharan Africa. In most of the regions, losses have exceeded 50%. Exceptions include areas protected from direct population pressure or ease of access to the forest. Angola, Congo, Zaire, Gabon, and limited areas of Kenya, Tanzania, and Zambia are examples. Importantly, ease of access and extraction of timber have had strong influences on the extent of the threat to specific forests (IUCN/UNEP, 1986a,b). Remaining virgin forests protected by inaccessibility include only the central and northeast areas of the Zaire Basin in central Africa (World Commission, 1987; Conservation International, 1990).

To summarize, the case of sub-Saharan Africa illustrates a number of specific impacts of economic activity and development on biodiversity and the environment. These include 1) habitat alteration, 2) changes in harvesting rates and patterns, and 3) introduction of species and alteration of interspecies equilibria. To proceed, it is useful to consider each of these effects in further detail.

Habitat alteration

While deforestation directly impacts habitat, its indirect effects go beyond the site of deforestation because deforestation typically results in fragmentation of habitat (Soule & Wilcox, 1980). Excellent examples of this fragmentation are increasingly apparent in sub-Saharan Africa as land clearing for timber and agriculture proceeds into savanna and forest areas. Habitat is also altered through changes induced in the local microclimate as solar exposure an surface water flows are changed. Water and air pollution are excellent examples of habitat change that may result from development. Water, air, and soil pollution can be expected to result from intensification of agricultural and industrial activities. In each case, food sources, breeding sanctuaries, and living spaces may be substantially altered. As sensitive species are impacted, interdependency among species carries the initial impact through the environment, leaving biodiversity altered (Raven, 1988).

Changes in harvesting

The role of dynamics in biodiversity is particularly apparent when the effects of changes in man's interaction with the resource is considered. Through any changes in harvesting rates of a particular species, the equilibrium within the impacted habitat can be expected to ripple through other species (Raven, 1988). In sub-Saharan Africa, substantial changes in harvesting rates of savanna brush have occurred over the past decade as fallow periods in slash and burn agricultural systems have decreased (Vitousek et al., 1986). Similarly, changes in harvesting rates of lake and river fish have resulted in changes in the balance of opportunity for survival of other species (Miller, 1989).

Introduction of species and alteration of interspecies equilibria

Introduced species of plants for any purpose may by design replace native varieties. For example, in some of the Rift Valley lakes, introduced species of fish have threatened many of the native species with extinction (Miller, 1989). Introduction of other plants may through added dominance lead to displacement of native species. Clearly, one of the substantial costs of widespread adoption of new agricultural varieties will be loss of biodiversity available in native land races (Prescott-Allen & Prescott-Allen, 1982). Despite the presence of these processes that pose threats to sub-Saharan biodiversity, substantial resources remain, suggesting reason for action.

A search for the roots of the paradox

The discussion to this point has provided some background on the nature and the extent of the impacts of economic activity and development in sub-Saharan Africa on biodiversity and the environment. Suppose we accept the premise that a strong imperative for economic development originates in declining standards of living, growing population, and the natural desire for improved standards of living within even a stable population.

It is important to understand what it is about decentralized, market-driven economies that lead human economic activity and development to have negative impacts on biodiversity and the environment. Some authors have argued these negative impacts are of such magnitude that they threaten the existence of these resources and human life itself. An understanding of two important characteristics will take us a long way to understanding why market systems lead to these malfunctions and how that performance can be corrected. These two characteristics have been labeled by economists as *externalities* and *public goods* (Nicholson, 1990).

Since the 1960s, social and political attention has increasingly focused on managing what economists have labeled *externalities* (Nicholson, 1990). These peculiar economic beasts involve the effects of an individual's consumption or production activities, which impact others (Fig. 16.9). Externalities have been of interest to economists for close to a century as an important source of failure in the ability of decentralized markets to allocate scarce resources in a way that leads to the greatest possible welfare for society. Early on, economists recognized that, in a decentralized economy, individuals make decisions about consumption or production based solely on the benefits or profits that can be personally realized. Some might say, incredibly, economists were able to show that under certain fairly reasonable conditions, such behavior would lead the economy "as if guided by an invisible hand" to produce the greatest possible welfare from the limited resources available. This result, if true, rationalizes the enthusiastic support for decentralized, market economies as a mechanism for best satisfying society's needs. By analogous argument, the result suggests that decentralized, market-driven economic activity offers an effective mechanism for developing countries to most effectively utilize their economic resources in development processes that do not negatively impact biodiversity or the environment.

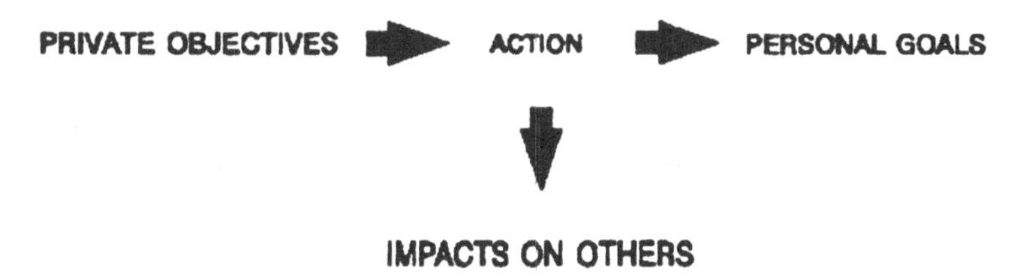

Fig. 16.9. Externalities.

Economists have been concerned with externalities for the simple reason that when they exist, and no compensating strategy is implemented, many of the attractive features of decentralized, market-driven economic activities are comprised (Nicholson, 1990). The immediate implication of the presence of externalities is that decentralized, market-driven economies will malfunction – resources will not be effectively utilized to produce the greatest possible level of social welfare. Stated differently, resources will be inefficiently used to produce the wrong products with the wrong inputs and technologies, and consumers will find themselves consuming too much of some goods and services and not enough of others.

Biodiversity and other features of the environment are often viewed as being negatively impacted by the externalities from economic activity and development (Wilkes, 1994). Concerns for the impacts development has on biodiversity and the environment are founded on another concern of economists that influences the performance of decentralized, market economies. The attractiveness of market economies holds when goods and services can be effectively exchanged between consenting consumers and producers. In order for this to be possible, those goods and services must be *appropriable* and *exclusively consumable.* In other words, the producer can control access to the good, forcing consumers to pay for it. Importantly, consumers must pay for it, since the benefits of consumption of the good are exclusively available to the individual consumer who pays for the good. Goods and services that satisfy these conditions are called *private goods* by economists.

In contrast, *public goods* are nonappropriable and nonexclusive in consumption. For this reason, producers of public goods cannot control access and consumers do not have to pay for access to the benefits of their consumption. Economists have shown that when public goods are introduced, decentralized, market-driven economies fail to perform. Decentralized, market-driven economies fail to produce the right amount of services associated with public goods or to utilize available supplies in a way that results in the greatest possible social welfare. Unfortunately, biodiversity and many other aspects of the environment have many features of public goods (Weaver, 1994). To conclude, *externalities and public good characteristics* represent two important bases through which market-based economic activity and development may lead to undesirable impacts on biodiversity and the environment. Economists have searched for solutions to correct the impacts of externalities and public goods as performance by trying to understand what it is about the presence of externalities or public goods characteristics that causes decentralized, market-driven economies to malfunction. In brief, three important implications of externalities and public goods should be noted (Starrett, 1988; Krutilla & Fisher, 1985; Conrad & Clark, 1991).

First, in the presence of externalities and public goods, markets establish prices of goods that fail to reflect their value to society (Nicholson, 1990). Economists interpret prices as the signals that direct economic activity in decentralized economies. They argue that the presence of externalities and public goods lead decentralized, market-driven economies to generate the wrong price signals. Consider the case of biodiversity. The centers of biodiversity are concentrated globally. Consider the case where valuable germplasm occurs within a forest that is exposed to demand for clearing for agricultural use. First, the agricultural-use value of the land would fail to reflect the total social-use value of the land. Second, current users of resources fail to pay the full social cost of use, transferring excess costs to current and future members of society. Because discount rates may fail to reflect the value of saving for the future, part of these costs are passed on to future generations. Finally, given prices that fail to measure social value, a number of other implications follow immediately. Social benefits and costs of utilization of the resources are not apparent. In their absence, social priorities are unclear and other social and political means must be used to guide decisions. Resource use is inefficient, and in the face of externalities and public goods, maximum social welfare cannot be achieved by a purely decentralized, market-based economic system. In the longer term, both private and public sector investment is misdirected.

Economic solutions to the paradox

The observations made above raise an important question. If externalities and public goods cause decentralized, market-driven economies to malfunction, could such malfunctions place market-based economic activity and development in direct conflict with biodiversity and the environment? Further, if such malfunctions result in conflict between market-based economic activity and development and biodiversity, does this imply that we must choose between such economic activity and development on the one hand, and biodiversity and the environment, on the other? Going back to popular imagery, is there credibility in the dramatic image of the conflict inherent between bulldozer and rainforest? Is this conflict an absolute, or merely a humanly resolvable situation? To shed some light on these questions, and to bring this argument to a close, it is useful to consider a number of approaches that have evolved over the past decade to reduce, manage, and in some cases, eliminate the basis of this conflict. Two general approaches exist – country-based and multinational.

Country-focused tools and approaches

Income growth

Increasing evidence appears to argue that public goods such as aspects of biodiversity and the environment are increasingly valued as income and education increase (World Bank, 1992; Krutilla and Fisher, 1985). In part, this may be due to increasing recognition of the cost of these effects, and the need for individuals to personally bear part of these costs in decreased health and direct avoidance costs. Recognition of these costs results in political and social pressure of management of the externalities and use of the public goods. If this is true, then an important solution occurs as a product of the development process itself. By increasing income, capacity is achieved to finance both private and public investments to alter the impacts of economic activity and development.

Key elements of income growth strategies will be the targeting of economic policy in order to stimulate sustainable growth (Dixon & Fallon, 1989). This must include macroeconomic, trade, and fiscal policies as well as specific sectoral policies. Important among these policies are those that eliminate distortions of market incentives directed at specific special-interest goals. For example, subsidies for particular inputs or economic activities may expand the scale of externalities or use of public goods, exacerbating impacts on biodiversity and the environment. Further elements of progrowth strategies would include: 1) stable and predictable macroeconomic policy, 2) financing of deficits by savings and toleration of them only within the context of business cycles rather than as perpetual disequilibria, 3) implementation of social goals such as income and wealth distribution through the use of lump-sum payments rather than price distorting subsidies or taxes, 4) minimization of the role of government in resource-allocation decisions, and 5) elimination of tariffs designed to protect domestic economic enterprises or consumers.

Slowing population growth

As noted above, an important force affecting incomes in sub-Saharan Africa is population growth. In the absence of adequate and efficient technologies and capital, the region's expanding population has strong survival imperatives to increase land clearing for agriculture, forestry, and housing. This land expansion typically occurs on marginal lands. Given meager resource bases, these populations also have very limited capacity to make even the simplest investments to preserve soil, water, or forest resources for sustainable use. Clearly, an important element of achieving improved impacts of development on the biodiversity and the environment will be reduced population growth rates (World Bank, 1992). Education and provision of economically, socially, and politically attractive alternatives will be important elements of any strategy to achieve such a goal.

Institutional changes and innovations

An important role of government in decentralized, market-driven economies is to provide sanction to institutions that are critical to the operation of such an economy (Nicholson, 1990). In many cases, evolution of institutions lags behind problem recognition and such evolution is not appreciated as an important means of government action to facilitate solution of the problem. Numerous important examples can be cited for the case under consideration. Consider *property rights* for agricultural land, for irrigation systems, for sanitary water supplies, or for fuel wood forest areas (Feder and Feeny, 1991). In each case, property rights assigned to interested users and beneficiaries of the resource can enhance the economic basis for investment and management of the resource over the long term. Importantly, property rights must be designed and implemented to solve the problems, rather than to create them. As a counterexample, consider the impact of property rights based on 1) land clearing, 2) continuous use, or 3) use of land for a particular type of crop.

A wide variety of property rights may be useful, ranging from full ownership to limited-use rights (Barzel, 1993; Alchian, 1965; Alchian and Demsetz, 1972; Demsetz, 1967). Zoning can be useful to establish limits on use rights and has proven effective as a means of ensuring compatible uses of resources where conflict might otherwise result (World Bank, 1992). In many cases, property rights need not be designed that conflict with indigenous cultural practices (Hopcraft, 1981). Where collective decision making is a tradition, such collective rights arrangements can often be useful. At the extreme, governmental organizations can be formed and enfranchised with management rights over the resource. Where intergenerational management of the resource is critical, property rights must be transferable even across generations.

Public institution development

While public institutions or other organizations may offer substantial potential, the efficacy of their operation depends on their ability to facilitate private sector behavior that is congruent with social goals. Such capacity cannot be expected to emerge instantaneously with the formation of new organizations. Instead, substantial resources must be invested in their design and development. An important function of these public institutions must be to retain participation of the users and beneficiaries of the resources within the functioning of the institutions (Mueller, 1993; Grindle and Thomas, 1991). Further, these institutions must provide an effective context for interest groups to mediate conflicts.

Regulation and use restriction

A common means for interest groups or governments to alter private sector behavior is through regulation (Mueller, 1993). Regulation can take several forms though two forms are most common: 1) "command-and-control" forms and 2) "market-directed" forms. In the first form, the government or collective claims the right to allocate resources. For example, input and/or output levels might be specified, or production or consumption practices or timing might be specified. Rationing systems are an example of "command-and-control" regulation. Similarly, government-directed cropping decisions are another example that remain in effect to various extents in sub-Saharan Africa. While market-oriented reforms during the 1980's have eliminated many of these policies (World Bank, 1991), many sub-Saharan countries still have in place "marketing campaigns" that specify the dates during which particular products can be marketed or planting directives that continue colonial practices of requiring that farmers in particular areas grow particular crops (e.g., coffee in Rwanda). Technology choices have not typically been regulated in sub-Saharan countries. Examples which would benefit the environment might include: 1) impoundment of animals to allow collection of manure and protection of wild vegetation, 2) restriction of fuel wood gathering to specified areas, or 3) prohibition of planting of erosive crops such as cassava, peanuts, or potatoes on sloped soils.

"Market-directed" regulations include use of taxes or subsidies that do not distort relative prices to provide incentives for desired behavior (Hahn and Noll, 1982; Hahn and Stavins, 1991). For example, deposit/refund-based recycling systems have been implemented for glass in many countries; similar approaches could be implemented with plastics and batteries that pose a serious threat to the environment in Africa. Experience has shown these tactics result in the emergence of private-sector recycling efforts (World Bank, 1991). Point-source pollution of water, air, or soil can be effectively managed by joint systems of rights and charges. Other examples include the use of in-kind subsidies such as access to public services as incentives for management of local public resources for a wider social base.

Market development to enhance value

Marketing represents an important though often forgotten element of the "technology" necessary to create economic value. By enhancing the value of agricultural and forest products, the productivity of those resources in generating income and satisfying consumer needs is enhanced. Substantial opportunities exist in sub-Saharan countries for marketing-system improvements in post-harvest handling, storage, efficient distribution (both geographically and over time), and for regional as well as global export. Importantly, many of these opportunities may be well known though left unexploited because of financial, cultural, or legal impediments. Clearly, an efficient market system must be viewed as a prerequisite for managing the impacts of development on biodiversity and the environment.

Education

Education can play an important role on several levels. Provision of information is a critical element of any strategy. By providing an appreciation of the value of common-property resources such as biodiversity and the environment, education can create and empower constituencies that have interest in those resources. In the longer term, education can develop the basis for the political and social processes necessary for coalition formation and resolution of conflicts over use of the resources. Education can also serve as a tool to facilitate diffusion and adoption of new technologies in production and marketing that will improve the efficiency of use of the common and open property resources. Importantly, education can enhance the productivity of human resources and facilitate substitution of labor and other resources for the common and open access property resources.

Investment in research and development of appropriate technologies

Research and development in appropriate technologies is often viewed as the first solution to managing the impacts of market-based economic development on biodiversity and the environment. While last in the options reviewed, technology does provide an important opportunity for dissipating the conflict between development and biodiversity. Throughout sub-Saharan countries, examples can be cited of technological innovations that have substantially reduced the use of common-property resources or the production and incidence of externalities. Often these innovations have been simple and required only local inputs. An example is the fuel-efficient cook stove made of clay that exploits the Venturi effect of carburation to intensify heat produced from charcoal. Diffusion of energy efficient food sources has also played an important role. Use of legumes in crop rotations and effective management of crop residues have done much to preserve soil productivity. Nonetheless, the potential of technology as an opportunity is dependent upon existence of sufficient human and financial capital to finance adoption and transition to new methods. As already noted, expanding population and declining per-capita GNP experiences in much of sub-Saharan threatens the potential of technology as a solution.

Multinational approaches

The strategy of forming collectives or government organizations to manage common property resources or externalities is based on the idea of internalizing the use conflict involved (Mueller, 1993). By creating a group of interested parties, an institutional context is created for the articulation of the nature of the conflict and for development of means for its resolution. In order for such approaches to be effective, all interested parties must be brought together. In the case of common property resources such as biodiversity or aspects of the environment, this requirement of inclusion may force the formation of groups that go beyond national and even regional boundaries. Plant and animal germplasm provide excellent examples. As with any element of value in biodiversity, germplasm extends the scope of interest in the management of biodiversity well beyond national borders to the global community and from people alive today to future generations. Necessarily, in order to develop effective institutions at such a global level, national sovereignty must be sacrificed to the extent that nations that are the sites of biodiversity must participate in solutions to its management even when these solutions go beyond their own national interests.

Substantial effort has been made in the development of such global institutions since the 1960's. It is clear from experience in multinational negotiations over the past decades that biodiversity and environmental resources and externalities are not distributed smoothly across the globe (most recently, see IUCN/UNEP, 1991). Instead, a surprising disparity is obvious across developed versus developing countries and Northern versus Southern Hemisphere countries. Faced by the immediate demands of impoverished rural populations, the opportunity cost of foregoing land clearing is very high for many developing countries. However, it must be recognized that these developing countries will not be the sole beneficiaries of sound land management practices. In fact, the entire global community stands to gain in the long run. To the extent that such

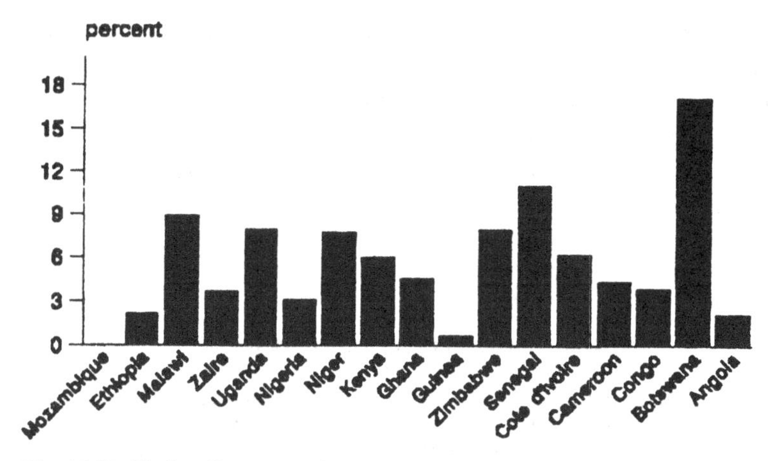

Fig. 16.10. Nationally protected areas as a percentage of total land area. Source: World Bank, 1992.

benefits occur in other countries, those countries must be expected to help finance the costs of such actions taken by other countries.

Conclusions

The first conclusion that is clear from the past three decades is that market-based economic activity and development and biodiversity and environmental management are not incompatible processes. To the contrary, given that actions are taken to repair the natural inadequacies of market-based economies, economic activity and development can strengthen the management of biodiversity and environmental resources and these resources can become critical elements in the growth process.

A second conclusion that is clearly warranted is that there exist substantive means for easing the impacts of economic activity on biodiversity and the environment. These means require innovation in national and transnational institutions and technologies in creating wealth from limited resources. They also require understanding of the subtle nature of the role of government as a means of strengthening rather than constraining economic potential. In this sense, numerous solutions exist that complement and reinforce economic activity and development processes, rather than stifle them.

Another conclusion that is clear from experience is that, at their core, many solutions to the apparent paradox of the relation between humanity and biodiversity will force change in the sovereignty of nations as multinational solutions are pursued. The changes will be of two types. First, solutions will require greater international cooperation and expansion of international institutions. It would seem inevitable that the necessary changes will include globalization of institutions to multinational scales. Second, solutions will require greater democratic participation and conflict resolution that will reduce the power of central governments and place the resolution of many conflicts in hands of subgroups of the national population.

References

Alchian, A. A. (1965). Some economics of property rights. *Il Politico*, **30**(4), 816–829.

Alchian, A. A., & Demsetz, H. (1972). Production, information costs, and economic organization. *American Economic Review*, **62**(5), 777–795.

Barzel, Y. (1989). *Economic Analysis of Property Rights.* New York: Cambridge University Press.

Conrad, J. M., & Clark, C. W. (1987). *Natural Resource Economics.* New York: Cambridge University Press.

Conservation International. (1990). *The Rain Forest Imperative.* Washington, D.C.: Conservation International.

Demsetz, H. (1967). Toward a theory of property rights. *American Economic Review*, **57**(2), 347–359.

FAO. (1981). *Tropical Forest Resources Assessment Project (GEMS): Tropical Africa, Tropical Asia, Tropical America* (4 Vols). Rome: FAO/UNEP.

Grindle, M. S., & Thomas, J. W. (1991). *Public Choices and Policy Change.* Baltimore: The Johns Hopkins University Press.

Goodson, J. (1988). *Conservation and Management of Tropical Forests and Biodiversity in Zaire.* Unpublished document, USAID, Zaire.

Hopcraft, P. (1981). Economic institutions and pastoral resource management: Considerations for a development strategy. In *The Future of Pastoral Peoples*, ed. John G. Galaty, D. Aronson, P. Saltzman, & A. Chovinard. Proceedings of a conference in Nairobi, August 1980. Ottawa: IDRC.

IUCN/UNEP. (1986a). *Review of the Protected Areas System in the Afrotropical Realm.* Gland, Switzerland: IUCN.

IUCN/UNEP. (1986b). *Review of the Protected Areas System in the Indo-Malayan Realm.* Gland, Switzerland: IUCN.

IUCN/UNEP (United Nations Environment Programme), and WWF (World Wide Fund for Nature). (1991). *Caring for the Earth, a Strategy for Sustainable Living.* Gland, Switzerland: IUCN.

Krutilla, J. V., & Fisher, A. C. (1985). *The Economics of Natural Environments.* Baltimore: The Johns Hopkins University Press.

Kim, K. C., & Weaver, R. D. (1994). Biodiversity and humanity: toward a new paradigm. In *Biodiversity and Landscapes*, ed. K. C. Kim & R. D. Weaver, pp. 393–423. New York: Cambridge University Press.

Miller, Daniel J. (1989). Introductions and extinction of fish in the African great lakes. *Trends in Ecology and Evaluation*, **4**(2), 56–59.

Mueller, D. (1993). *Public Choice II.* New York: Cambridge University Press.

Myers, N. (1994). We do not want to become extinct: the question of human survival. In *Biodiversity and Landscapes*, ed. K. C. Kim & R. D. Weaver, pp. 133–150. New York: Cambridge University Press.

McNeely, J. A., Miller, K. R., Reid, W. V., Mittermeir, R. A., & Werner, T. B. (1990). *Conserving the World's Biological Diversity*. Washington, D.C.: The World Bank.

Myers, N. (1988). Threatened biotas: Hotspots in tropical forests. *Environmentalist*, **8**(3), 1–20.

Nicholson, W. (1990). *Intermediate Economics and its Applications*, 5th ed. Chicago: The Dryden Press.

Olson, M. (1965). *The Logic of Collective Action.* Cambridge, MA: Harvard University Press.

Pearce, D. W., & Warford, J. J. Forthcoming. *World Without End: Economics, Environment, and Sustainable Development.* New York: Oxford University Press.

Prescott-Allen, R., & Prescott-Allen, C. (1982). *What's Wildlife Worth? Economic Contributions of Wild Plants and Animals to Developing Countries.* London: International Institute for Environment and Development (Earthscan).

Raven, P. H. (1988). Biological resources and global stability. In *Evolution and Coadaptation in Biotic Communities*, ed. S. Kawano, J. M. Connell, & T. Midaka, pp. 3–27. Tokyo: University of Tokyo Press.

Starrett, D. A. (1988). *Foundations of Public Economics.* New York: Cambridge University Press.

Schramm, G., & Warford, J. J. (1989). *Environmental Management and Economic Development.* Baltimore, MD: The Johns Hopkins University Press.

Soule, M. E., & Wilcox, B. A. (1980). *Conservation Biology: An Evolutionary-Ecological Approach.* Sunderland, MA: Sinauer Associates.

Vitousek, P. M., Ehrlich, P. R., & Matson, P. A. (1986). Human Appropriation of the products of photosynthesis. *BioScience*, **36**, 368–373.

Weaver, R. D. (1994). Economic valuation of biodiversity. In *Biodiversity and Landscapes*, ed. K. C. Kim & R. D. Weaver, pp. 255–269

Wilkes, G. (1994). Germplasm conservation and agriculture. In *Biodiversity and Landscapes*, ed. K. C. Kim & R. D. Weaver, pp. 151–170

Wilson, E. O. (1988). *Biodiversity.* Washington, D.C.: National Academy Press.

World Bank. (1991). *The Challenge of Development.* World Development Report 1991. New York: Oxford University Press.

World Bank. (1992). *Development and the Environment.* World Development Report 1992. New York: Oxford University Press.

World Commission on Environment and Development. (1987). *Our Common Future.* Oxford: Oxford University Press.

17

Technology and biodiversity conservation: are they incompatible?

JOHN CAIRNS, JR.

Introduction

The major question considered in this chapter is whether we can develop a civilization that allows *wildness* to persist. By this I do not mean the wildness that sends big city teenagers on a rampage mutilating and sometimes murdering innocent persons nearby. I am using wildness in the sense that Thoreau used it; a natural system unadjusted by nature trails, parking lots, and concession stands to satisfy the needs of human society. Environmentalists have often been accused of not being open to compromise that is so characteristic of reasonable people. Unfortunately, all the major environmental compromises possible have been made. These have resulted in less than 3% of the globe's land mass remaining in what Thoreau would have accepted in his most charitable mood as wildness. A world that runs itself is fading from memory. In its place is a "managed" environment supervised by governments that cannot balance budgets. If technology and biodiversity are to coexist, it must be a technology that wildness can endure. Ecologists perpetually talk about the interdependence of things and lip service is given to this notion on Earth Day, but, in practice, we approach environmental problems one fragment at a time, not as a complex, multivariate, interdependent landscape. The coexistence of technology and biodiversity depends on switching from a fragmented to a landscape view.

The reintroduction of the wolf to a portion of its former range in the United States (even when this is designated a wilderness area) is an excellent illustration of the point just made. Despite the fact that the wolf is a relatively shy creature, fears of attacks on tourists, livestock, etc. may preclude the reintroduction or greatly restrict the number of sites for reintroduction. The fact that the wolf belongs there and that, even after the reintroduction, the truly wild areas will be geographically small and ecologically widely spaced is not persuasive to much of the general public or to decision makers.

The case of the spotted owl in the Pacific Northwest is an even better illustration of the intolerance for wildness. The spotted owl cannot injure people or livestock or damage anything of commercial value. Unfortunately for the spotted owl, it requires old growth forests of substantial size (i.e., wildness) and is, therefore, an obstacle to the lumbering industry and the unions that wish to preserve a few lumbering jobs for a short period of time. The wildness will lose and so, ultimately, will human society, which favors a short-term economic gain against the survival of a fellow species.

McKibben (1989) suggests in his book *The End of Nature* that humans are no longer required to adapt to nature and, therefore, nature, as a force high in the human consciousness, no longer exists. McKibben is, of course, not referring to tornadoes, volcanoes, hurricanes, floods, and the like, but to the habits of the buffalo, migratory fish species, medicinal plants, and the like. Humans no longer perceive their survival as dependent on understanding the habits and characteristics of the biological portion of natural systems (Weiner, 1990). Instead, man shapes these to his own ends and, as a species, appears confident that he is in control. McKibben's point is well worth attention because he is really discussing nature as a system rather than an array of individual components. Man's perception of being in control of nature is a serious error. As the former United States Surgeon General C. Everett Koop made abundantly clear, humans must either modify their sexual behavior in recognition of the rapid spread of the AIDS virus or die. Both individuals and political systems have been slow to respond. If, as Ehrlich and Ehrlich (1990) speculate, the AIDS virus mutates so that it is much more transmissible, the adjustments by both individuals and political systems will necessarily be phenomenal if society is to survive.

Furthermore, while McKibben's perception of human attitudes toward nature is on target for the majority of individuals, it is an unwarranted assumption that man can ignore with impunity the workings of nature. For example, changes of only a few degrees in the average global temperature, accompanied by changes in rainfall (not only amount but timing), may have enormous effects upon the world's agricultural systems (Schneider, 1990). Furthermore, as Ehrlich and Ehrlich (1990) note, human society exists today in both total number and quality of life as a result of the exploitation of the one-time bonanza of fossil fuels, fossil water, world forests, biodiversity, accumulated topsoil, and the like, all of which are being depleted at rates many orders of magnitude faster than replacement rates. We are told by growth-oriented economists that when one resource ends another will replace it. One cannot help wondering if the replacement will be as attractive as the original, or even provide the same service. The overall rate of global change and the high probability of specific drastic events, such as increasing the rate of global warming, indicate that the economists may be looking at fragments of the system rather than the entire biosphere. However, lost resources must not only be replaced, but the *quantity* must nearly double early in the next century just to maintain present per-capita levels. In the six seconds it takes to read this sentence aloud, 18 people will have been born. Clearly, shipping people into space, even if the technology for emigration of billions were available at a bearable cost, is not a feasible option. If scientists and engineers can make other less-hospitable planets habitable, why can they not seek to keep this one habitable? Neither of these can be accomplished if we continue on our present course.

Societal dependence on technology

The world's present population is heavily dependent upon existing environmentally unfriendly technology and should be rapidly developing new, less environmentally damaging technologies such as solar power. During the transition period, waste minimization (source reduction), recycling, and consumer restraints will buy time to develop new lifestyles and new technologies in a more cost-effective way. Since well over half the world's population lives in cities, the technology of transportation is essential to deliver food and building materials to them. Countries such as the Netherlands are, in this regard, more like big cities since they are heavily dependent on imports of foodstuffs; when the present Dutch gas fields are depleted, energy will also be imported as well. *In short, the life-support system for most human societies is both technological and natural*! On the other hand, the present use of technology is a definite threat to biodiversity, since the world's species are being depleted at an alarming, almost certainly unprecedented, rate (Wilson, 1988). In fact, it is quite possible that, unless dramatic changes occur in the use of technology, as much as half the species inhabiting the planet may be gone during the lifetime of some younger people now alive (Cairns, 1988). Few dispute the estimate of a world population of 10 billion sometime in the first half of the next century, barring some catastrophic event such as an epidemic disease. Man is dependent upon the world's biota for genetic stocks of agricultural cereals and the like, for templates for developing new medicines and drugs, for maintaining the equilibrium of ecosystems that permits reliable delivery of their services, and, of course, for aesthetic (and a variety of other) reasons. The book *Biodiversity*, edited by E. O. Wilson (1988), provides more detail on the benefits of maintaining global biodiversity. The benefits of technology have been expounded endlessly for most of this century. The question here is the compatibility of present and future technologies with the maintenance of global biodiversity.

The fragmentation problem

In human societies

Some members of society deplore any barriers to the use of technology for economic gain. Many people use technology in their daily lives but are increasingly uncomfortable with both the disastrous environmental effects and the control over their personal lives. Theoretical ecologists studying pristine systems who loudly proclaim the academic value of their studies and the ecological value of the systems being studied unhesitatingly use word processors, complex computers, satellite imagery, and sophisticated measuring instruments. Even "earth first"ers unashamedly use electronic amplification for their guitars and songs deploring earth's technologies. Finally, the newly developing field of restoration ecology regularly uses, as a matter of necessity, various types of technology with the intention of restoring damaged ecosystems to some semblance of their predisturbance condition. This includes the use of bulldozers, front-end loaders, and the like to restructure a stream or river to its predisturbance contours, elimination or reduction of hazardous materials stored in ecosystems, or the cleanup of spilled oil.

The lack of a holistic view is apparent everywhere. For example, the residents of La Trinidad de Dota (Kim Wilkinson, personal communication), a community of 30 houses in the Talamanca Mountains of Costa Rica, have fragmented the original forests of the area and are now putting such pressure on the remaining fragments as a source of fuel wood that even the fragmented system is almost certainly in jeopardy. This case is particularly poignant because villagers recognize the beauty of the forest fragments, and the older villagers want to see that their descendants are able to share this appreciation. One might excuse the villagers of Costa Rica who are destroying their forests for fuel wood since they feel they cannot afford alternative fuels and it is a traditional way of life. However, the same lack of a holistic perspective is evident in comparatively very wealthy societies, such as the United States. A stand of very ancient trees in Bowen Gulch near Rocky Mountain Biological Laboratory in Colorado, which represents one of the few such mature stands globally as well as in the United States, may well be sacrificed for short-term economic gains (High County Citizens' Alliance, 1989). Ironically, these trees are on federal lands and could easily be included in the Never Summer Wilderness or in the Rocky Mountain National Park or be used as an ecological connection between the two.

The lumbering industry claims that logging jobs will be lost if older forests on federal lands cannot be cut. Ironically, most lumbering jobs will be lost anyway because of increased mechanization in logging, a fact that lumber companies carefully conceal. More importantly, most logging companies are heavily subsidized by government funding and are far from being an economic asset. It is estimated (Robert Zahner, personal communication) that the logging companies using federally managed forests pay only 10 cents on the dollar of the actual cost, and taxpayers subsidize logging companies for the other 90 cents. At the very least, the federal government should be breaking even on the cost incurred by logging companies using federal lands; ideally, it should make a profit. Better yet, forests should be preserved for the use of the citizens of the country who are supposedly represented by Congress and the U.S. Forest Service rather than a special interest group whose power will decline once the primeval forests have been eliminated. Issues are so fragmented that citizens rarely realize they are subsidizing (via the U.S. Government) technologies that have a severe environmental impact. I have no objection to lumbering companies managing the forest lands they own as they choose. However, these areas are government lands belonging to all citizens who are heavily subsidizing lumbering technology without generally being aware of this. Until environmental management integrates all the issues, valuable long-term ecological assets will be lost for perceived (short-term) economic gain for technology based industries subsidized by tax dollars.

In the educational system

Most would like to enjoy the benefits of technology and the benefits of natural ecosystems fully. This should be possible if the human population explosion is brought under control in time, and our educational system is restructured so that those responsible for the technology and those responsible for restoring and maintaining the integrity of natural ecosystems are well aware of each group's activities and thoughts. Presently, the educational system effectively isolates the disciplines from each other so that engineers rarely understand ecologists and vice versa. When exchanges do occur, they are usually hostile, each side maintaining that the other side does not understand its views or problems. They are right! The reason for this is that the educational system (geared toward increased specialization as one progresses through it to the Ph.D. level) has set up a series of very effective isolating mechanisms, including prerequisites or rights of passage for obtaining degrees, that are carefully controlled by the tribal members and are so rigorous and time-consuming that they prevent the interactions so necessary for developing a more harmonious relationship between technology and natural systems. A report entitled "The University in the 21st Century" produced by the Commission on Higher Education in the Commonwealth of Virginia (Virginia Commission on the University in the 21st Century, 1989) calls this "The Tyranny of the Disciplines":

> The "tyranny of the disciplines" in American higher education is an extremely perplexing problem. We understand this phrase to mean that the academic disciplines and departments that support them define acceptable methods of inquiry and what it means "to know" something about ourselves and about the world. Discipline-based departments set the criteria by which research, scholarships, and teaching are evaluated and, as a result, how rewards are meted out to faculty members (promotion, tenure, salary increases, teaching schedules, research space, and so on). Membership in a discipline and the corresponding department, rather than in a particular college or university community, is the basis for many faculty members' professional lives.
>
> We think that this disciplinary-based system must change. The president of one Virginia university has observed that much exciting teaching and research called "interdisciplinary" is really a mark of shame: the present disciplines are no longer adequate to what we know and the problems we must solve. Nevertheless, they exert too much control in our colleges and universities. As a result, the rewards for working outside the established boundaries of the disciplines are limited. For junior faculty, unprotected by tenure, the sanctions often are fatal.
>
> The culture of higher education is broader and stronger than the colleges and universities of Virginia, which cannot change the disciplines by themselves.

Orr (1990) points out that those who lived sustainably in the Amazon rainforests and elsewhere on Earth could not read (but were not uneducated in the broad sense). In contrast, those whose decisions are destroying the planet are often very well educated. Clearly, it is the type of education (highly specialized, technology oriented) that is at fault, not education itself. Education should take students beyond analysis to synthesis. It should not permit compartmentalization of disciplines that isolates those responsible for development and utilization of technology from those interested in preserving and protecting Earth's biodiversity. But even this is not enough; both groups should base their decisions on a guiding set of beliefs (which the Greeks called "ethos") and a sense of equity or fairness. This would involve yet another set of disciplines in the humanities. The present system of "higher education" is shamefully inadequate in this regard. I am proud that the Commission on Higher Education in the Commonwealth of Virginia has condemned the isolation of the disciplines from each other and the factors that permit this dangerous practice to continue.

In the political system

Although the genesis of fragmentation in the functioning of a larger system is clearly due to the present organization of the educational system, the political system has raised fragmentation to a fine art. Lobbyists of special interest groups dominate the political system in this country, and, although opposing forces may reluctantly compromise, they do so predominantly to achieve some of their goals rather than to maintain the well-being of the larger system. In fact, short-term gain is the name of the game, whether it means a profit for a particular industry or getting re-elected. But the well-being of complex multivariate systems, such as Planet Earth, so that long-term sustained beneficial use is possible, requires a perspective measured in centuries, not months or years. This is evident in such situations as global warming. In this scenario, the probability of rates of temperature increase and climatic change capable of disrupting the agricultural system and the lives of people living in coastal areas justifies designing plans to reduce the probable rate of change as much as possible. Rather than offend special interest groups in the short-term, major gambles are being taken with long-term prospects. A harmonious relationship between technology and the maintenance of global biodiversity and integrity is simply not possible when a short-range view dominates and long-range planning is given either lip service or no attention whatsoever. The fragmentation of the political system into special interest groups is one of the major obstacles to developing a strategy for using technology without destroying the integrity of natural systems. Since nations seem unable to control the effects of fragmentation or the forces affecting global biodiversity, it comes as no surprise that the United Nations is no better at this, although some heartening efforts are underway. Just as the federal government of the United States must occasionally override states that are not acting in the long-term enlightened self interest of their own citizens or the nation to which they belong, so the United Nations may serve a similar function when nations are not acting to preserve the integrity of the life-support functions of the global ecosystem.

In the regulatory community

In the United States (and although I have not investigated each and every country, circumstantial evidence indicates that the U.S. is not exceptional in this regard), Congressional or state legislators fund agencies line item by line item. No substantive funding is given for integration, and, although sometimes interagency efforts are highly publicized, behind the scenes these intragency efforts generally represent turf battles between warring groups. Leopold noted in his superb Abel Wolman Distinguished Lecture, February, 1990, to the National Research Council:

> . . . the proliferation of public agencies dealing with water has led to a disassociation of their policies, their procedures, and their outlook from the operational health of the hydrologic system. Everything one entity does affects many other entities, yet each entity operates as if it alone is the flower facing the sun. There is no guiding belief, no

ethos involving the natural world. There is no concern for the common – as Garrett Hardin expressed it – no overriding responsibility for the whole.

Again, fragmentation is fatal to effective integrated environmental management, which requires that interactions of technology and natural systems be carefully documented and the impact of technology on global biodiversity and integrity of natural systems not be deleterious on a short- or long-term basis. Additionally, maintenance of the integrity of natural systems requires that they be fragmented as little as possible by highways, forestry practices, urban and regional developments, and the like (Soule et al., 1988).

The need for a national environmental agency

An agency is needed that would be responsible for the integrity and long-term well being of the nation's natural life-support systems. Fortunately, the formation of a national institute for the environment (NIE) is already underway (Saxton, 1990). There are other possibilities such as an organization from NOAA (National Oceanographic and Atmospheric Administration). Whatever organization is chosen, it should employ professionals whose interests and internal sense of responsibility transcend disciplinary boundaries. For example, Luna Leopold (1990) put ethos and equity first in his Abel Wolman lecture "Ethos, Equity, and the Water Resource." Leopold clearly realizes that the contributions of his discipline must fit into a larger system, including nonscientific values.

The frightening loss of the nation's wetlands is a clear indication of the undesirable consequences of a fragmented approach to environmental management. Estimates vary of the original wetland acreage in the United States when it was settled, since information is scattered and largely incomplete. However, a reliable account is 215 million acres for the conterminous United States. Thus, today's wetland resources in the lower 48 states probably represents less than 46% of our original wetlands (Tiner, 1984). Moreover, the wetlands have been drained, filled, and covered by airport runways, housing developments, and other artifacts of a technological society. Because wetlands have not been viewed as a crucial part of a hydrologic system, which includes groundwater, streams, rivers, and lakes, additional technology must be employed to manage floods and, in some cases, recharge groundwater to replace services nature was providing at no cost. President Bush's no-net-loss-of-wetlands announcement is essential to preserve the few remaining wetlands that might be used as models for restoring or replacing wetlands lost through short-sighted practices. One hopes that reports of a retreat from this statement because of circumstances in Alaska are unfounded. However, even if they are unfounded, no-net-loss assumes the present acreage and distribution of wetlands is adequate, which it almost certainly is not. Furthermore, an NIE or equivalent should target optimal, not marginally acceptable, quantity and quality of natural systems.

Because past events indicate federal agencies are unlikely to create a holistic, integrated environmental management system, the effects of fragmentation should be offset by the formation of an agency with powerful system-level and long-term views (e.g., Saxton, 1990). This nation is unlikely to optimize the benefits of technological and natural systems in any other way.

Communications technology and science superb – the skill of communications miserable

People are deluged daily with information transmitted by electronic means. However, recent surveys have shown that even the VCRs present in virtually every U.S. household appear bewilderingly complex to nearly half the people who have them. In contrast, communication, both visual and audio, seems simple. Unfortunately, this is far from the truth since most problems "communicated" by the mass media are multifaceted, multivariate, and much more difficult to analyze than the technology that transmits them. At a recent meeting of the American Chemical Society in Washington, D.C. (September 27, 1990), I participated in a special symposium on chemophobia. At this meeting, the curious fact emerged that while the general public was perceived by the chemists to have an inordinate fear of chemicals, numerous polls have shown that chemists ranked high in the public opinion polls in terms of trust, respect, and credibility. Yet the chemists' view of the hazards and risks associated with the production and use of chemicals was light years away from that of the general public. Not only chemists, but other scientists, members of the medical profession, and engineers, tend to be dissatisfied with the view of research and technology presented by the mass media. In contrast, notable contributions have been made by publicity to public perceptions of risk, such as the campaign by former U.S. Surgeon General C. Everett Koop describing risks associated with cigarette smoke (both active and passive) and the dangers of AIDS.

On the other hand, specialists often tend to provide evidence for the news media that strengthens the pet theory or private belief of the moment and, in many instances (for example, the banning of cyclamates or the scare that certain common foodstuffs might cause rheumatoid arthritis), is not scientifically justified (to be most charitable). Possibly, the evidence is even counter to the public interest, since laypersons generally fail to realize that science works best by disproving hypotheses, not by proving them. The use of fluoride in public water supplies has been vigorously, sometimes fanatically, opposed by the con fraction and nearly as strongly supported by elements of the pro side. Publicity in itself does not always lead to public understanding. A specialist may, when faced with questioning by the news media, shy away from it because it is difficult to deliver "media bites" that reflect a complex multivariate problem adequately. Clearly, one cannot place sole reliance for value judgments about safety and harm with scientists. Although hazard or risk assessment is a probabilistic exercise based on scientific evidence, perceptions of safety or lack thereof are value judgments based on individual perceptions of cost–benefit ratios.

Not only does science confuse the general public, but often itself. Professors are most likely to teach what is known rather than what is not. This blinds everyone to the areas of ignorance in which science is immersed. Global restoration of ecosystems damaged by anthropocentric activities is badly needed and, moreover, the need, although not universally accepted, is widely recognized. Nevertheless, ecologists have not developed the methodology to reconstruct damaged ecosystems. Most articles in ecological journals actually focus on populations rather than systems. We know how components of ecosystems function in many cases but not how the components function in the aggregate. Not only are the answers to the question unsatisfactory, but the question itself (i.e., how the whole system really works) is rarely asked. Furthermore, the isolation of the disciplines from each other into what amounts to tribal enclaves (each with its own language, territory, rites of passage, and other attributes), developed so that one tribal member can recognize another and, to a degree, exclude aliens. This has hampered the integration of knowledge on problems that transcend the capabilities of a single discipline, which is practically all of the world's major problems at present. Communication of science and engineering with the general public must be preceded by the capability and intent to foster communications among disciplines. This is probably the key to developing a harmonious relationship between technology and biodiversity, and evidence abounds globally that the relationship is worsening, not improving!

Recommendations

1. That a powerful national institute for the environment (whatever the title) be formed with power to integrate the various federal agency activities so that the integrity of the nation's ecosystems is not further destroyed.
2. That the number of ecosystems being restored to some semblance of predisturbance condition exceed by a wide margin the number of ecosystems being deleteriously affected or severely impaired by various types of tech-

nology; as a brief intermediate step while this adjustment takes place, that no-net-loss of all ecosystem services, area, and quality be enforced without qualifications.
3. That, since, as the report "The University in the 21st Century" notes, the disciplines are unlikely to integrate in such a way that students now in the system interact (e.g., engineers and ecologists), outside pressures must be applied to the educational system to enable integrative science and thinking to at least be on a par with reductionist science and specialization.
4. That substantive funds be allocated for integrating the probable effects of a particular technology with existing technologies and, most importantly, determining the relationship of the particular technology to the well-being of natural systems.
5. That a set of quality control conditions, explicitly stated, be developed for the nation's ecosystems; any technology that forces the attributes being measured outside acceptable boundaries must immediately adjust so that the deleterious effects are eliminated regardless of the economic consequences to the proponents of this technology.

Acknowledgments

I am indebted to Teresa Moody for transcribing the dictation for this manuscript and to Darla Donald for editorial assistance in preparing the manuscript. Michael Soule furnished some valuable information on the NIE and on fragmentation, as did Rob Atkinson for wetlands.

References

Cairns, J., Jr. (1988). Can the global loss of species be stopped? *Spec. Sci. Technol.*, **11**(3), 189–196.

Ehrlich, P. R. & Ehrlich, A. H. (1990). *The Population Explosion.* New York: Simon & Shuster.

High County Citizens' Alliance. (1989). *New Forest Plan Threatens West Slope Economies.* Flyer, published jointly by the High County Citizens' Alliance, P.O. Box 1066, Crested Butte, CO 81224 and Western Colorado Congress, 7 North Cascade, P.O. Box 472, Montrose, CO 81402.

Leopold, L. (February 1990). First Abel Wolman Distinguished Lecture, National Academy of Sciences, Washington, DC.

McKibben, B. (1989). *The End of Nature.* New York: Random House.

Orr, D. W. (1990). Is conservation education an oxymoron? *Conserv. Biol.*, **4**(2), 119–121.

Saxton, H. J. (1990). Earth Lab 2000 – a Congressional Initiative for a National Environmental Institute. U.S. House of Representatives, 324 Cannon House Office Building, Washington, DC.

Schneider, S. H. (1990). Global warming: Causes, effects and implications. In *On Global Warming*, ed. J. Cairns, Jr. & P. F. Zweifel, pp. 33–51. Blacksburg, Virginia: Polytechnic Institute and State University.

Soule, M. E., Bolger, D. T., Alberts, A. C., Sauvajot, R., Wright, J., Sorice, M. & Hill, S. (1988). Reconstructed dynamics of rapid extinctions of chaparral-requiring birds in urban habitat islands. *Conserv. Biol.*, **2**, 75–92.

Tiner, R. W., Jr. (1984). *Wetlands of the United States: Current Status and Recent Trends.* Newton Corner, Massachusetts: U.S. Fish and Wildlife Service, Habitat Resources.

Virginia Commission on the University in the 21st Century. (1989). The academic organization. In *The Case for Change*, Section VIII. Richmond, Virginia.

Weiner, J. (1990). *The Next One Hundred Years: Shaping the Fate of Our Living Earth.* New York: Bantam Books.

Wilson, E. O. (1988). *Biodiversity.* Washington, D.C.: National Academy Press.

18

"Emergy" evaluation of biodiversity for ecological engineering

HOWARD T. ODUM

Introduction

Older information of life (genetic biodiversity) and the newer information of human technological society are being combined in a unified, self-organizing process that Vernadsky called the "noosphere" (Vernadsky, 1929, 1944). To adapt human economy into the geobiosphere harmoniously requires reorganization of new and old information for symbiosis of whole-earth ecological engineering (Fig. 18.1). The term "ecological engineering" is becoming popular for designs of the new landscape that use information to unify ecosystems and human technology (Mitsch 1987; Mitsch & Jorgensen, 1989).

Principles of self-organization for maximum performance

The maximum power principle predicts that those designs of human economy which fit with nature for mutual survival will prosper. The main mechanism allows that the producers and the consumer-servicers reinforce each other (Fig. 18.1). This system develops efficiency, hierarchical roles, and division of labor. The patterns that succeed are saved as information in the form of genetic and learned biodiversity. By trial and error and sometimes by intelligent prediction, the human-aided self-organization of the earth and its economy move towards maximum, useful, resource utilization.

If human choices among alternatives are to be useful, the ongoing self-organization of human economy and environment requires quantitative measurement of contributions. This chapter explains the way "emergy" (spelled with an "m" – the work, measured in emjoules, contributed by the environment and human economy on a common basis; the available energy of one kind required to produce, copy, or maintain that information) and transformity quantitatively evaluate information of biodiversity and landscapes. Measuring the contributions of information and biodiversity in emergy units provides a

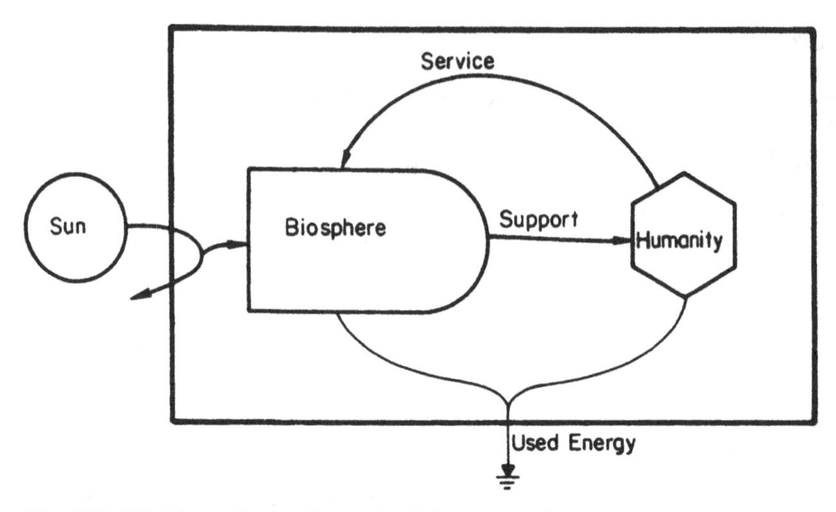

Fig. 18.1. World symbiosis of mutual reinforcement of economy and nature (Odum, 1986b).

scale of value for choices in the further organization of the biosphere consistent with the principle of maximum performance.

Unfortunately, the concept of maximizing profits that applies to the smaller scale of human individuals and business during growth periods is being incorrectly applied to the larger scale of environment and to the times close at hand when growth is not possible. Maximizing a person's income or a business's profit often pulls resources away from its greater contribution to the public economy, not recognized by market evaluation.

Money is only paid to humans for their service on a market price basis. The interface between environmental work processes and the economy is shown in Figure 18.2. The dashed line shows the pathway of money received for sales of environmental products returning to the economy (on the right) to pay for labor, goods, and services. None is ever paid to environment.

When resources, such as wood, fiber, food, and fisheries are abundant (from the left in Fig. 18.2), standards of living are high but prices are low. When resources are scarce, there are shortages, and standards of living are low. But prices are high because costs of effort to obtain and process resources are great. Market values cannot be used to evaluate environmental contributions, impacts, or policies. To evaluate the performance of the whole system of humans and environment requires a common measure, which can evaluate all the inputs of nature and the economy on a similar basis.

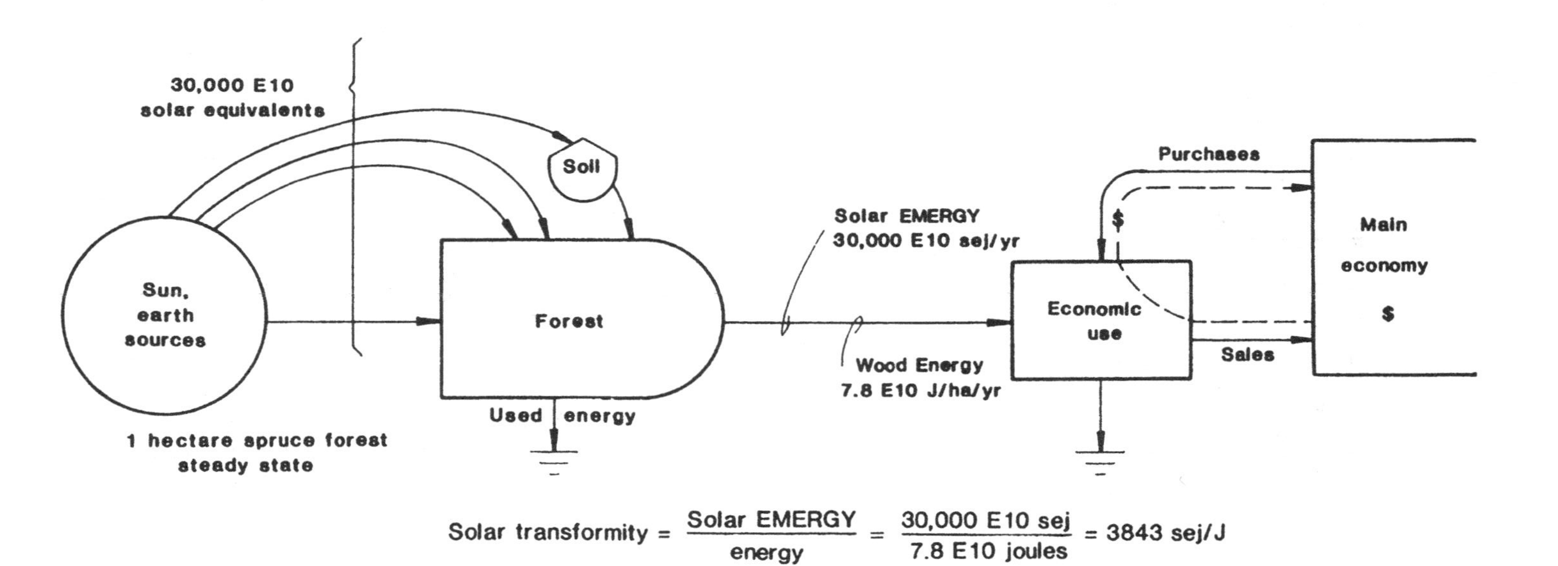

Fig. 18.2. Environmental–economic interface, forest example.

Emergy, transformity, and Em$

Emergy is the prior work in making a product or service, including contributions of resources, environment, and human labor. It is a more general measure of work than money. Emergy is the sum of direct and indirect inputs required for a product expressed in equivalents of one kind of energy. Here we use solar energy equivalents and express everything as solar emergy. The definitions of emergy and related indices introduced previously are presented in Table 18.1 (Odum, 1986a, 1987a, 1987b, 1988; Scienceman, 1987).

Solar transformity is the solar emergy per unit energy. It is a measure of position in energy hierarchy. Solar transformity increases with each successive energy transformation. The more stages of work required to make a product, the more emergy has been used and the less energy remains. For example, bears have higher solar transformities than grasses.

Having evaluated the total annual solar emergy budget for many countries,

Table 18.1. Definitions

Emergy = energy of one type required directly and indirectly to produce a product or service – expressed as emjoules

Solar emergy = solar energy required directly and indirectly to produce a product or service – expressed as solar emjoules

Transformity = emergy per unit energy – expressed in emjoules/joule

Solar Transformity = solar emergy per unit energy expressed in solar emjoules/joule; abbreviated sej/J

Em$ of a product = gross economic product $ multiplied by the proportion of national emergy contributed by that product

Net emergy yield ratio = emergy yield divided by emergy from the economy

Emergy investment ratio = emergy purchased from the economy divided by the emergy free from environment

Emergy exchange ratio = emergy received with a purchased product or service divided by the emergy of the buying power in the money paid for the product

Emergy of information
(1) of a single copy = emergy to make a copy
(2) of shared information = emergy to make and install duplicate copies so that there is a population with the same information
(3) to sustain information = emergy to maintain a population with shared information, including the necessary cycle of extracting, duplicating, testing, and selecting, which eliminates error and sustains adaptiveness
(4) to originate information = emergy required to sustain a population of the size and for the time necessary to evolve the new information from the precursor

we calculate the solar emergy per unit currency for each country. After converting the currency to international dollars, we calculate the solar emergy per dollar (Table 18.2). This is an index of the ability of that currency to purchase wealth.

Solar emergy contributions to a product can be expressed in "macroeconomic dollars" by dividing by solar emergy/$ ratio. The macroeconomic value of a product is the part of the gross economic product that is dependent directly and indirectly on that product (measured in Em$).

Also given in Table 18.1 is the net emergy yield ratio, an index for evaluating the contribution of primary sources of the economy (the fuels). Fossil fuels generally average about 6/1, which means that only 1/6 is used in processing the emergy, leaving 5/6 to operate other aspects of the economy. Because this ratio decreases as reserves are used up, this ratio will decline in the future, causing us to depend more on the renewable resources that require biodiversity for efficiency.

The investment ratio (Table 18.1) indicates whether a process can compete economically. It is the ratio of purchased emergy to free environmental emergy. To be economical the ratio has to be that of the regional economy or less. In other words, when the free inputs of environment are high, costs are less, and the economic use competes well. In the U.S. the ratio in 1992 was about 7, and it was less than 1 in the least developed areas of the Earth.

The investment ratio is also a measure of the environmental loading. The higher the ratio the more the impact of the economic development. Projects

Table 18.2. Emergy buying power for different countries in the 1980's.

Country	Solar emergy per $ (sejJ/$)
China	8.7
Ecuador	8.7
Brazil	8.7
India	6.4
World	3.8
Italy	2.7
United States	1.8
Japan	0.9
Switzerland	0.7

Modified from Odum & Odum (1983); Pillet & Odum (1984).

planned with high investment ratios are environmentally destructive and not economical. Controlling investment ratios keeps economies regionally uniform and protects their environmental basis. For example, a low investment ratio was calculated for the Everglades National Park (Debellevue, Odum, Browder, & Gardner, 1979), which suggested that it was underutilized, and thus subject to pressures for more development. On the other hand, plans for development in mangrove areas in the Cape Coral area of southwest Florida had very high investment ratios, suggesting that the density was too high and it would not be economical. Recent bankruptcy of companies involved on that coast tend to bear out the prediction.

Protecting the contributions of environment by ecological engineering symbiosis maximizes the economic vitality. The choice is not between environment or jobs. Maximizing environmental function maximizes jobs.

Energy hierarchy and whales

The geobiosphere runs on energies of solar insolation, the deep heat driving geologic convection, the tides, and the nonrenewable fossil fuels now being processed by society. Each of these is intimately coupled to the others. In Figure 18.3 a systems diagram shows the way these energies drive a successive chain of energy transformations of the energy hierarchy of the earth.

The quantity of energy flux diminishes with successive transformations. The emergy used increases, and the energy remaining is less. Consequently, the solar transformity increases. In Figure 18.3 items to the left have small territories, turn over rapidly, and have abundant, if low quality energy. Items to the right have larger territories, slow turnover times, and small quantities of high-transformity energy. Transformity increases up the energy hierarchy from the left to the right on the diagrams. Transformity indicates position in energy hierarchy of the universe. For example, baleen whales of the southern hemisphere oceans are major consumers of the krill (small pelagic shrimp), which are dependent on a food-chain hierarchy from sunlight to phytoplankton clusters to zooplankton and krill (Fig. 18.4). Data from papers by George Knox (1984) estimating original stock of these whales were used to assign a part of the earth's annual emergy budget to each whale (Table 18.3). Based on an average age of 15 years, an average emergy value was obtained with a macroeconomic dollar equivalent of 2.6 million 1990 U.S. dollars each. This is the share of the world's wealth used in maintaining each whale. Maintaining the whale population maintains the genetic information.

If, however, the species is threatened and only a few are left, what also has to be evaluated is the genetic information of evolving the first (or the last) copy

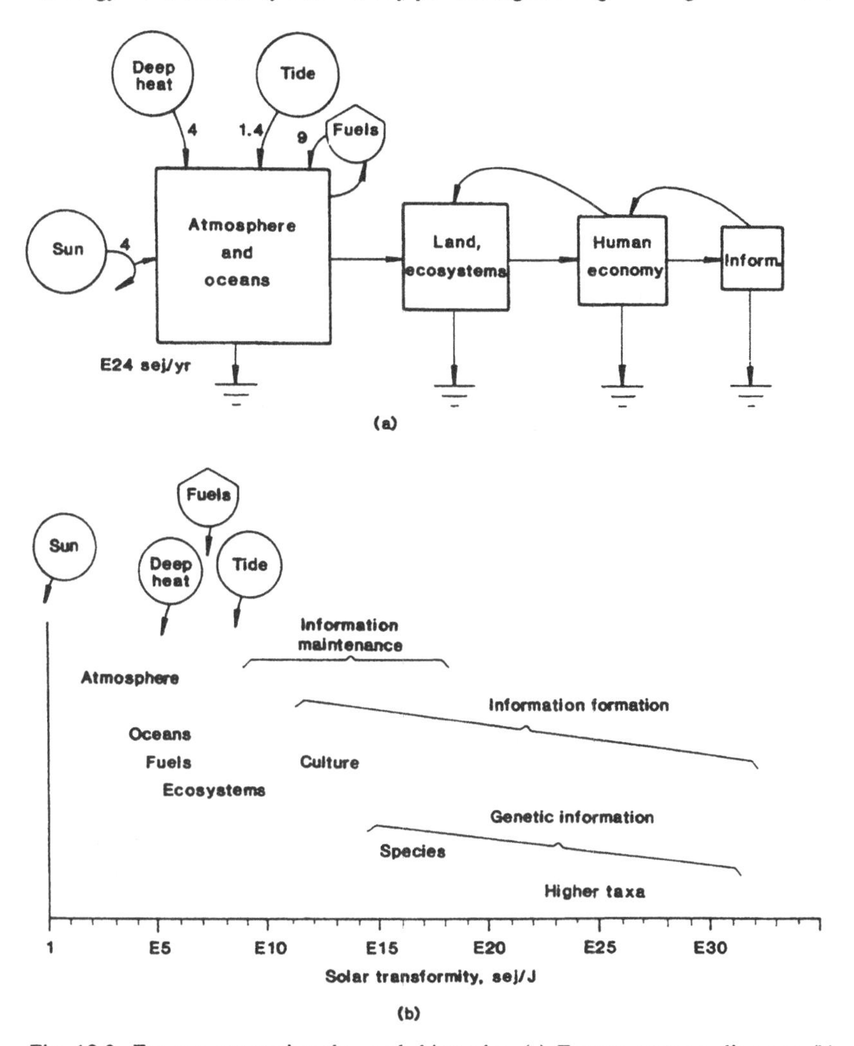

Fig. 18.3. Emergy supporting the earth hierarchy. (a) Energy systems diagram; (b) representation of energy hierarchy with a graph of energy types and solar transformity.

from its precursor. Evaluating biodiversity is a question of evaluating information, which has several aspects.

Emergy and information

Emergy and transformity provide a new scale for quantitatively measuring information and its utility. Emergy and transformity evaluations put informa-

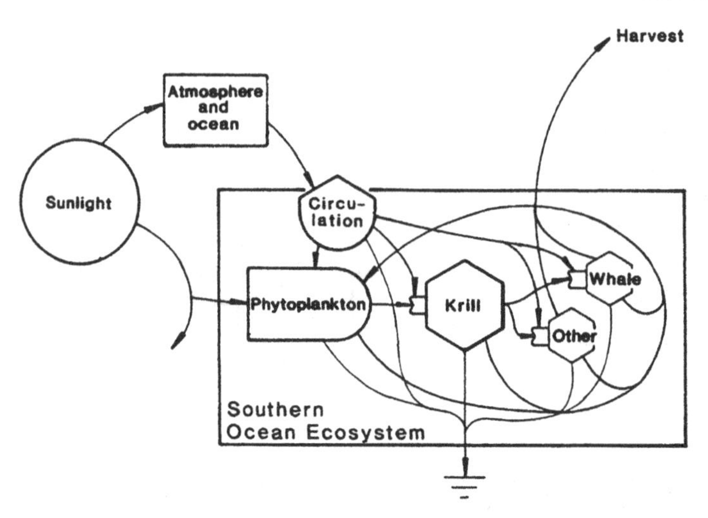

Fig. 18.4. Energy overview for evaluation of a southern hemisphere baleen whale (Odum, 1987b).

tion and biodiversity in perspective with other aspects of ecologic and economic systems, such as energy and materials.

Information of a system may be defined as the identification of parts and connections of a system. Information is useful if it can make a system operate. Examples of useful information are the genetic codes of living organisms, the complex organization of an ecological system, and the cultural information of human societies. Information may be extracted from the system it operates and stored or transported in compact form. Examples are the genetic information in seeds, information in books, messages by telephone, and archeological artifacts

Table 18.3. Emergy evaluation of one southern hemisphere whale.*

Area of ocean per whale[a]	3.8 E7 m^2
Annual solar emergy supporting each whale[b]	3.5 E17 sej/yr
Emergy stored in average whale[c]	52.5 E17 sej
Em$ (solar emergy/2 E12 sej/U.S. 1992$)[d]	3.8 E6 $

*Modified from Odum (1987b).
sej = solar emjoule; E7 = 10^7.
[a]Area of 3.8 E13 m^2 of area between antarctic convergence and the south polar ice which originally supported one million baleen whales (Knox, 1984).
[b]Half of the solar emergy per whale territory = annual solar emergy of the earth, 9.44 E24 sej/yr, times the fraction of the earth's area per whale (3.8 E7 m^2/5.1 E14 m^2).
[c]Average lifetime, 15 years times the annual emergy supporting one whale.
[d]Emergy per whale, 52.5 E17 sej divided by the 1992 emergy/$ ratio, 1.4 E12 sej/$.

in ancient deposits. Knowledge is information which may be utilized and understood by human minds.

Information in a system may be saved, copied, improved, transmitted, and used to make and operate new systems, whereas emergy may be used to evaluate each phase of information processing. Several information-related transformities are defined and illustrated in Tables 18.1 and 18.4 and in Figure 18.5.

Emergy for making a copy of a system that contains information is not large (Fig. 18.5a). For example, some emergy is required to copy (reproduce) a tree. Emergy to extract the information from a system is not large (Fig. 5b) For example fruits are made by a tree. Emergy to duplicate extracted information is not large. For example, many seeds may be made in a fruit. Emergy to transmit information may not be large. For example, tree seeds are dispersed in the forest by wind or birds.

More emergy is required to make enough copies, transmit them and establish them in a large population, so that the same information is widely shared, with larger impact. For example, a species of tree becomes widely established in a region. Even more emergy may be required to sustain that information, through the cycle of extraction, testing, and reapplication necessary to keep the information adapted. Maintaining information and its utility requires an information cycle (Fig. 18.5e).

Much more emergy is required to generate useful information from scratch than to maintain information. This is why systems store useful information once it is developed. Information is continuously lost by the dispersal and depreciation processes usually described as the "second laws of thermodynamics." Loss of the last copy is extinction.

Even when isolated in compact form, information requires some form of energy as a carrier such as that in the DNA of seeds, the paper of books, or the electromagnetic waves of radio transmission. The emergy requirements for developing information for a functional, competitively surviving system is large, but the energy of its carrier is small. Consequently, the transformities of information (emergy per unit energy) are very large. Shared information has large territory, slow replacement time, and high transformity. The highest transformities are for the replacement of the last copy, since this represents the formation process (Fig. 18.3b). Transformities of endangered species may be useful in putting their values in perspective.

Power and efficiency are maximized by maintaining and evolving information, which includes extracting, reproducing (copying), transmitting, reapplying, sharing, testing, and thus repairing or revising information by selection. Ecosystems and biogeochemical processes of the earth are now configured and

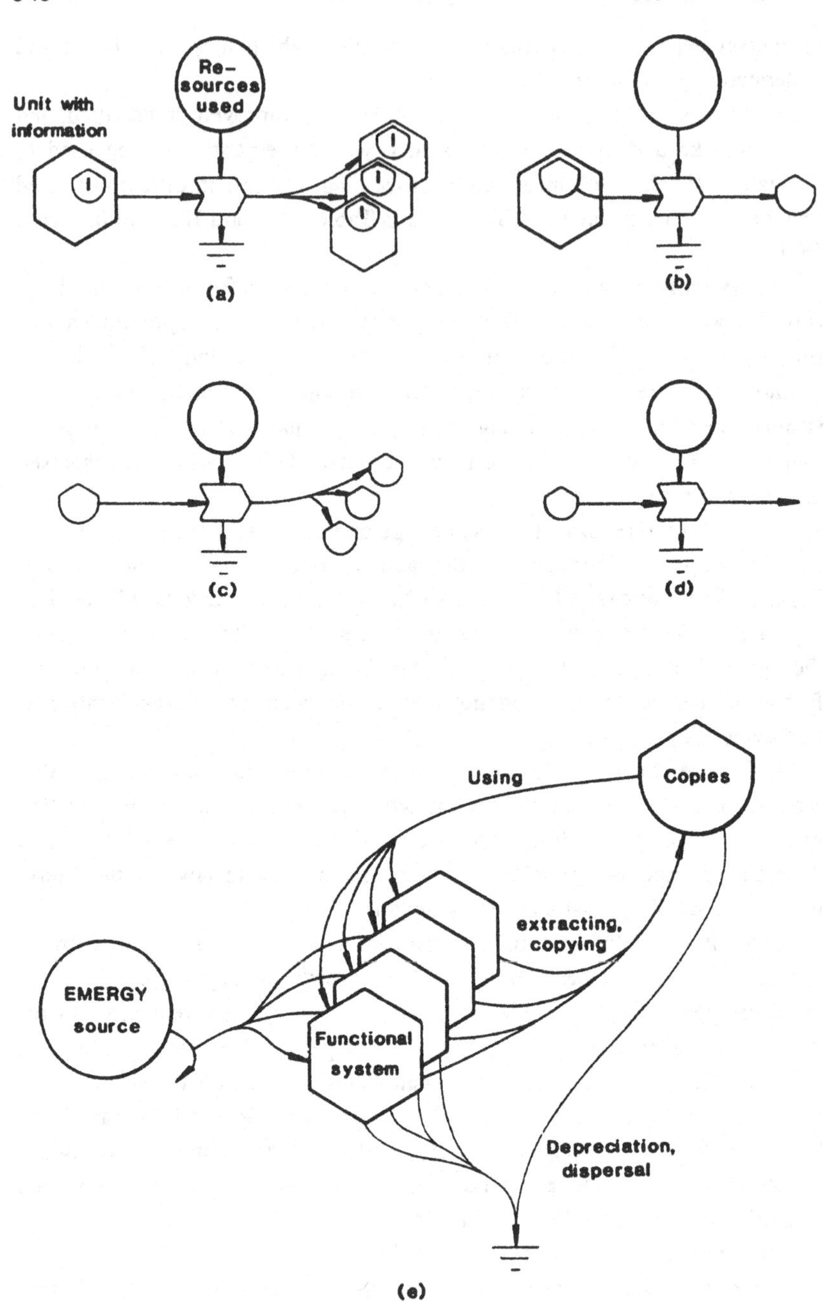

Fig. 18.5. Information processing. (a) Copying units that contain information; (b) extracting information from its operational system; (c) copying information; (d) transmitting information; (e) information cycle for sustaining and evolving information.

Table 18.4. Kinds of transformity for information with rainforest examples.*

Name	Calculation	Example	Solar transformity (sej/J)
Copy of a unit containing information[a]	Solar emergy to make a copy	Growing a tree	3.2 E7
Extracting information from unit[b]	Solar emergy to isolate information	Growing seeds	1.13 E9
Maintaining sustainable population[c]	Solar emergy per year to support cycle copying, testing, etc.	Supporting tree species population	7.26 E11
Formation of species information[d]	Solar emergy for years of evolution	Supporting tree species for time of its evolution	4.8 E15

*Data from El Verde, Puerto Rico (Odum, 1970); DNA data from Canoy (1970). Sej = solar emjoule; E7 = 10^7.

[a]Annual emergy support to the forest, 6.0 E14 sej/hectare/yr, divided by the estimate annual flux of DNA reproduction based on 1.5 turnover times of leaves and other tissues per year. Using DNA in El Verde forest from Canoy (1970), (600 mg DNA/m^2) (0.001 g/mg) (1E4 m^2/ha) (5 kcal/g) (4186 J/kcal) = 1.25 E8 J/ha; annual DNA flux is (1.5) (1.25 E8 J/ha) = 1.88 E8 J/ha/yr.

[b]DNA flux in seedfall is (292 J/m^2/d) (5 E-4 J DNA/J seed) (1 E4 m^2/ha) (365 d/yr) = 5.33 E5 J DNA/ha/yr solar emergy 6.0 E14 sej/ha/jr divided by DNA flux.

[c]One tree species out of 153 species occupying 19,648 ha of Luquillo forest. (6 E14 sej/ha/yr) (0.0065 for one species) (19,648 ha) = 7.7 E16 sej/yr/species. One tree out of 3140 trees per 4 hectares, larger than 10 cm diameter, occupies (4 E4 m^2/ha)/3140 trees = 12.7 m^2/tree. DNA flux per tree: (12.7 m^2/tree) (0.6 g DNA/m^2) (5 kcal/g DNA) (4186 J/kcal) (1.5/yr turnover) = 1.06 E5 J/tree/yr. Emergy supporting population divided by DNA in one tree: (7.7 E16 sej/yr/species)/(1.06 E5 J/tree/yr) = 7.26 E11 sej/J DNA.

[d]Species formation taken as 10,000 years with sustenance emergy from c (7.7 E16 sej/yr/population) (10,000 yr) = 7.7 E20 sej/population. Using DNA content of one individual, solar emergy to evolve one individual: (7.7 E20 sej/pop.)/(1.59 E5 J DNA/ind.) = 4.8 E15.

controlled with living information. Large storages of information from earlier evolution are the basis for a stable and efficient biosphere.

Rainforest examples

Data from the Luquillo rainforest of eastern Puerto Rico were used to illustrate definitions of the solar transformities of several kinds of information products (Tables 18.4 and 18.5). Solar transformities increase with copying, extracting, transmitting, sharing, sustaining, and evolving. The solar transformity and macroeconomic dollar value of the last (or first) copy of stored genetic information are very large. These figures reinforce the qualitative feelings of those appraising environmental systems in other ways.

The useful information in any system is that which was seeded from the past plus that which was developed in the self-organizational processes. For ex-

Table 18.5. Emergy and macroeconomic value of rainforest trees at El Verde, P. R. (Annual solar emergy per hectare, 6.0 E14 sej/ha/yr)*

Item	Emergy (sej)	Macroeconomic value† (1990 US $)
sej = solar emjoule; E7 = 10^7.		
Plantation monoculture, 10 years, 1 ha[a]	1.5 E15	750.
Mature forest, 300 years old, 1 ha[b]	1.53 E17	90,000.
Annual support of all 153 tree sp., 19,648 ha[c]	1.17 E19	5.9 E6
Mature forest, 300 years old, 19,538 ha[d]	3.51 E21	1.755 E9
Sustain 1 species, 128 ha, 1 year[e]	7.7 E16	38,500.
Evolve 1 species, 128 ha, 10,000 yr[f]	7.68 E20	3.8 E8
Evolve 153 tree species, 10,000 yr, 19,648 ha[g]	1.18 E23	5.9 E10
Average tree, 12.7 m^2 area, age 50 years[h]	3.8 E13	19.
Dominant climax tree, 300 yr, 500 m^2 crown[i]	4.5 E15	2,250.
Emergy of species diversity per bit[j]	4.9 E13	24.7

*Preponderance of solar emergy that rain used: (2,140 g/m^2 water transpired/day) (5 J free energy/g) (365 d/yr) (1 E4 m^2/Ha) = 3.9 E10 J/ha/yr; solar transformity of rain 1.54 E4 sej/J (3.9 E10 J/ha/yr) (1.54 E4 sej/J) = 6.0 E14 sej/ha/yr.

†Solar emergy divided by 2 E12 sej/U.S. 1990 $.

[a]Formation in 10 years with average solar emergy half of that used at the end of growth which is half the metabolism of the mature forest: (6.0 E14 sej/ha/yr) (0.5) (0.5) (10 yr) = 1.5 E15 sej/ha

[b]Average emergy used in formation taken as that after 100 years. (6.0 E14 sej/ha/yr) (300) = 1.8 E17 sej/ha.

[c](6 E14 sej/ha/yr) (19,648 ha) = 1.17 E19 sej/yr/forest.

[d](6 E14 sej/ha/yr) (19,648 ha) (300 yr) = 3.5 E21 sej/forest.

[e](6 E14 sej/ha/yr) (128 ha) = 7.7 E16 sej/species.

[f](6 E14 sej/ha/yr) (128 ha) (10,000 yr) = 7.68 E20 sej/species.

[g](6 E14 sej/ha/yr) (19,648) (10,000 yr) = 1.17 E 23 sej/forest species.

[h]assumed average tree over 10 cm as 50 years: (6 E14 sej/ha/yr) (12.7 E-4 Ha) (50 yr) = 3.8 E13 sej/tree.

[i](6 E14 sej/ha/yr) (0.05 ha) (300 yr) (0.5) = 4.5 E15 sej/tree.

[j]Shannon diversity 4.6 Bits/individual (N. Sollins, 1970); 785 individual trees per hectare. The total tree bits in whole sustaining forest is (785 ind./ha) (19,648 ha/forest) (4.6 bits/ind.) = 7.09 E7 bits/forest solar emergy for whole forest from item d divided by forest bits: (3.5 E21 sej/forest)/(7.09 E7 bits/forest) = 4.9 E13 sej/bit of trees.

ample, ecological succession represents the influx of the seeded information plus that in the pattern of species developed by the selective processes during succession. These include some self-selection by competitors at the start, followed by reinforcement selection by cooperative parts of the system in later succession. The pool of genetic diversity as a storage of information reinforces productive processes through its many reinforcing feedbacks from right to left (Fig. 18.6). Maintaining a pool of high diversity apparently is a protection against interruption of productivity during changed conditions in regional economics as well as within smaller ecosystems (Zucchetto, 1981).

Emergy and diversity

In ecology, as in many other fields, complexity has been evaluated as the logarithm of the possible combinations and called "information." The field of "information theory" has evaluated letters in messages, occupations in a city, structures in a molecule, etc., expressing the result in bits (logarithm to the base 2), nits (logarithm to the base e), or Hartleys (logarithm to the base 10). One of the most commonly applied formulas, the Shannon-Weaver-Wiener index (Shannon & Weaver, 1949) evaluates the bits per individual in the distribution of units and classes of units. Information indexes have also been used to evaluate the complexity involved in their interconnectedness in ecosystems organization.

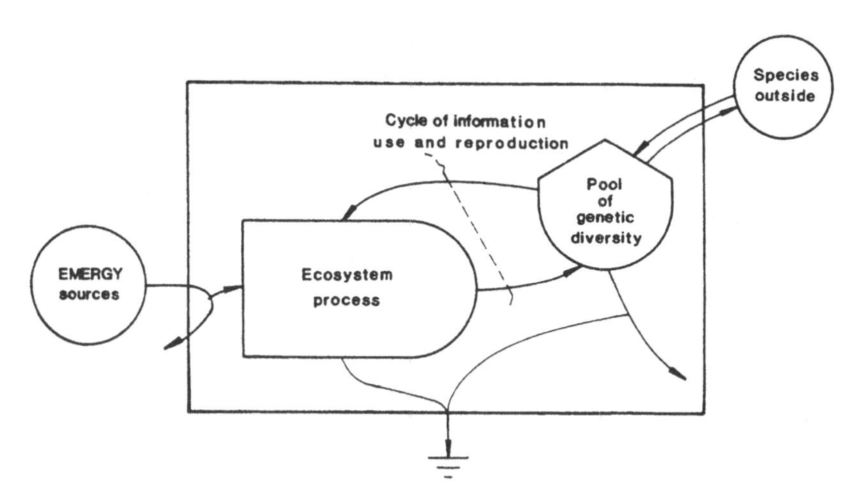

Fig. 18.6. Emergy basis for the ecological system of biotic functions, mutually symbiotic with the pool of genetic information and diversity.

These "information" measures do not indicate whether the units are organized or whether they are yet to be organized, but the bits that are organized and useful can be expressed in these units and subtracted from the bits of total complexity to find the part that is not organized or specified.

Margalef (1956) initiated the application of these information measures to ecological species diversity, and there is now a large literature and abundant data on the bits involved in the complexity of species diversity. There is still a schism among ecological scientists about the extent to which a high species complexity represents useful organization, or whether it represents disorderly lack of organization.

The solar emergy required to sustain diversity in an ecosystem may be inferred from field data comparing high and low information ecosystems in similar environmental circumstances. In Table 18.5 a moderate respiration of a forest plantation is compared with the higher metabolism of the complex naturally organized forest nearby at El Verde Puerto Rico. The difference in emergy between the two is what accompanies the higher diversity. Hence the calculation of emergy/bit for the trees of that forest (Table 18.4).

Transformity for comparing information bits of different scales

At the molecular level, complexity measured with the information measure is entropy (Boltzmann's equation), (Boltzmann, 1886), increasing with the number of molecular possible states (N) and thus with temperature:

$$S = K \log N$$

Information, in bits, etc., can be calculated for molecular scale and any other scale of size in natural hierarchy such as enzymes, cells, organisms, species, landscape units, stars, etc. A means of relating this one mathematical measure across levels of size has been needed. As we suggested recently (Odum, 1987a, 1987b, 1988), the emergy required to sustain a bit of information provides a scaling factor for comparisons. Much less emergy is required to maintain a bit in a small system in a bowl compared to the same bits of complexity for units of landscape size. Keitt (1991), using data from remote sensing of the Luquillo Rain Forest, showed the increasing emergy per bit required in changing to a larger-scale view.

Placing an emergy evaluation on bits of complexity also provides a value measure for complexity. A set of information that has been through a process of becoming selected to operate a useful system will have a much higher emergy than a happenstance "random" set of characters with the same bits.

Human culture and genetic information

Late in evolution the human has emerged as the earth's information-processing specialist. Humans have highly developed information-processing organs (brains) and stored programs for joining individual capabilities through social mechanisms and institutions. The worldwide sharing of language and symbols provides very large territory and fast turnover times for human information, genetic and learned.

The system of humans, their genetic inheritance and their learned information can be represented hierarchically (left to right in Figure 18.7). As a start on the process of evaluating culture, a preliminary emergy calculation was made in Table 18.6 and compared with that for genetic information. The emergy in copying genetic information of humans is no larger than that in learned information, but the emergy of formation (first or last copy) is very large. Stored genetic information has a replacement time on a geological time scale.

Culture is an information storage of shared information among a group of people. In Figure 18.7 culture is the learned information part of the system of native people. The total emergy supporting the system (Table 18.6) is primarily that from the transpired rain, the earth cycle, and contributions from marine resources. Assuming a steady state, an estimate of human metabolic energy can be related to the solar emergy basis to provide solar transformity of people. A smaller fraction of the metabolism may be identified with the informational processing between people. A smaller part of the metabolism maintains the genetic information. For each of these there is a solar transformity (Table 18.6).

Multiplying the annual flows times the time required to develop the storages provides the emergy of stored information (genetic fraction and cultural assets

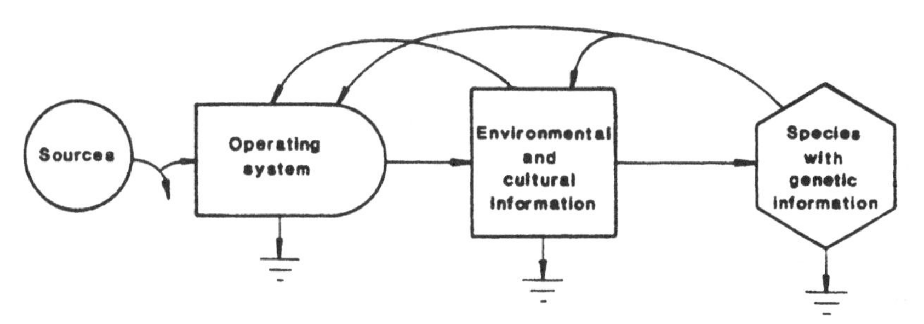

Fig. 18.7. Energy systems overview of the hierarchical interaction of biotic process, human culture, and genetic heritages; arranged from left to right in order of solar transformity.

Table 18.6. Emergy of native culture of New Guinea based on renewable emergy (H. T. Odum, S. Doherty, and M. T. Brown).

Item	Emergy flux (sej/yr)	Energy flux (J/yr)	Solar transformity (sej/J)	Em$* (E9 $/yr)
Annual Flux:				
Renewable sources[a]	1052 E20			52.6
Human metabolism[b]	1052 E20	1.57 E16	6.7 E6	
Information flux[c]	1052 E20	5.73 E15	1.8 E7	
Genetic flux[d]	1052 E20	6.3 E10	1.7 E12	
Steady-state storage:	sej	J	sej/J	E12 $
Human population, 3.5 E6[e]	3.74 E24	9.98 E4	3.5 E9	1.74
Cultural information[f]	3.47 E25	2.29 E13	1.51 E12	17.4
Human DNA (genetic info.)[g]	1.05 E28	2.1 E12	5 E15	5240

*Emergy divided by 2 E12 sej/U.S. 1990$.
sej = solar emjoule; E7 = 10^7.
[a]Renewable emergy for Papua New Guinea (Doherty, 1990), 1052 E20 sej/yr.
[b]Metabolism of population: (3.5 E6 people) (2927 kcal/day) (4186 J/kcal) (365 d/yr) = 1.57 E16 J/yr
[c]36.5% of the human metabolism for social interaction and learned information: (0.365) (1.57 E16 J/yr) = 5.73 E15 J/yr.
[d]49 year life expectancy; average birthing age, 16 yr; generation: 49–16 yr; 33 yr turnover time (0.03/yr); human DNA from item g: (0.03) (2.1 E12 J DNA) = 6.3 E10 J/yr.
[e]Energy storage in population: (0.2 dry) (454 g/lb) (3.5 E6 people) (150 lb ea) (5 kcal/g) (4186 J/kcal) = 9.98 E14 J. Emergy storage: (1052 E20 sej/yr) (33 yr) = 3.47 E24 sej.
[f]Storage of learned population information (culture) based on 10 generations with social information flux from item d. (1052 E20 sej/yr) (10 generations) (33 yrs per generation) = 3.47 E24 sej. Information carrier storing information as 2.3% of biomass 9.98 E14 J (item e).
[g]Human DNA: (2.1 mg DNA/g dry) (0.2 dry) (454 g/lb) (3.5 E6 people) (150 lb ea) (0.001 g/mg) (5 kcal/g) (4186 J/kcal) = 2.1 E12 J human DNA in population: Genetic differences from precursor stocks generated in 100,000 years and 3 generations per year: (1052 E20 sej/yr) (100,000 yr) = 1.05 E28 sej.

fraction). Very large solar emergy and equivalent macroeconomic dollars result (Table 18.6).

Mitigation and wetlands

To facilitate economic development, laws in some areas call for mitigation, the substitution of one area for another based on judgments about equivalent values. Giampetro and Pimentel (1991) suggest comparing landscapes by expressing their stored attributes as the product of their area and time of development, thus in units of hectare-years. The work of nature on one hectare for 100 years is equal with this procedure to the work on 100 hectares for one year. This simple procedure assumes that all areas of the earth have the same work done per unit time, which is not very accurate, since energy flows in nature are distributed hierarchically, being high on beaches and in active volcanoes.

Emergy evaluation provides a more refined way to quantitatively relate one area to another according to the work done by nature and humans in making that landscape (Tables 18.4 and 18.5). For example, an emergy evaluation was made of Florida salt marsh by Odum and Hornbeck (1992). A comparison of salt marsh with rainforest is made in Table 18.7. Whereas the salt marsh has more annual emergy value because of the large tidal contribution, the rainforest has much the larger stored emergy. Mitigation has to consider the storages which are involved in the short run as well as the annual values for evaluating the long-term contributions.

Emergy evaluation of foreign trade of environmental products

Most environmental products contribute more wealth (emergy) to the buyer than the seller receives in the emergy buying power of money received in payment (Fig. 18.2). This is because payment is only made for human services, not for nature's work. Thus when fish, timber, or bananas are sold in international trade, much more emergy goes abroad than is received in buying power. New Zealand ships the emergy of its water power to Japan as aluminum ingots at a fraction of their value to the public economy (Odum, 1984). Figure 18.8 shows the poor emergy exchange for shrimp from ponds in Ecuador shipped to the U.S. and also the bad balance of emergy in all the other exports as well (Odum & Arding, 1991).

If the resources were not exported but kept at home, wealth at home would

Table 18.7. Emergy evaluation of ecosystems to mitigation. E14 solar emjoules/hectare.

	Plantation forest *	Rain forest*	Salt marsh[†]
Annual contribution	3	6	13
Stored in ecosystem:	15	1530	467

*See Table 18.5.
[†]Data from Odum & Hornbeck (1992).

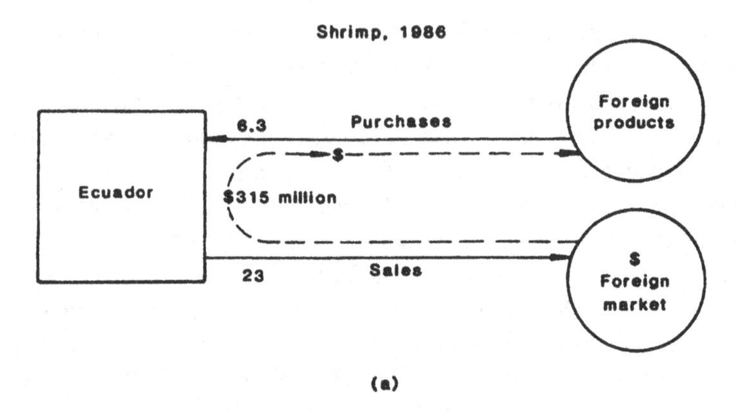

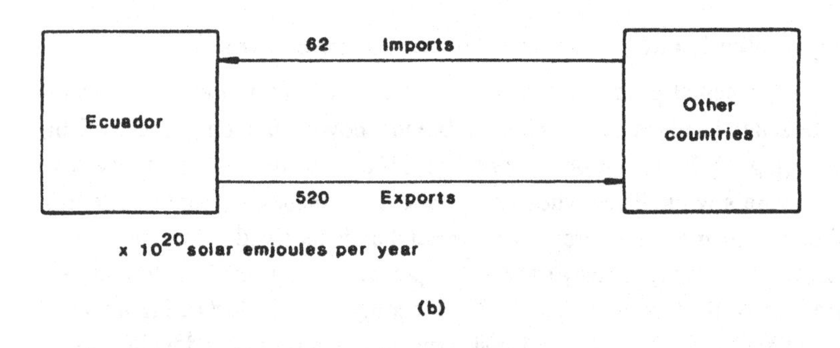

EMERGY EXCHANGE

Fig. 18.8. Emergy exchange between Ecuador and developed countries in 1986. (a) Sale of shrimp; (b) all export sales (Odum & Arding, 1991).

be larger, standards of living would rise, more money could be printed without inflation, and their currency would become stronger on world markets.

When emergy evaluations were made for different countries (Table 18.2) large differences were found. The rural (less developed) countries circulated less money, its people receiving more of nature's services direct from the environment without any money being paid to anyone. Labor by a person there is more environmentally subsidized, allowing them to work for less cash. If people there working for less cash export a manufactured product, they are sending more emergy value abroad.

If a country with a high emergy/$ ratio such as China, Brazil, or Ecuador borrows a million dollars from one with a lower ratio such as U.S., Japan, or Switzerland, it borrows a million dollars of foreign buying power. Later, it pays it back with its own buying power, which is much larger. Thus, the payback may really be 300 to 800% interest, quite enough to destroy the economy of the undeveloped country and enough to cause wasteful affluence in the developed country. In the struggle to pay back the absurd usury, lending institutions call for stripping of natural resources and destructive environmental projects that take emergy away from the local public use and send it abroad with only a handful of local people profiting. Displaced local people have higher emergy investment ratios than the local economy, so that the products are too costly to sell at home. The local business that sells the products abroad becomes a part of the foreign economy, an isolated enclave of the developed economy in an undeveloped country.

Economic policy and biodiversity

Emergy evaluation shows that the stripping of tropical forests, coral reefs, and the rest of the world's remaining resources is being driven by loaning and investing from the developed areas. Peoples in both lending and borrowing countries are being tragically misled by the erroneous concept that maximizing individual profit helps the public good.

The situation is potentially fatal for the lending institutions and the political regimes of the countries as well. Everyone involved becomes committed to unsustainable stripping of the rest of the world. This inequitable neocolonialism will soon stop, partly because these countries get wise and partly because the resource reserves (forests, soils, minerals, fisheries, and aesthetic tourist attractions) are finally exhausted.

In the U.S., the last of the great virgin forests of the northwest (spotted owl habitat) being shipped to Japan also involves major imbalance of emergy exchange. There would be many more jobs in the U.S., lower costs of housing, and a conservation of biodiversity if the forest wealth were used in the U.S. on a renewal basis. With the lower emergy/$ ratio of Japan compared to the U.S., any transaction between U.S. and Japan drains the U.S. economy. Similarly transactions between Ecuador and the U.S. drains the wealth of Ecuador without equitable compensation.

For maximizing world process and productivity, emergy equity is required between countries. Treaties concerning trade are needed, but not along the lines of those at GATT in 1994, trying to force underdeveloped countries to follow free-market policy, which the above analysis shows is wasteful exploitation.

Instead, international equity can be established with exchanges made according to balance of emergy. This would allow reasonable economic development with uniform investment ratios, maximum sustainable prosperity, and the joining of countries in mutualistic trade partnerships that prevent war. All countries with foreign debts could recalculate their obligations using the emergy basis (Scienceman, 1987). It is likely that many of these debts considered on a real wealth basis have already been paid off.

Emergy evaluation of Taiwan shows how the best economic development comes from using resources at home, importing only information, and not borrowing. Taiwan went from an undeveloped country to one with emergy indices equivalent to the U.S. in 25 years (Huang & Odum, 1990).

Until we can get a new paradigm established substituting planetary welfare for personal profit and growth, the future of world biodiversity is in the hands of money-oriented, profit-maximizing, policy makers, who knowingly or unknowingly are destroying the earth's heritage and their own futures. Emergy evaluation of environments and their biodiversity is a means for developing more international equity and a more prosperous fit of environment and economy.

References

Boltzmann, L. (1886). Der Zweite hauptsatz der mechanischen Warmtheorie. *Almanach Det K. Acad. Wiss. Mechanische Wien,* **36,** 225–259.

Canoy, M. (1970). Desoxyribonucleic acid in rainforest leaves. in *A Tropical Rainforest* ed. H. T. Odum & R. F. Pigeon, pp. G69–G70. Division of Technical Information, U.S. Atomic Energy Commission, TID2470.

Debellevue, E., Odum, H. T., Browder J., & Gardner G. (1979). Energy analysis of the Everglades National Park. In *Proceedings of the First Conference on Scientific Research in National Parks,* vol. 1, ed. R. M. Linn, pp. 31–43. Washington, D.C.: National Park Service.

Doherty, S. J. (1990). *Policy perspectives on resource utilization in Papua New Guinea using techniques of EMERGY Analysis.* Masters Project Paper, Department of Urban and Regional Planning, Univ. of Florida. Center for Wetlands, Gainesville, FL.

Giampietro, M., & Pimentel, D. (1991). Energy efficiency: assessing the interaction between humans and their environment. *Ecol. Econ.,* **4**(2), 117–144.

Huang, S. L., & Odum, H. T. (1990). Ecology and economy: EMERGY synthesis and public policy in Taiwan. *J of Env. Mgt.* (1991) **32,** 313–333.

Keitt, T. (1991). Hierarchical Organization of Information and Energy in a Tropical Rainforest System. Masters Thesis, Environmental Engineering Sciences, Univ. of Florida, Gainesville, FL.

Knox, G. (1984). The key role of krill in the ecosystem of the southern ocean with special reference to the convention on the conservation of Antarctic marine living resources. *Ocean Mgt.,* **9,** 113–156.

Margalef, R. (1956) Information Theory in Ecology. *Mem. R. Acad. Cienc. Artes Barcelona,* **23,** 373–449 (translation in *Soc. Gen. Syst. Yearbook* **3,** 36–71).

Martinez-Alier, J. (1987). *Ecological Economics.* Oxford: Basil Blackwell.
Mitsch, W., & Jorgensen, S. (eds.). (1989). *Ecological Engineering, an Introduction to Ecotechnology.* New York: Wiley.
Odum, H. T. (1970) Summary: An emerging view of the ecological system at El Verde. In *A Tropical Rainforest,* ed. H. T. Odum & R. F. Pigeon, pp. I191–I289. Division of Technical Information, U.S. Atomic Energy Commission, TID2470.
Odum, H. T., & Odum, E. C. (eds.). (1983). *Energy Analysis Overview of Nations.* Working paper WP-83-82. Laxenburg, Austria: International Institute for Applied Systems Analysis.
Odum, H. T. (1984). Embodied energy, foreign trade, and welfare of nations. In *Integrations of Economy and Ecology, an Outlook for the Eighties,* ed. A. M. Jansson, pp. 185–200. Stockholm, Sweden: Asko Laboratory.
Odum, H. T. (1986a) Emergy in Ecosystems. In *Ecosystem Theory and Application,* ed. N. Polunin, pp. 337–369. New York: Wiley.
Odum, H. T. (1986b). Las nuevas energia y los ecologistas: razonar o sentir. Prospectiva del Ano 2000. In *El Hombre, la Sociedad, el Mundo,* pp. 201–222. Barcelona, Spain: Fundacion Caja de Pensiones.
Odum, H. T. (1987a) Living with Complexity. In *Royal Swedish Academy of Sciences, Crafoord Prize in the Biosciences,* pp. 19–85. Stockholm, Sweden.
Odum, H. T. (1987b). What is a whale worth? *The Siren – News from UNEP's Oceans and Coastal Areas Programme,* No. 33, May 1987, 31–35.
Odum, H. T. (1988). Self organization, transformity, and information. *Science,* **242,** 1132–1139.
Odum, H. T., & Arding, J. E. (1991). EMERGY Analysis of Shrimp Mariculture in Ecuador. Working Paper prepared for the Coastal Resources Center, University of Rhode Island, Narragansett, RI.
Odum, H. T., & Hornbeck, D. (1994). EMERGY evaluation of Florida salt marsh and its contribution to economic wealth. In *Proceedings of Salt Marsh Conference, 1989, Florida A & M University, Tallahassee* (ed. C. Coultas). Gainesville, FL: Univ. of Florida Press.
Pillet, G., & Odum, H. T. (1984). Energy externality and the economy of Switzerland. *Schweiz Zeitschrift for Volkswirtschaft und Statistik,* **120**(3), 409–435.
Scienceman, D. (1987). Energy and Emergy. In *Environmental Economics,* ed. G. Pillet & T. Murota, pp. 257–276. Geneva, Switzerland: Roland Leimgruber.
Shannon, C. E., & Weaver, W. (1949). *Mathematical Theory of Communication.* Urbana, IL: Univ. of Illinois Press.
Sollins, N. (1970). Comparison of species diversity at El Verde with some other Ecosystems. In *A Tropical Rainforest,* ed. H. T. Odum & R. F. Pigeon, pp. I245–I247. Division of Technical Information, U. S. Atomic Energy Commission, TID2470.
Vernadsky, V. (1929). *The Biosphere.* Abridged and reprinted (1986), London: Synergetic Press.
Vernadsky, W. I. (1944). Problems of Biogeochemistry II. The fundamental matter-energy difference between the living and inert natural bodies of the Biosphere. *Trans. Conn. Acad. Arts & Sci.,* **33,** 483–517.
Zucchetto, J. (1981). Energy diversity of regional economies. In *Energy and Ecological Modelling,* ed. W. J. Mitsch, R. W. Bosserman, & J. M. Klopatek, pp. 543–548. Amsterdam: Elsevier.

19

Urban horticulture: a part of the biodiversity picture

HAROLD B. TUKEY, JR.

Introduction

Urban horticulture is plants for cities – functional uses of plants to improve urban environments for the benefit of the people who live there. Functional uses mean screens against unpleasant views and against headlights. It means essential food and special nutrition in the human diet – fruit and vegetable production in cities is important. It means effects upon noise and air pollution and on the amelioration of climates in urban areas. It also means plants to improve the human psyche – plants and gardens are good for people.

Why should we talk about plants in cities? Because most of us live in cities and as world populations expand, more and more people are gathered in tighter and tighter clusters. It includes us all. We get most of our information about city plants from observations and anecdotes, from nurserymen, from landscape architects, and from people who manage properties in the city. For example, the civil engineer in the city knows more about root growth of tree species than does a research scientist because he has to repair the clogged sewer lines and broken pavement caused by urban trees.

Our most substantive information comes from agriculture and forestry. And although the principles of plant growth can be demonstrated equally well with an apple tree, a Douglas fir, or a chrysanthemum, strategies developed from these plants don't apply well when we try to manage rhododendrons in gardens and parks, or a grouping of sweetgum trees along a city street, or the collection of common plants in a backyard, or the rushes and cattails of an urban wetland. Indeed, the whole concept of commercial agriculture is different from urban horticulture. Agriculture talks about large acreages and single cultivars, uniformity, and precision. A poinsettia plant must come into flower before Christmas, not on December 26. The azalea must be ready for Mother's Day. The tomatoes must all be ripe at the time the big harvesting machines are available. Commercial forestry has similar requirements.

But that's not the way we do things in our urban area. The *Acer palmatum* in my front yard on the south side is very different from the *Acer palmatum* you have in your yard on the east side, and your's is very different from the one just down the street. We don't talk about stands of trees as the foresters do; we talk about single plants and assemblages of plants, entirely different concepts in management from commercial agriculture and forests.

Urban horticulture

Urban horticulture is an integrated science utilizing the expertise of other people and disciplines. It is based on good botany, soils, ecology, chemistry, and physics. It's based on the artistic concepts of perception and space, color and texture. It's also based on social considerations of business management, law, personnel, and even conflict resolution. All of those things are integrated into the profession which we call urban horticulture, which includes research, graduate and undergraduate teaching, cooperative extension, and management.

What are some of the specific aspects of research in urban horticulture? There is environmental horticulture, which means good plant husbandry; how do plants grow in cities? The professor of urban environmental horticulture and his students measure rates of photosynthesis of street trees that grow in the city, an environment that is much different from the commercial nursery area where many street trees are grown. What rates of photosynthesis and respiration are there in a tree that receives its light from eleven to one in the middle of the day in the concrete canyon in the city? Not surprisingly, we learn that those photosynthetic levels are different from what we expected; the leaf number and chloroplast number are different. In fact, the tree growing on a city street is entirely different from the same plant grown in a commercial nursery. Nursery-grown trees should be acclimated to the city, much as foliage plants from the tropics must be acclimated to indoor environments. Such simple facts save thousands of dollars for street-tree managers. The people we work with are very clever, and anything we learn is adapted quickly into their management techniques.

Next is basic plant physiology, especially stress physiology. The urban physiologist and her students are interested in communications in plants. How do the stems and leaves know what is going on in the root zone? Root growth is a serious limitation in the city. There are problems of aeration, flooding, poor soil, or no soil at all, human vandalism, and designers who try to put large trees into tiny holes in city pavement with underground utilities. Roots sense that there is a problem and a signal is transmitted to the top that says "don't grow." These are apparently chemical signals, in very minute amounts, which require the help of biochemists to identify and quantify, and perhaps genetic engineering to correct. There are similar reactions and signals caused by attacks by pests, by extremes in temperatures, and moisture gradients. When we know what these signals are, our management friends will be able to manipulate city plants to great advantage, just as our colleagues in commercial agriculture maximize production through plant and environment manipulation.

Plants are for beauty, for enjoyment, and for function. Plant systematics brings order to plants and their naming and classification, which is critically important in marketing of plants by the multimillion dollar nursery industry. But horticultural taxonomy is more – it is plant exploration and plant introduction, not the "bring 'em back alive" experiences of former generations, but the careful selection of germplasm introduction only after careful evaluation of invasiveness, and proper selection. Selection must include knowledge of the urban environments, not just because a plant looks nice on a Korean hillside or on a Himalayan slope.

Pest management with city plants is entirely different from pest management in agriculture. The number of different plants is much greater in urban horticulture, with 125 or more species in a typical Seattle backyard. Each plant flowers at a different time and is susceptible to specific pests about which relatively little is known. Chemicals aren't the total answer; they are hazardous, and often too expensive, and in some locations, are prohibited by law. Solutions sometimes proposed by groups that call themselves environmentalists are emotionally satisfying, but don't manage pest problems or urban plants effectively.

This suggests that integrated pest management (IPM) is the best solution. The first step in integrated pest management is plant selection – the right plant in the right location for the right functional requirement. The first person in an IPM scheme should be a horticulturist. The second step is a change in attitude, one of toleration of some pest problems, rather than absolute cleanliness and control. These two considerations, proper selection and toleration, will be of immense value in developing effective approaches to pest management in cities.

Another discipline in urban horticulture is landscape ecology. This includes restored landscapes, land reclamation, assemblages of plants in gardens and botanical gardens, wetlands, plants along transportation rights-of-way, including city streets, and home and institutional plantings. The principles of community and restoration ecology apply equally well to city situations, but relatively few scientists have taken an interest in the urban area. Yet, the management of urban areas, big and small, benefits from information derived from ecological studies, as in water usage, vegetation management, and species diversity. In addition, we are interested in how plants grow together and how they should be arranged. The landscape architect on the faculty of the University of Washington, Seattle, helps to design the research plots and teaching gardens so that assemblages of plants demonstrate management solutions as well as an attractive setting. A landscape architect is a fully accredited member of a faculty in urban horticulture, and mixes easily with the biological scientists and helps the teachers and educators to interpret plans visually. Other areas of research include soils and nutrition, genetics, plant breeding, education, communication, and business management and administration.

Of course the best research goes for naught if the results are not communicated. Continuing education and cooperative extension are absolutely essential. The Center for Urban Horticulture at the University of Washington, Seattle, has seventy faculty, staff, and students. Yet, more than 50,000 people a year come to our buildings in programs of education in horticulture. That is in part because we have good programs. But mostly it is because people want to learn about horticultural plants, and we can provide a convenient and supportive location. We facilitate, we cooperate, and we are now the largest outreach effort at the University.

Values of plants and gardens

The philosophers, poets, artists, and some knowledgeable horticulturists have extolled the enriching, ennobling, and therapeutic values of plants and gardens. It's an old saying that gardeners make good friends. Yet we continue to talk about ornamental plants and amenity plants as if these are not essential to the people who use them. Let me assure you that there is very little point in talking ornamental anything to municipal and state leaders, or even campus administrators whose daily decisions include public health, the homeless, regional growth and planning, and drugs and crime. Therefore, I talk about functional plants and what they do in the environment. But in many cases, we are not sure. Why are plants good for people? Is there an advantage to placing indoor plants in an office, as some studies show? Do hospital patients recover more quickly with views of plants and nature around them? Why is it that when I turn off the freeway and into the Arboretum, I can feel it. My shoulders relax, my breathing is easier. Is blood pressure reduced, serotonin levels affected? These are questions not answered by professional horticulturists, but rather by the medical profession, including clinicians, rehabilitative and educational psychologists, and physiotherapists. It is time to demonstrate conclusively in scientific terms the essentiality of plants and gardening. Today there is much interest in this newly developing field of plant/human interactions, and organizations such as the American Association of Horticultural Therapy and the People–Plant Council are fully accredited and have well attended annual meetings and workshops.

Urban horticulture is unique in another way – in the diversity and excellence of the audiences it talks with. Some of these audiences have never worked with academic university personnel before. Some have never worked with each other, and in fact may have an antipathy for some allied professionals.

We deal with design professionals – landscape architects, city and regional planners, developers, and those in transportation. Landscape architects need help on the horticultural aspects of their designs – plants and plant specifications and performance. Regional planners understand the value of greenbelts, parks, and open spaces, but they do not understand the management of the plants that make up these spaces.

We talk with plant managers, those who manage landscape plantings around institutions and business establishments – a very large and very lucrative profession. These people have never dealt with a horticultural department to find out about fertilizers, water, pruning, and other kinds of plant husbandry activities.

We talk with parks departments, estate gardeners, and city arborists. The plant managers with the biggest budgets are in public utility companies, companies with strength in electrical engineering, accounting, and law. But these large firms know almost nothing about plants, especially trees, although the effect of trees on their operating costs is very large.

We deal with regulatory organizations, such as the U.S. Corps of Engineers, departments of natural resources and agriculture, the Environmental Protection Agency, departments of transportation and engineering, all of which have mandates for managing plants without very much knowledge about plants. So-called environmental groups (the Audubon Society, Nature Conservancy) are part of urban horticulture; they are interested in management of biological resources, especially plants, and require information on good management systems which we in urban horticulture can and should provide.

We also talk to politicians and other government leaders, few of whom have extensive knowledge of plants but who understand the value of parks and gardens, open space, property improvement, and support of an enlightened electorate that insists on environmental sensitivity and an improved quality of city life.

A most interesting and influential audience is made up of the thousands of dedicated horticultural hobbyists in hundreds of plant societies. No one from these groups has attended the symposium on biodiversity upon which this book is based. Yet these people know more about their particular plants than anyone else I know of. The local chapter in Seattle of the American Rhododendron

Society is arguably the best informed group on rhododendrons in the country. Similarly the Native Plant Society, the societies interested in ferns, bonsai, dahlias, begonias, fuchsias, and the local support groups for arboreta and botanical gardens have their own programs of education through international symposia and publications, extensive gardens, and plant exchange.

Urban horticulture and biodiversity conservation

Now what does all this have to deal with a discussion of biodiversity? Urban horticulture brings a diversity of people and organizations who are fully aware of problems of the environment, including biodiversity, and are already playing a role, but are sometimes not recognized or appreciated by professionals in the field.

Every major city and many smaller towns in America and around the world has a botanical garden. It is supported by lay people who realize that one of the cultural amenities of any community is a botanical garden. We have one in Seattle, a very famous one, the Washington Park Arboretum. All botanical gardens have four major roles – conservation, education, research, and display (recreation). In the past, botanical gardens were sometimes like Noah's Ark with two of everything. That may still work in some places, but it isn't appropriate in most, and it doesn't work in Seattle where we can grow a great variety of temperate zone plants but have only 200 acres and a limited budget to work with. Today, responsible botanical gardens temper their enthusiasm for exotic plants from other places with conservation of the native flora in their area. We are learning how to propagate native species, grow them, and introduce them into local landscape plantings, sometimes with important savings of water and fertilizers, and avoidance of costly methods of pest control. Botanic gardens also have a large role to play in biodiversity. There are still plant explorers; our people go to the winter-rainfall region of Chile, which hasn't been explored extensively, looking for plants that will fit niches in our program of responsible plant introduction.

Maintenance of germplasm is an important consideration. As discussed in this volume, critical efforts are being made at maintenance of diversity in commercial agricultural and horticultural plants. However, the germplasm for city plants is also very narrow, and no one is providing any funds for gene banks of city plants. There are no deep freezes, no government subsidies. Our germplasm maintenance programs are in the formal living collections in botanical gardens and arboreta, the informal collections in conservancies and preserves, and in the wild. We who work with plants for cities have many more plants to be concerned about than our colleagues in agriculture, yet there are few people who are making a specialty of conserving plants for functional uses in cities.

There are other groups of people who have a role in biodiversity. When the plant explorers brought plants back to Europe, they were supported by owners of estate gardens. Some of these gardens still exist with important living collections. There are people of means who are interested in plants, sometimes rare and unusual plants. One such person in Seattle has tried to recreate a forest floor. Now the ecologist might laugh, but this woman has learned her horticulture the hard way by growing plants in her own yard, bringing plants together from different parts of the world to create an absolute masterpiece. They are all labeled, described, and catalogued. She is not alone; there are similar garden gems throughout the world, all with important implications for maintenance of biological diversity, but completely unknown to the policy formulators and the discussion leaders.

Plant societies are very active within their specialty. For example, the Rhododendron Species Foundation of Tacoma, a private organization, has collected all the species of rhododendron that can be grown in the Puget Sound region, a very large representation of the genus. The collections have been assembled in a fine display garden; species are propagated and offered for public sale, and the Foundation conducts its own programs of education and research. Many private organizations have similar efforts in many locations in the world, with excellent programs focused on one plant or small groups of plants.

Urban horticulture also brings together different assemblages of people, often groups and individuals who have never met and never worked effectively together. Take a local forestry forum as an example. We invited all the managers of urban trees to come together in one place to share mutual problems. We were not sure who would come; we knew only a few names, and had never worked professionally with most. But they came – municipal and state arborists, landscape and park maintenance personnel, national tree services, landscape architects and other design personnel, educators, commercial arborists, researchers, representatives of the U.S. Forest Service and the Washington Department of Natural Resources, garden writers, local politicians and State legislators, and most important, tree managers from public and private utilities. All of these people did not know each other, some had territory they did not want to share easily, and none had worked effectively with academics in a university department of research and teaching.

But they talked, and found that all share some common concerns and problems. For example, a major utility company will provide street trees free of charge or at a low cost if municipalities and individuals will plant species selected by the company to be compatible with utility lines. The initial cost of proper trees is much less than the later costs of maintenance of improper selections. Today we have an active Forestry Forum which shares ideas and resources, supports research and education on tree selection and industry management, and is providing State leadership in tree planting initiatives. Here is a new resource, gathered around urban horticulture and urban trees, which brings knowledge, influence, finances, and enthusiasm to environmental concerns.

This kind of mutual support is very important, especially with the current emphasis on world environment. The United States has reacted to environmental concerns with nationwide initiatives of tree planting. Even the President mentions tree planting, and the White House staff is leading one of the major efforts. Yet all this enthusiasm, the tremendous amounts of money that are being raised, new mandates for government and state agencies is based on very little horticultural information about urban trees – selection, planting, pests, and plant husbandry. Fortunately, the national leaders of tree initiatives recognize that they do not have the expertise and must look elsewhere for technical assistance. Thank goodness, for there is no point to all the excitement and effort if we are going to do the job badly. Here is a chance for making a real impact on environmental problems, biodiversity, germplasm maintenance,and quality of urban life, a place in which urban horticulture and urban forestry is taking a leadership role.

Urban horticulture brings plants to the discussion of biodiversity, not the handful of species that is the basis of economic agriculture, but plants in great numbers. At the Washington Park Arboretum alone we have 5500 taxa. We talk more than plant numbers, we deal with assemblages of plants – botanical gardens, residential and institutional landscapes, along streets and highways, abandoned areas and landfills, greenbelts, and wetlands. I had management responsibility for 250 acres of choice land bordering Lake Washington in the middle of Seattle – 200 acres at the Arboretum and fifty acres of naturalized grassland on top of Seattle's old garbage dump. These are now multi-use areas, each with its particular style of research, education, and public outreach. The rushes and invading species on the grassland are very different from the Douglas fir and red cedar that form the overstory of the Arboretum, and the management of the purple loosestrife that threatens the ephemeral ponds is different from the ornamental cherries that attract thousands of visitors each spring. These are all urban plants, requiring urban plant management.

Habitats and landscapes

Urban horticulture also addresses differences in habitat. In the city, habitat is often in very small areas. On the naturalized grassland on top of a landfill and in the adjacent wetlands are plants and animals that were not there before – mostly exotic species. There is a particular group of pheasants, there are eagles and great blue herons, and swans. There are coyotes, yes, coyotes in the middle of the city, apparently adapted to urban life on a major university campus.

The ephemeral ponds in the grassland are beneath a major migratory flyway, and have attracted waterfowl that aren't commonly found elsewhere. It is easy to know when something new and interesting has flown in because looking out my office window, one notices groups of people, all with binoculars and cameras raised.

All habitats are not native, or naturalized. The landscape of a major city is a habitat management opportunity. When you fly over the Seattle area you see trees and greenery, which means birds and animals tucked away in small habitat zones. We in urban horticulture don't know how to manage it well, wildlife scientists and agencies don't know how, and ecologists don't know either. Sadly, many professionals don't consider city habitats worthy of study and understanding. But urban spaces and habitats should be added to the discussion of biodiversity.

Urban horticulture for people

Perhaps the most important advantage that urban horticulture can offer is people. We deal with everyone, because almost everyone lives in an urban area. These are good people, smart people, people who have contributed dollars and have been taxed to support things that they think are good. They've built our Center and Arboretum with $15 million of private funds. They are voters who write to their legislators, attend public hearings, make political contributions, and lobby. Thus, they have access to decision makers. That is different from some groups who support concerns such as biodiversity, who are not always effective because their position may seem to be extreme, or too self-centered. We have friends who provide us with clout, access, and dollars, and the will and perseverance to make substantive changes.

Another important segment of urban horticulture audiences is young people. I have two grandsons in the new generation, just starting school, and just becoming aware of the world around them. They are smart and they know many more things than we ever knew at the same age. And they can do many things that we couldn't do. They are environmentally aware and committed; they recycle; they plant trees; they save whales. When a five-year-old gives up his beloved French fries in response to an environmental call, that is truly commitment. They balance opposing options better than their elders – they know we must plant trees, but they also know the products we depend on that come from trees. These are the young people that come to the Arboretum, who are interested in plants, conservation, and the environment. These youngsters are an important audience for urban horticulture.

All of this is what we talk about in urban horticulture. In the pressure-cooker situations of modern cities where people are put together in closer and closer continuity, plants apparently provide a civilizing influence. It means a geranium on the back stoop, a plant in the pilot house of the dirty, crowded ferry boat crossing the Pearl River in Canton, it is the ivy growing from a window in a 1000-year-old wall overlooking the Coliseum in Rome, and a tree for a schoolgirl on Arbor Day. It means a pea-patch vegetable garden, a kitchen garden, a rooftop planting, or a shopping mall display. It means the great estate gardens, parks and landfills, native plants, city planning, and open space. These are all things we talk about in plants for cities.

There is a place for urban horticulture and its audiences in discussions about biodiversity. When we talk about natural systems, don't forget the systems that people are working in and live in – they are natural too. People cause the problems, but they also provide the opportunity for improvement through their numbers, influence, interest and commitment, and enthusiasm. Plants in cities are a part of biodiversity.

Acknowledgment

The financial support of the Department of Horticulture of The Pennsylvania Sate University, Dr. S. J. Wallner, Chair, is acknowledged gratefully.

20

The watchdog role of nongovernmental environmental organizations

M. RUPERT CUTLER

From personal experience I know that, with the President's support, U.S. federal political appointees can help save the biosphere. But I've also learned that, without the help of private environmental groups throughout the political process, from the nomination of candidates for elective office through the legislative program-authorization and budget-appropriation processes, government officials' hands are tied in this regard.

Initiatives taken by nongovernmental environmental organizations (NGOs) have been and will be critical to the success of the biodiversity conservation movement. When I was President Carter's point person on forestry and soil conservation policy (Assistant Secretary of Agriculture for Natural Resources and Environment, from 1977 to 1980) I saw to it that the National Forest Management Act of 1976 was interpreted to require protection of National Forest biodiversity. With the President's support I initiated the second roadless area review and evaluation of the 191-million-acre National Forest System (RARE II) which identified some 3,000 National Forest wilderness-protection opportunities throughout that system in forests from Puerto Rico to Alaska. The result: President Carter asked Congress to double (from 12 million acres to 24 million acres) the amount of National Forest System land in the National Wilderness Preservation System, and Congress eventually adopted most of those recommendations. But without the help of the environmental NGO community throughout the Forest Service administered RARE II inventory and evaluation phases and the follow-on legislative process, those executive-branch efforts would have been fruitless. The NGOs took the recommendations made by the Carter Administration, modified them, and won from the Congress statutory protection for millions of acres of intact ecosystems (e.g., Allin, 1982; Roth, 1984).

Experience has shown that environmental protection requires endless political pressure, endlessly applied. That's why environmental NGOs are essential: to institutionalize that source of pressure so the work load can be shared and passed from one generation of volunteers to another. The successes of the first generation of American environmental organizations have been described by many authors, including Roderick Nash (regarding Yellowstone National Park and New York's Adirondack Forest Preserve) (Nash, 1982), Henry Clepper (regarding the National Forest System) (Clepper, 1971), Ira Gabrielson (regarding the National Wildlife Refuge System) (Gabrielson, 1985), and Horace Albright (regarding the National Park System) (Albright, 1985). An essential ingredient in these political/environmental success stories was the supportive involvement of a diverse array of interest groups. Agricultural and railroad representatives desiring stable water flows from western mountains (less flooding from logged-off mountains) supported the creation of a public forest system, women in the Audubon movement protesting the use of bird feathers on women's hats were behind creation of the first wildlife refuges, and big game hunters represented by the Boone and Crockett Club saw the need to save the last American bison and pronghorn in Yellowstone Park.

We've heard a lot recently about predicted changes in global weather patterns. Let's consider another kind of climate change – a change in the political climate. Needed, if global biodiversity is to be protected, is a political climate in which elected and appointed government officials feel free – ideally, feel obliged – to support the policy changes needed to protect ecosystems for the long term, because of constituent pressure. That's the special role and responsibility of private environmental-protection advocacy organizations. In the United States some of them specialize in the protection of a particular geographic area such as the Greater Yellowstone ecosystem, the wilderness "canoe country" of northern Minnesota, and the Florida Everglades. Others are national or international in their focus and actions agendas. They have in common the goal of saving our genetic heritage.

Using house organs (magazines and newsletters), legislative alerts, the mass media, the upraised voices of their individual members, and the threat of litigation, environmental NGOs inform and involve their members and the general public. Their goal is to encourage legislators to pass and agencies to implement new laws and rules to better protect our biological diversity heritage. All such groups – international, national, and local in focus – make important contributions to the global campaign. Locally focused NGOs represent examples of the well-known dictum to think globally and act locally.

Without political help from nongovernment environmental organizations, even the most dedicated public administrator cannot withstand the predictable pressure to exploit commodities and pursue other economic-development opportunities associated with natural areas. Countervailing pressure has always been the name of the environmental-protection game. The environmental movement is a protest movement, organized to protest damage to our life support system. Its credo is, fight for biodiversity or lose it. It's as simple as that (see, e.g., "The Sierra Club Century," *Sierra,* May/June 1992).

All biodiversity conservation is local. Local political support is a sine qua non for every habitat set-aside by government or a private land trust, even though a good national policy framework may be in place. The NGO's job is easier, of course, when public employees, both elected and appointed, dedicate themselves to helping rather than hindering this biodiversity-protection process. I've often asked employees of public natural resources management agencies these questions: "Are you responding positively to the opportunity posed by the global movement to protect biodiversity? Are you providing this movement with informed leadership, or by failing to act are you asking to be told by a legislative body or a judge how to manage the lands and resources for which you're responsible? Do you want Congress and the courts to do your thinking and your land-use planning for you?" For that's what will happen if agencies drag their feet in this regard: NGOs will put the agencies' feet to the fire through litigation and the passage of prescriptive new laws.

What seems to me to be absent today is an explicit commitment by our federal and state governments to participate in joint, coordinated land use planning and decision making at ecosystem and bioregional scales. If such policy direction was given by the president and the states' governors, or even by agency administrators, it would pave the way for real progress in preventing the further endangerment of species in the United States, and it would provide a model for the rest of the world.

To avoid the endangerment of more species in the future, I am urging all environmental NGOs to adopt biological diversity conservation as one of their central missions for the 1990s. One way they can help create a strong political constituency to support the policy changes needed to achieve the goal of healthy populations of *all* native species is to support *nongame* wildlife management programs. These programs address the needs of the appreciative, rather than the consumptive, users of wildlife. They often are called "watchable wildlife" programs (Vickerman, 1989).

Their elements include the designation of wildlife viewing areas open to the public. Each viewing area is to be identified by a standard binoculars-symbol sign, a design approved for this purpose by the U.S. Federal Highway Administration. It would be helpful if every state and comparable jurisdiction published a wildlife viewing guidebook to these areas like those printed by Falcon Press of Helena, Montana for Idaho, Montana, North Carolina, Oregon, and Utah. These colorful books help wildlife watchers and photographers find and enjoy diverse public and private habitats. Furthermore, they provide an economic incentive for private landowners to save habitat, because they can charge interested visitors a fee for access to their land.

What do watchable-wildlife guidebooks have to do with creating political support for the conservation of global biodiversity? Promoting nonconsumptive wildlife recreation and ecotourism are the most obvious functions of these guides, but wildlife viewing areas also constitute part of a global network of biodiversity maintenance areas. Areas too sensitive or inaccessible to be included in a list of sites likely to be visited by thousands of wildlife enthusiasts, such as most research natural areas, will not be included in these networks and books, of course.

The point is that there is a close relationship between publicizing wildlife viewing areas and conserving biodiversity. The wildlife viewing guides help create a large, well-informed, vocal, grassroots constituency to lobby for and monitor enforcement of laws and land use plans that require ecosystem-based resource planning and management and protect diverse habitats and linkages between them. Viewing-guide users swell the ranks of the political constituency for biodiversity protection.

Let me return to the observation that public agency land managers cannot withstand the pressure to enter and exploit the commodities found in now-natural areas unless they receive political support from volunteer environmental activist groups. The political scientist's "iron triangle" concept comes into play here: No government program survives without three kinds of active advocates: (1) a pressure group made up of dedicated private citizens demanding the program; (2) enthusiastic legislators willing to authorize, fund, and monitor the program; and (3) a competent agency; able to carry out the program. Those three elements constitute the three sides of the "iron triangle." Conclusion: Government officials may plan to protect biodiversity, but to stay in business they must be responsive to citizen members of NGOs, who pay the bills and elect the congressmen, and win these groups' active political support.

For those wondering how to define biodiversity or how to go about making the case for its conservation, I have some recommended reading for you. Good examples of how to carry out biodiversity education include the National Park Service's handbook for interpreting biological diversity called *Makes a World of Difference* and the Office of Technology Assessment's 1987 report, *Technologies to Maintain Biological Diversity.* Other examples of excellent publications in this field include The Wilderness Society's *Conserving Biological Diversity in Our National Forests,* the World Resources Institute's *Keeping Options Alive: The Scientific Basis for Conserving Biodiversity,* and a comprehensive report prepared jointly by WRI, the World Wildlife Fund, Conservation International, the World Bank, and the International Union for the Conservation of Nature called *Conserving the World's Biological Diversity.* Then there are the Defenders of Wildlife publications, *In Defense of Wildlife: Preserving Communities and Corridors,* featuring a lead paper by Larry Harris of the University of Florida and sold by Island Press, and the Defenders Magazine Special Report entitled *The Biodiversity Challenge,* based on a workshop held in Denver. Island Press published in 1991 a useful book based on that workshop called *Landscape Linkages and Biodiversity* edited by Wendy Hudson.

More public education on this issue is badly needed, as is positive national media coverage, to make a vote for biodiversity conservation a political winner for a majority of the members of Congress. I served on a team assembled by the Keystone Center in Colorado to prepare a consensus document with industry representatives on how to protect biodiversity on American public lands (Keystone Center, 1991). I can tell you that, when some representatives of the extractive industries now hear "ecosystem management" or "bioregional planning" being advocated, they think its proponents are simply out to enlarge the National Wilderness Preservation System, and they've transmitted their fears to their representatives in Congress.

These skeptics may well be able to prevent the adoption of strong biodiversity-protection policies until conservation NGOs have done three things: (1) clarified the fact that human economic activities (such as domestic livestock grazing) and biodiversity maintenance actually are *compatible* in some instances (there will be some situations in which plant communities can withstand various degrees of fairly intense human use); (2) documented the fact that biodiversity maintenance, with its attendant high quality of life, provides a positive economic return to an area because people like to live in such places; and (3) organized a strong political coalition to win the adoption of biodiversity-protection policies despite the predictable opposition.

Gifford Pinchot won the enlargement of the National Forest System over considerable resistance. Sigurd Olson fought relentlessly for the airspace reservation over the Superior Roadless Area until President Truman signed the order. Howard Zahniser never let opposition to passage of the Wilderness Act stop him from his successful campaign. The Alaska Coalition hung in there and won passage of the Alaska National Interest Lands Act. Those who favor stronger, more explicit biodiversity-protection policies and programs shouldn't let anyone deter them, either.

The ideal geographic basis on which to plan for biodiversity conservation is the bioregion. But the most *practical* administrative or political unit within which to begin coordinated land-use planning to achieve biodiversity conservation in the U.S. may well be the state. Why? Because the habitat gap analysis work initiated by Dr. Michael Scott in Hawaii, Idaho, and Oregon is being done state by state. Natural Heritage Inventory databases have been set up state by state. These state inventories of plant communities and rare species provide a good starting place. And because many environmental NGOs (including the state affiliates of national groups) are organized along state lines they can adopt a state biodiversity-protection plan as their action goal.

The identification of biosphere reserves, buffers, and resource-development sites then can begin at the political level where people are used to dealing with each other (the state level), with opportunities for trans-state-boundary cooperation taken advantage of where possible. A good example of the latter is the multi-state/multi-agency cooperation being pursued in the Yellowstone ecoregion.

Strong and effective public agency action to protect biodiversity is long overdue. Since the Endangered Species Act was adopted, six listed species have become extinct. Five hundred and seventy-four U.S. species remain on the endangered species list. Some 3,700 more species are backlisted or await consideration. Only 308 of the 574 listed species have recovery plans. Few of the plans can be fully implemented with existing funding. The Endangered Species Act (ESA) is a safety net sagging under the strain of the sheer magnitude of species needing attention. As it is currently being implemented, the ESA can be compared to emergency room medical care. The tendency has been to do too little too late, with a strong emphasis on high-tech solutions to problems in a crisis atmosphere. We have waited until the patients are dying before we administer aid. Too few of the patients ever make it to the operating table. Too little has been done to prevent the crisis in the first place.

I'm not critical of the efforts of those who have fought to bring individual species back from the brink of extinction but simply recommend a complementary approach – the broader opportunity to protect entire habitats. Only adoption of a cooperative strategy, with this goal, will prevent massive extinctions of plant and animal life, as well as recover individual species. The bottom line is not how many species we save from extinction in the next decade but how many species will survive the next hundred years or more.

Habitat fragmentation probably is the single most threatening activity affecting the viability of wildlife populations in the United States. Already our national parks – virtually the only federal lands besides wilderness areas that are administered primarily to perpetuate their "natural" values – are losing species, possibly because animal populations exist there in wildland remnants too small to sustain those species over time. We have established preserves as if there will be no biological changes over time. We have focused on individual parks and preserves instead of landscapes or bioregions. We have focused on individual populations and species, instead of protecting the system in which they live. We have been too oriented toward species diversity instead of native diversity. Some wildlife managers have been engaged in the process of introducing exotic animals and propagating them artificially, often at the expense of native fish and wildlife.

After all these years of applied wildlife management, the United States still has no coherent, systematic approach to wildlife conservation. We can't hope to prevent species extinction until we inventory our stock of all our native plant and animal communities, define the habitat needs of all our native wildlife, and take steps to protect the lands and processes necessary for their survival. We haven't taken these obvious, basic steps yet. Our state and federal agencies must go beyond their traditional efforts to manage for only game species and endangered species.

A new goal is called for: the maintenance of viable ecosystems for all species, slimy as well as "charismatic." All species – "all the cogs and wheels," as Aldo Leopold called them – are essential to the maintenance of viable systems. In short, we need to manage distinct biological communities, ecosystems, and their processes to maintain representative examples of all habitats and linkages of communities across regional landscapes (e.g., Little, 1991). It will be easier and more cost-effective to protect ecosystems while they are still functioning and intact than to wait until crisis management has to be invoked for one endangered species after another.

Fortunately, conservation biologists and environmental activists are beginning to coalesce around more holistic approaches to wildlife conservation such as that used by Michael Scott, leader of the U.S. Fish and Wildlife Service Cooperative Fish and Wildlife Research Unit in Moscow, Idaho, and his colleagues, called biological diversity inventory or "gap analysis." Gap analysis begins with maps of vegetation types found in each region. Computerized distribution maps highlight areas of greatest animal diversity. When overlain by maps of land ownership and managed areas, these maps identify "gaps" in the network of biodiversity maintenance areas. Once the data from such statewide biodiversity inventories are in hand, we can determine which lands and resident wildlife populations are at greatest risk of degradation or extinction.

These reports represent a special, once-in-a-lifetime opportunity for private environmental NGOs to make a noteworthy contribution to biodiversity conservation. The work products from the Scott-type habitat gap analyses cry out for private citizens' groups to take these research results and run with them. Local volunteers in each state can take the gap analysis findings as to which plant communities are totally without dedicated protection and see that protection of some kind is given them. Whether it's a conservation easement or a research natural area or a private land trust or a wilderness or refuge designation, whatever fills the bill in each instance should be pursued until more than one good example of every plant community in this country is protected within an American Biodiversity Network.

Clearly, a mechanism is needed to coordinate land use planning between agencies and across political boundaries, to take into account the entire ranges of every species. Private landowners will be important to the process also, especially to help provide landscape linkages between species-rich preserves on the basis of conservation easement agreements that provide the owner with an economic incentive. Given the number of environmental agencies and organizations in the United States, the U.S. should be able to serve as an international model in the field of comprehensive wildlife preservation. We should show by example what can be done to save biodiversity. An integrated public/private strategy is needed, but there are many reasons why such a plan does not exist.

One is that no single authority exists to draft and implement such a plan. Virtually all existing resource agencies have operated as though they have been given mandates to produce something of monetary value such as timber, forage, minerals, ducks, deer, facility-oriented recreation, and so forth. This is based more on tradition than on law. Healthy shifts *away from* this traditional single-output-dominated mentality are surfacing, however. The second reason such a plan doesn't exist is funding. Inadequate funding exists for the purpose of protecting wildlife habitat on a broad scale.

I advocate getting ahead of the endangerment curve by protecting biological diversity. Thus I recommend the following 10-point biodiversity protection political action platform.

1. Enactment of a National Biological Diversity Conservation and Research Act to ensure the maintenance of biological diversity on public lands. It must be adequately funded and identify a lead agency to implement a proactive plan for identifying and protecting native ecosystems by means of a national system of biodiversity conservation areas.
2. Full funding for the habitat gap analyses being conducted by co-op research units of the U.S. Fish and Wildlife Service. Approximately three million dollars per year are needed for six years to complete the inventory nationwide.
3. Passage of the American Heritage Trust Fund Act. It would establish a $24 billion fund for land acquisition. Although $900 million is authorized annually now for spending from the Land and Water Conservation Fund,

only about one quarter of that is spent each year. If the trust fund is established, a billion dollars each year would be available for land conservation.

4. Establish of a trust fund or other new funding scheme by every state. Such measures could include everything from general obligation bonds to earmarked real-estate transfer taxes. In 1989, Oregon established a Resource Conservation Trust Fund to finance the acquisition of parks, natural areas, interpretive facilities and recycling programs. The Oregon law requires the development of an interagency habitat conservation plan. It will be based on the biological diversity inventory or gap analysis being conducted by the U.S. Fish and Wildlife Service.
5. Fully fund the Environmental Protection Agency's work on biological diversity and global warming.
6. Give universities more incentives to maintain plant and animal collections through traineeships and fellowships in systematic botany and zoology.
7. Create more effective interpretive programs to explain the importance of preserving biological diversity. That will be one of the roles of the Environmental Education Center of Virginia at Explore Park in Roanoke, which I direct. We will focus on the need to conserve the amazing diversity of the plants and animals of the Southern Appalachian Bioregion.
8. Increase efforts to protect significant lands.
9. Find ways for resource agencies at the state and federal levels to work more closely together. They must incorporate explicit biological diversity goals into their existing land use plans, to create an effective network of protected wildlife habitat linked together to allow animal and plant populations to disperse from one preserve to another.
10. Set a good example for the rest of the world by U.S. adaption of the "golden rule" – do unto others as you would have them do unto you – to other countries' wildlife resources. This means we apply Endangered Species Act-based constraints on U.S. agencies *abroad* as well as at home, so that we will not be responsible, through U.S.-funded and/or directed habitat destruction, poisoning, etc., for harm to other nations' biodiversity.

Time is running out. As with the parallel global problems of acid rain, climate change, and ocean pollution, we cannot afford the luxury of just studying biological diversity and debating the merits of landscape ecology versus single-species management any longer. Wild land and wild species are disappearing at an alarming rate. We need to put into place, by means of the collective, grass roots-based political muscle of our environmental NGOs, a comprehensive, integrated approach to protecting biological diversity – one that emphasizes habitat protection over individual species management and bioregion-based planning over individual administrative unit-based planning.

Ann Ronald, in her 1982 book, *The New West of Edward Abbey,* distills this marvelous quote from *Abbey's Road:*

> Dreams. We live, as Dr. Johnson said, from hope to hope. Our hope is for a new beginning. A new beginning based not on the destruction of the old but on its reevaluation. If lucky, we may succeed in making America not the master of the earth (a trivial goal), but rather an example to other nations of what is possible and beautiful. Was that not, after all, the whole point and purpose of the American adventure?

And in her 1974 biography of Aldo Leopold, Susan Flader concluded, "We must encourage the greatest diversity to preserve the widest possible realm where natural processes can achieve their own equilibrium." Only if we do encourage the greatest diversity will the quality of life of all living organisms on this fragile planet, including our own species, be assured.

Members of grass-roots-based private nongovernment environmental organizations everywhere can play a pivotal role by creating political support for the policies, programs and appropriations needed to get the job done.

References

Albright, H. (1985). *The Birth of the National Park Service.* Salt Lake City: Howe Brothers.

Allin, C. W. (1982). *The Politics of Wilderness Preservation.* Westport, Conn.: Greenwood Press.

Clepper, H. (1971). *Professional Forestry in the United States.* Baltimore: Johns Hopkins University Press.

Gabrielson, I. (1943). *Wildfire Refuges.* New York: Macmillan.

Keystone Center. (1991) *Biological Diversity on Federal Lands: Report of a Keystone Policy Dialogue.* Keystone, Colorado: Keystone Center.

Little, C. W. (1991). *Greenways for America.* Baltimore: Johns Hopkins University Press.

Nash, R. (1982). *Wilderness and the American Mind.* New Haven: Yale University Press.

Roth, D. M. (1984). *The Wilderness Movement and the National Forests: 1964–1980.* USDA Forest Service, FS 391, December 1984.

Vickerman, S. E. (1989). State Wildlife Protection Efforts: The Nongame Programs. In *Defense of Wildlife: Preserving Communities and Corridors,* (pp. 67-84). Washington, D.C.: Defenders of Wildlife.

21

Legislative and public agency initiatives in ecosystem and biodiversity conservation

MICHAEL J. BEAN

Introduction

Many of the chapters in this volume have explored either the scientific or the ethical dimensions of the biodiversity issue. Others have examined some of the practical techniques of conserving biological diversity, from germplasm preservation to ecological restoration. The ambition of this chapter is to survey the field of public policy, which attempts to meld all of these and more. Its task is to explore what policy makers in Congress, guided by an often very superficial understanding of the scientific issues, torn by seemingly conflicting ethical imperatives, and utterly without any practical experience in the techniques of conservation, have recently propounded as the nation's public policy with respect to the conservation of biological diversity. This chapter will also examine the recent initiatives of those in the federal agencies charged with implementing congressional policies. The aim is to convey a better understanding of both the limits and the opportunities for advancing the conservation of biodiversity through the recent initiatives of public policy makers.

An evolving awareness of the importance of biological diversity

The origins of many of today's federal policies concerning the conservation of biological diversity can be traced to about a century ago. Three initiatives stand out prominently in the early part of that history. The first was the creation of the national parks, starting with Yellowstone in 1872. The second was the establishment of national forests in the final decade of the Nineteenth Century. The last was enactment of the first significant federal wildlife legislation, the Lacey Act of 1900. Many of the impulses behind these original conservation initiatives can be perceived to be still at work in current public policies.

Yellowstone was preserved as the world's first national park largely because it was a curiosity. Its geysers, steam vents, boiling mud pots, and other geological features sparked wonder and amazement. Its biological resources were significant also, but in 1872 they were not much different from those that could be found over vast areas outside the park. Conserving biological diversity was a purpose of the first national park, but it was clearly secondary to preserving the geological wonders that set Yellowstone apart from anywhere else.

Two decades later, the country's first national forests were established. Until that time, the nation had pursued a policy of disposing of federally owned lands to homesteaders, miners, and others eager to take what the government had in such seeming abundance. The policy of land disposal, however, led to much abuse. In particular, the forests of the West were being plundered without regard to impacts on watersheds or the need for reforestation. The National Forest System was a reaction to this pillage; large areas of federal forest land were "withdrawn" from the disposal laws. They were to be managed intelligently, with a view to their long-term value as renewable resources. Their establishment, in short, recognized that our appetite for short-term gain had to be constrained in order to meet our long-term interests.

With the turn of the new century, the Congress enacted the first major national wildlife conservation law, called the Lacey Act. In one sense, this first federal step into the realm of wildlife conservation policy was a halting one. For the most part, the Lacey Act merely provided federal law enforcement and federal penalties against those who violated state wildlife laws and then transported their illegal booty across state lines. But implicit in the "federalization" of state law offenses was the recognition that the problem of overexploitation of wildlife to supply distant markets was one that states alone could not handle. National action was needed to address a problem that was at least national in scope.

One other facet of the Lacey Act is worth noting. Because its scope was determined by the scope of the underlying state conservation laws, the Lacey Act applied to only a relatively few species. The nearly exclusive focus of state conservation laws at the time was the small number of species valued for recreational or commercial purposes. Thus, the Lacey Act, like the state laws it embraced, was limited to what was then considered "useful" wildlife; the vast majority of species for which there were neither recreational nor commercial uses remained outside its ambit.

At the same time that these early conservation initiatives signalled an awakening of public policy makers to a growing conservation problem, other policy initiatives of the same era only compounded that problem. The Reclamation Act of 1902 provided authority for numerous major "reclamation" projects, aimed at "reclaiming" the arid West, and sought to turn the deserts green with irrigated agriculture. Similarly, the Swamp Lands Acts of 1849, 1850, and 1860 were the means by which the federal government sought to encourage the draining and filling of "useless" swamps and bogs so that they might be put to productive use. Today, we recognize the enormous adverse biological impact that many of these activities had, from the endangerment of much of the native fauna of Western aquatic systems to the plummeting of continental waterfowl populations, and the poisoning of wildlife as a result of the leaching of trace chemicals from desert soils by irrigation waters. At the time, however, the results of these laws were almost universally viewed as socially desirable. The connections between these purposeful alterations of the natural landscape and declining wildlife populations were simply not appreciated. Nor was the importance of predatory animals to the functioning of biological communities. In 1909, Congress appropriated funds to the Agriculture Department to promote the destruction of "noxious animals," the first of many such measures (Bean, 1983).

By the early years of the present century, therefore, the seeds of many modern conservation policies were planted – the need to preserve unique resources, the need to restrain our utilization of natural resources, and the recognition that intelligent conservation would require coordinated action at the state and federal levels. But these seeds were planted in a garden in which many other contrary and competing public policies had already taken root. Today, in the waning years of the same century, conservation efforts have blossomed and borne fruit, but they still compete for light and space in an even more crowded public policy garden.

The modern conservation dilemma

Making public policy choices is relatively easy when only a few things have value and everything else can be ignored. Unfortunately, however, today's public policy makers no longer have the luxury of such simplemindedness. The truth of Ralph Waldo Emerson's observation that a weed is but "a plant whose virtues have not yet been discovered" is now apparent (Emerson, 1878). The world of living things can no longer be divided into those useful and those not. Instead, the world must now be divided – as Emerson foresaw – into those things whose utility we have already discovered, and those others whose utility we have yet to discover. That vastly complicates the task of restraining our short-term appetites to preserve our long-term interests, for now we can see that our long-term interests may depend upon conserving many more different things than we ever thought of before.

Likewise, saving unique and awe-inspiring things – the inspiration for creating Yellowstone National Park – is a manageable task when there are only a few of them. But what happens when the diversity of life itself becomes the source of such sentiments? When the sense of awe and wonder that inspired our ancestors to preserve Yellowstone inspires us (at least some of us) to want to preserve sea turtles, spotted owls, even snail darters?

These dramatic changes in our awareness – awareness of the uniqueness and potential usefulness of all living things – create the modern conservation dilemma for public policy makers: what to conserve and how best to conserve it while meeting other important societal objectives. That dilemma is reflected in the ebb and flow of political support for the protection of endangered species.

The Endangered Species Act

In 1966, Congress enacted the first of several federal laws aimed at slowing the accelerating loss of species. This first law and a successor enacted in 1969 were limited in scope and consequence. They set the stage, however, for enactment in 1973 of a more sweeping, comprehensive measure with truly grand aims. The Endangered Species Act of 1973 sought to establish a program for the conservation of species threatened with extinction as a result of "natural or manmade factors." That program was to encompass not only vertebrates, to which virtually all earlier conservation legislation had been limited, but also invertebrates and even plants. Further, the scope of the program encompassed species both here and abroad. Most significantly, the new law embodied a novel and unqualified commitment that no agency of the federal government would be responsible for any action that jeopardized the continued existence of any species.

The Endangered Species Act was passed without great controversy and with widespread, bipartisan support. For at least a few years, accommodating the Act's requirements was relatively painless. Then came the controversy over Tellico Dam and the snail darter. Tellico Dam was a Tennessee Valley Authority project designed to bring industrial growth and increased recreational opportunities to a largely rural area of eastern Tennessee. The snail darter was an endangered species of fish that was discovered in the area to be impounded by the dam only after work on the dam had begun.

Congressional (and public) enthusiasm for the noble and sweeping principles embodied in the Endangered Species Act waned noticeably when those principles collided with the other societal interests that Tellico Dam was intended to promote. Ideas that had seemed unobjectionable when the survival of the bald eagle or the panda was at stake suddenly sounded ludicrous when nothing more important than a three-inch minnow was involved.

Fortunately, Tellico Dam was not without its warts and blemishes as well.

A dam designed to produce no hydroelectric energy and with dubious economic benefits, Tellico was – in the eyes of those willing to look beyond the comical "little fish versus big dam" characterization of the conflict – simply another pork barrel project designed to pour federal taxpayer dollars into the home state of influential Congressmen. As a result, despite the initial widespread outrage that a mere fish could stop a nearly finished federal project, and suggestions that the Endangered Species Act should either be repealed or so weakened that it would never stand in the way of another major development project again, Congress decided to leave the Act virtually intact. In response to the Tellico Dam controversy, Congress in 1978 enacted amendments to the Act creating a process for exempting qualified federal projects from the Act's requirements. The standards that had to be met to qualify for such an exemption, however, were high: no project could be exempted unless it was of national or regional significance, and then only if its benefits clearly outweighed the benefits of alternatives that would not imperil an endangered species (Bean, 1983). These standards were so high that only a single contested project has ever been exempted.

Today, the Endangered Species Act is again at the center of major public controversy. The listing of the northern spotted owl as a threatened species has called into question the future of logging the remaining "old growth" forests of the Pacific Northwest. The Bush Administration called for the invocation of the exemption process added by Congress in 1978 and for amendments to the Endangered Species Act that would assertedly expand the authority of the committee empowered to grant exemptions. Many members of Congress have called for an overhaul of the Act, to ensure "flexibility" and the consideration of economic factors when decisions about the listing of species are made; others have called for revisiting the issue of protecting subspecies and geographic populations under the Act. Still others want to emphasize captive propagation and species relocation as a way of escaping rigorous regulatory controls over activities affecting species in their natural habitats. As a result, Congress seems likely to take a more careful and critical look at the whole notion of endangered species protection in the year ahead.

Congressional interest in reexamining the Endangered Species Act comes at a time when many in the biodiversity community are questioning the adequacy of the Act as well. Just as, in decades past, myopic notions of "useful wildlife" gave way to a broader recognition of the potential importance of all species, conservation efforts focused on species are today being challenged by the view that broader needs must be addressed. Most people writing about biological diversity today are at pains to emphasize that it is a concept much broader than just the preservation of endangered species. They point out that much biological diversity (in the form of genetic diversity, local community diversity, etc.) can be lost without losing species themselves. In addition, they argue that because many endangered species are already so close to extinction, conservation resources devoted to their last-minute rescue may only divert resources from other endeavors with bigger payoffs (Westman, 1990). Some authors also liken endangered species conservation efforts to "retail" conservation, when what is truly needed is "wholesale" conservation aimed at protecting many species simultaneously. This can best be done "by shifting some of our focus on individual endangered species to a more land-based ecosystem approach aimed at conserving biological diversity" (Scott, 1990). For these reasons, and because the Endangered Species Act has allegedly become "bogged down in a bureaucratic listing process" and "has created a bureaucratic quagmire that diverts attention from larger goals," at least some conservationists have questioned the wisdom of regarding the Endangered Species Act as a central focus of biodiversity efforts (Karr, 1990).

This congruence of congressional and conservationist interest in reexamining the Endangered Species Act might be a welcome thing if Congress were likely to want to embrace more substantial and far-reaching initiatives for protecting biological diversity. Unfortunately, however, the most intense pressure in Congress today is not from those who want to advance the cause of conserving biological diversity, but from those who want to retreat from it. They would weaken and undermine the protection afforded to endangered species without providing any offsetting benefits for biodiversity conservation generally.

The ongoing debate over the Endangered Species Act does at least illuminate the perceptions of federal policy makers regarding some of the issues that concern biodiversity interests. While it is true that biodiversity conservation encompasses more than just species preservation, it is also true that species loss represents a fundamental and irreparable diminution of biological diversity. It is, if not the most basic, at least the most tangible, readily understood and appreciated aspect of biodiversity conservation. If conservationists are unable to persuade policy makers to support this "bottom line" of biological diversity, one cannot be particularly optimistic about success in garnering support for more amorphous and abstruse concepts.

The dichotomy often drawn between protecting endangered species and protecting endangered ecosystems also proves, upon closer examination, to be not nearly so clear as is often argued. In the first place, except when a species has been reduced to the point at which it survives only in captivity, effective endangered species conservation generally means habitat conservation. This was evident to the drafters of the Endangered Species Act who, in its opening section, declared the Act's purpose to be "to provide a means whereby the ecosystems upon which endangered species and threatened species depend may be conserved." Nevertheless, endangered species advocates are often accused of using endangered species as "surrogates" for their real objective of protecting habitats. The northern spotted owl, these accusers say, is merely the tool by which environmentalists want to save old growth forest from the chain saw; the snail darter, they said, was just a pretense for leaving the Little Tennessee River undammed. These accusations seek to shift attention away from conservation needs and toward conservationists' motives. Of course, spotted owl preservation requires old growth forest protection; by focusing on the motives of environmentalists, the accusers seek to divert attention from the inescapable fact that endangered species cannot be preserved in the wild without preserving their habitats.

To those who suggest that what we really need is an "Endangered Ecosystems Act," the lesson to be drawn from this is simply that protecting endangered ecosystems is not likely to be any easier, politically, then protecting the individual species that depend upon them. The spotted owl controversy, if recast as an "endangered old growth forest ecosystem" controversy, involves the same difficult trade-offs between short-term economic interests and long-term conservation interests. Indeed, the tradeoffs are almost certainly greater, since the areas identified for spotted owl conservation in the recent report of the Interagency Spotted Owl Scientific Committee (Thomas et al., 1990) do not suffice to assure the conservation of a broader array of old-growth dependent species. Neither is the recasting of such a controversy from its species focus to an ecosystem focus likely to avoid the interminable skirmishing about whether that ecosystem is truly endangered (there is already as much debate over how much old growth forest remains as over how many spotted owls remain), or for that matter how that ecosystem should be defined and demarcated. In short, the "bureaucratic quagmire" that has characterized the Endangered Species Act is likely to characterize any Endangered Ecosystems Act as well; there is simply too much at stake in controversies such as this for the affected interests and their lawyers not to fight over every issue on which they might secure some advantage.

Other legislation

The Endangered Species Act debate in Congress is the pivotal public policy debate over the conservation of biological diversity. There are other congressional initiatives, of course, but they are being overshadowed by the endangered species debate. One is a proposed "biodiversity bill" that was approved by the House Science, Space and Technology Committee. It has several features. First, it declares a national policy in favor of conserving biological diversity. Second, it requires an explicit discussion of impacts upon biological diversity of federal actions in environmental impact statements prepared under the National Environmental Policy Act. Third, it establishes a research-oriented national biological diversity center and authorizes funding for biodiversity research. Declarations of policy, discussions in environmental impact statements, and research do not add up to restrictions on activities that destroy biological diversity. For that reason, the proposed legislation has been not especially controversial, but neither has it galvanized particular enthusiasm among the ranks of conservation activists. As a result, the proposed bill failed to pass in two succeeding Congresses. However, in 1993, the creation within the Interior Department of a "National Biological Survey" was at least partially an outgrowth of this legislative initiative.

Many of the most effective policies for conserving biological diversity are reflected in laws and regulations that were intended to serve a much broader public purpose. The National Park System, for example, is intended to provide both public outdoor recreational opportunities and the conservation of scenic, historic, and natural resources. Units of the Park System are not uncommonly the last, or the best, remaining examples of unusual biological communities. Likewise, the laws relating to the management of National Forests and the lands of the Bureau of Land Management are intended to assure "multiple use" of the resources of such lands for commodity production, mineral extraction, recreation, wildlife conservation, and so forth. These lands too are often of critical value for biodiversity conservation. The regulations of the Forest Service under the National Forest Management Act require the maintenance of "viable populations" of native vertebrates on National Forest lands. It was this requirement that initially forced changes in the Service's management practices with respect to old growth forests in the Northwest.

Another important law of more general applicability is the National Environmental Policy Act ("NEPA"). Its main purpose is to force disclosure of the environmental impacts of federal actions before they occur, by requiring that agencies prepare environmental impact statements for their major actions. That disclosure, however, has served the valuable purpose of alerting federal agencies to less harmful alternatives or to ways of avoiding or minimizing the adverse environmental impacts of their proposed actions. Conserving biological diversity, while not an explicit objective of NEPA, has often been the result of the evaluation and disclosure that it compels.

Conclusion

At the same time that conservation biologists perceive an urgent and growing threat to biological diversity, the nation's public policy makers are contemplating a retreat from the flagship law that has the conservation of biological diversity as its exclusive purpose. Among congressional policy makers, the questioning of the nation's commitment to endangered species protection (and implicitly to biodiversity conservation) is more widespread today than at any time in the past decade. While other laws and programs will continue to be highly useful in protecting biological diversity, no other measure is likely to embody as clearly the fundamental commitment to the conservation of biological diversity. Thus, the storm clouds gathering around the Endangered Species Act portend a difficult and trying future for advancing a broader vision of, and a deeper commitment to, the conservation of biological diversity.

References

Bean, M. J. (1983). *The Evolution of National Wildlife Law.* 2d ed. New York: Praeger.

Emerson, R. W. (1878). *Fortune of the Republic,* quoted in E. Norse (1990). *Ancient Forests of the Pacific Northwest.* Washington, D.C.: Island Press.

Karr, J. R. (1990). Biological Integrity and the Goal of Environmental Legislation: Lessons for Conservation Biology. *Conservation Biology,* **4,** 244–250.

Scott, J. M. (1990). Preserving Life on Earth: We Need a New Approach. *Journal of Forestry,* **88,** 13–14.

Thomas, J. W., Forsman, E. D., Lint, J. B., Meslow, E. C., Noon, B. R., & Verner, J. (1990). *A Conservation Strategy for the Northern Spotted Owl.* Washington, D.C.: U.S. Dept. of Agriculture.

Westman, W. E. (1990). Managing for Biodiversity: Unresolved Science and Policy Questions. *Bioscience,* **40,** 26–33.

Part Seven

Biodiversity and landscapes: postscript

22

Biodiversity and humanity: toward a new praradigm

ROBERT D. WEAVER and KE CHUNG KIM

Introduction

The Rio Summit 1992 (United Nations Conference on Environment and Development), reflected the extent of global consensus that biodiversity, the environment, and the biosphere are in a perilous state and that the current state of these natural systems has been caused by human activity. The urgency for altering the relationship between humanity and biodiversity ultimately follows from the realization that current processes threaten the integrity of human habitat. Ozone depletion, global climate change, species extinction, and destruction of ecological functions are each calamities of major consequence to the biosphere. Moreover, they are anthropogenic and foretell human disasters because the fate of humankind and that of nature are inextricably intertwined.

The extent of the peril to humanity is such that fundamental changes in the concepts and approaches of our technological society must take place to manage our life-support system if the human species is to sustain the social, political, economic, and cultural capacities essential for its existence. Adding to this challenge of averting catastrophe is perhaps the more temporal challenge of finding ways to support these human systems, achieve acceptable standards of living for a global population of 5.5 billion that is annually increasing by 100 million, while simultaneously sustaining essential natural systems. These dual challenges are the most critical that humanity has ever faced. In view of the precarious state of both natural and human ecosystems the piecemeal approaches of the past will undoubtedly fail to address these challenges.

In this chapter we will argue that biodiversity conservation must take on a broad interpretation that includes the conservation of ecosystem and biosphere processes, in addition, to the preservation of endangered and threatened species or specific habitats or ecosystems. Preservation of the global atmosphere and protection of the biosphere cannot be limited to specific current threats such as automobile emissions or deforestation. Achieving this generalized notion of biodiversity conservation will require fundamental transformations of human cognition and values, and environmental values and ethics, as well as the social, political, and economic equilibria of our technological society. Such transformations will require expansion of the information bases of science and technology, education and communication of the implications of human impacts on biodiversity, and adoption of new social, political, and economic institutions and processes to facilitate and catalyze movement to new equilibria.

The transformation of human cognition to appreciate the role of these human life resources will require massive efforts in education throughout the world at every level of society. Given the current state and dynamics of biological systems, new science and technology will be required to halt current processes of degradation and to reclaim and rehabilitate essential systems. New science and technology essential for the restoration of biological systems will include ecological engineering (ecotechnology), including bioremediation and detoxification of contaminated soils (e.g., Cairns, 1988; Mitsch & Jorgensen, 1989; Mitsch, 1987; Odum, 1994). Equally essential to the realignment of the relationship between humanity and biodiversity will be the adoption of a sustainable economic approach to development and humanity's impact on the environment.

The outline of the argument to be presented is as follows. First, we will reconsider the essence of the paradox of humanity and biodiversity, clarifying a generalized interpretation of biodiversity, assessing the extent of the conflict, and reconsidering the bases of relevance of the paradox. Second, we will attempt to move beyond the symptoms of the conflict and consider the origins of the conflict between humanity and biodiversity. Here we will focus on the limits of human cognition and the nature of human values, including the notion of human dominion over nature. Despite the role of these factors, the crux of the paradox will be argued to lie in their balance or equilibrium across our highly heterogeneous human society. Next, based on these premises, the nature of the challenge for change in the relationship between humanity and biodiversity will be reconsidered focusing on opportunities and constraints, as well as the goal of that change. In the final section, the options for change will be discussed, highlighting means for shifting social, political, and economic consensus.

The essence of the paradox

Humanity, biodiversity, and landscapes: the nature of the conflict

The paradoxical conflict between humanity and biodiversity has been considered from a variety of perspectives in this volume. Stated succinctly, the paradox is simple. Human life depends for survival on biodiversity, nature, and the ecological systems which are subsumed in landscapes and the biosphere. Yet, at the same time, human activity during the past century apparently threatens the survival or existence of those very biological entities and systems. To review the nature of this conflict, it is useful to reconsider the dependence of humanity on biodiversity and landscapes, the necessity of that dependence, the extent of the burden placed on biodiversity and landscapes by human activity, and the extent to which current extinction rates and ecosystem degradation are the result of human impacts.

The conceptualization of biodiversity adopted in this chapter will be one which is operative within the context of prescriptions to manage or alter the impacts of human activity on biological systems. It is the composition of these systems that is of interest as a basis for defining the concept of biodiversity. Despite the apparent focus on species by many early government policy responses to human impacts on biological systems (e.g., the U.S. Endangered Species Act), it is now generally appreciated that the static and dynamic interrelatedness of species, as well as of species with the physical structures in ecological systems, requires that a more general focus be adopted in the conceptualization of biodiversity. Here we recognize that biodiversity, the diversity of biological entities, is inextricably linked to ecological systems and, more generally, interrelated to and an integral part of the biosphere. For this reason, we acknowledge the need to step beyond a focus that is limited only to biodiversity. In so doing, we will for brevity interpret biodiversity within this more general focus as encompassing ecological systems and the biosphere. Based on this generalized notion of biodiversity, the scope of preservation and conservation strategies emerges directly. Whether this focus is labeled a broad-based ecosystem approach (Scott et al., 1987), a total diversity approach (Norton, 1987), or a biodiversity and landscapes approach (as in the title of this volume), the intent is the same: to recognize the need for a scope and focus that goes well beyond species to admit the highly interactive relationships among species, species diversity, physical structure of habitat, landscapes, and the biological and physical structure of the biosphere. For simplification, we refer to this scope of focus hereafter simply with the word biodiversity.

Human dependence on biodiversity

The extent of human dependence on biodiversity, in an instrumental sense, is substantial and includes dependence on biological resources as consumption goods (e.g., food, fuel, and medicine), dependence on biosphere services (such as wind, moisture, energy, photosynthesis, and waste processing), and dependence on ecological services (e.g., biological decomposition and transformation). The dependence of humanity on the biosphere has been viewed as stretching beyond regard of the biosphere as simply a supply of instrumental values. From this perspective, the biosphere is home – a highly complex, dynamic, living shelter of humanity (e.g., Odum, 1969). In this sense, biological entities and systems cannot be characterized as resources, in the conventional sense (e.g., Randall, 1986). Consistent with this view is the recognition of human dependence on the biosphere for satisfaction of biophilia (Wilson, 1984; Kim & Weaver, 1994) as well as a source or basis for transcendence and transformation of human perspectives and knowledge (Thoreau, 1960; Sagoff, 1974; Emerson, 1982; Norton, 1987).

The utilization of biodiversity and biosphere products and services by humanity would seem to be a clear necessity. Both from a perspective of dependence on biological products for food, energy and health, or biological products and services, as well as from the perspective of dependence of human interaction with biological systems to satisfy biophilia and benefit from aesthetic inspiration, the dependence of humanity on biological systems for purely instrumental use would seem strong. This view of the dependence of humanity on biodiversity is not universally held (see McKibben, 1989; Weiner, 1990). A logical result of the process of abstraction of human life from nature has been the emergence of the view that humanity can survive without biodiversity. From this perspective, human intellect is viewed as capable of generating substitutes for biological products and services as well as biosphere and ecological services.

Human values: a categorization

The structure and extent of human dependence on biodiversity must be viewed as reflecting human values that underlie human behavior and decisions that impact biodiversity. Throughout this chapter, we will proceed by adopting a categorization of value that distinguishes *utilitarian* from *nonutilitarian* value. The categorization based on this dichotomy is by definition structurally similar to others (see, e.g., Norton, 1987). Utilitarian value will be defined [more generally than Norton (1986) or Kellert (1986)] as originating from consumption of a good or service or from an experience that is instrumental to satisfying human preferences. As defined, this generalized notion of consumption will be interpreted as including both appropriable, exhaustive consumption (e.g., eating an apple) as well as nonexhaustive, nonappropriable consumption (e.g., recreational, aesthetic, or learning experiences). Distinction of these sources of utilitarian value does not require adoption of other elements of utilitarian philosophy such as any particular form or structure of human preferences or a singular role of those preferences as the motivation for human behavior or the basis for human bliss. Nonetheless, consistent with this definition, utilitarian values are clearly anthropocentric.

We define nonutilitarian values as nonanthropocentric values that emerge from human adoption of otherwise natural values that are perceived or recognized by humans. For example, particular forms of animism could be a source of such values that emerge from common human beliefs. The Twa tribe, in Rwanda, believe their spirits inhabit the hill upon which they live and take their values from their perceptions of those spirits and their directives. Other theologies give rise to deities that are interpreted as directive through their establishment of values that must be adopted by believers. Finally, a nontheological origin for intrinsic value could be argued to follow from the evolution of each individual species and the interpretation of this evolution as a basis for admission of intrinsic value. In many cases, these value systems may generate what could be collectively labeled as *intrinsic values*. We choose to use nonutilitarian values as a label rather than intrinsic values to allow clarification of the source of these values as including both human perception and cognition of nonhuman conditions, events, or entities.

The burden of humanity on biodiversity

The extent of anthropogenic burdens on the biosphere are of such magnitude that a mass extinction of species has already begun and will continue over the next few decades (Myers, 1994; Kellert, 1986; Lovejoy, 1986; Mittermeir, 1988; Mooney, 1988). Human impacts on biodiversity include habitat alteration; overharvesting; cross-species impacts such as introduction of species, and catalyzing shifts in dominant species; waste disposal; and alteration of macrosystems such as solar radiation through the ozone layer or acid balance of precipitation and, at least at local levels if not at global levels, climate change (McNeely et al., 1990, Chapter III). Human demands for the goods and services of natural systems have evolved to an extreme where genetic resources are threatened to such an extent that an accepted strategy for their preservation is *ex situ* isolation in gene banks, abstracting the genetic material from its ecological context, and halting its natural evolution (Wilkes, 1994). While this strategy may enhance the convenience of access to germplasm, it also implicitly suggests that human pressure threatens continued existence of the germplasm, necessitating its isolation. The loss of biodiversity posed by human demands has been viewed as an important threat to agricultural productivity, as well as to the volume and stability of the food supply (Frankel & Hawkes, 1974;

Wilkes, 1977; Plucknett et al., 1987; Wilkes, 1994). Others have noted that such burdens of human demands may threaten entire ecosystems (see e.g., Ehrlich, 1968; Raven, 1987; Weiner, 1990) including human life itself.

The satisfaction of increased human demands for biodiversity and biosphere products and services resulting from these processes have been facilitated by revolutions in agricultural and industrial productivity, which have leveraged the productivity of human labor. However, an important fuel for this leveraged labor productivity has been the increase in intensity of consumption of biosphere resources per unit of human labor (see, e.g., Meadows et al., 1972). Western economists have labeled this phenomenon as input or factor biased technological change. This type of technological change achieves increases in productivity of limited or relatively highly valued resources (such as labor) through substitution of less limited or relatively less valuable inputs or resources such as manufactured capital and machinery (see, e.g., Nicholson, 1990).

Three manifestations of the impact and burden of humanity on biodiversity have been frequently cited: 1) the rapid extinction rate that has evolved during the past half decade, 2) rapid human alteration of landscapes involving alteration of either or both the physical and biological structure of habitat, and 3) rapid human exhaustion of biosphere capacities as a waste sink for nonconsumptive products of technological processes.

Species extinction

The current extinction process is of such magnitude that many conclude it may eventually threaten human life itself (Ehrlich & Ehrlich, 1981; Myers 1979b, 1983, 1989b, Raven, 1988; Myers, 1994; Lovejoy, 1980; Sayer & Whitmore, 1991). Extinction of species and alteration of natural landscapes is occurring at a substantially greater rate today relative to past periods (e.g., Myers, 1989b; Kim & Weaver, 1994). While this extinction process is often viewed at a global level, the processes of degradation of species are even more apparent at a micro level of specific ecologies or habitats (Morton, 1985). The implications of extinction have been cited as two-fold. First, the loss of a particular species involves both utilitarian and nonutilitarian anthropogenic losses. However, in addition, such a loss also implies an alteration of dynamic interaction of the remaining species, fundamentally altering their evolution and survival as well as the pool of information constituted in those dynamics. Similarly, each species lost carries with it a loss of information of that species' interactive facility and capacity in its habitat. In the case of plant germplasm, the loss of a species implies the loss of an information base of biological function, adaptation, and evolution that goes beyond the current functions of the lost germplasm (Weaver, 1990).

The magnitude of the threat of the extinction process is clarified if the interspecies dynamic implications of extinction are acknowledged. The interrelatedness of species and habitat implies that a single extinction can be expected to result in other extinctions as well as structural changes of the habitat. From this perspective, extinction is a spiraling or nonlinear process that accelerates (see Norton, 1987 for a summary of this argument). Logically, three alternative inferences concerning the role of humanity could be drawn from the spiraling nature of the extinction process. First, if the current extinction process can be viewed as a dynamic result of the evolution of earlier extinctions, then the past role of humanity in this process could be claimed to have been only one of catalyzing further acceleration of an ongoing process. Alternatively, the magnitude of the process of extinction could suggest to some that the potential for human actions to contribute to redirection of the process is small. The inference drawn most commonly has been that the observed negative spiral of reduction must be interpreted as a result of human activity and that a change in the relationship between humanity and biodiversity could fundamentally alter the direction of this spiral (e.g. Erhlich & Erhlich, 1981; Myers, 1988; Raven, 1988).

Landscape alteration

Alteration and loss of landscapes resulting from human activity results in three significant impacts on biodiversity: 1) displacement of species, 2) destruction or displacement of habitat, and 3) fragmentation of habitat (Terborgh and Winter, 1980; Lovejoy, 1986; Vermeij, 1986). The implications for species displacement through modern agricultural use of "improved varieties" or dominant, aggressive species has resulted in direct displacement of native species or in intermingling and, in some cases, substitution of artificial for natural selection processes (Wilkes, 1994). In either case, genetic erosion of the germplasm resource occurs statically, and the dynamic path of the remaining germplasm's evolution is definitionally altered. For example, when the genetic base is so narrowed, the continued existence of that base at risk (Wilkes, 1994). The impact of the destruction of tropical forests on the extinction process has been highlighted as an example of the interaction between human activity in our technological society and biodiversity (Raven, 1988; Silver and DeFrieds, 1990; Sayer & Whitmore, 1991). This process results in both displacement of or alteration of habitat as well as in fragmentation of habitat. Human alteration of landscapes leads to reduction and degradation of ecological systems and structures and often results both from and in evolution of landscapes through invasion by human grant species that may further impoverish indigenous biota through species exclusion (Mooney, 1988).

Fragmentation of habitat as a result of human occupation and manipulation of the landscape has accelerated with increased population and increased intensity of ecosystem and biosphere utilization made feasible by revolutions in agricultural and industrial productivity (Vitousek et al., 1986; Kristensen and Paluden, 1988; Western, 1989; McNeely et al., 1990). The fragmentation impacts of tropical forestry have often been cited (McNeely et al., 1990; Uhl et al., 1994). The biological consequences of such fragmentation often involve extinction as species mobility and interaction with the landscape is altered (Harris, 1984; Vermeij, 1986; Saunders et al., 1991; Robinson et al., 1992; Stern et al., 1992), while fragmentation of habitat clearly limits opportunity for the option of movement for large animals. Given that the potential for adaptation to change is finite and sometimes inadequate for a species, the option of movement is indeed an important one if extinction is to be avoided (Vermeij, 1986).

The biosphere as a waste sink

Exhaustion of biosphere capacities to absorb the waste products of our technological society has been recognized on several fronts (e.g., Ravera, 1979; Peters & Lovejoy, 1992). A critical element of the evolution of our technological society has been the use of the biosphere as a waste sink. Importantly, until recently, the long-term costs of such use of limited capacity has not been explicitly recognized in use decisions, suggesting that the cost of such use is not passed through the economic system as a signal to producers and consumers demanding products dependent on such services.

Humanity and biodiversity degradation

While some authors have related the conflict between humanity and biodiversity to postindustrial society, others argue that the conflict was endemic to preindustrial society just as it has been in the postindustrial revolutionary period (Weiskel. 1989; Sanders & Webster, 1994). The observation of numerous periods of extinction, even before the presence of humanity on the earth (see Vermeij, 1986; McNeely et al., 1990) would also seem to put in question the logic that postindustrial humanity has been the sole factor in the current extinction processes. Nonetheless, these authors argue that both the intensity and geographical scope with which the biosphere is impacted by human activity has increased during the postindustrial period, leading to the inference that postindustrial human society has played an important contributory role in catalyzing the current extinction processes. Other authors claim that human impacts have been the predominant cause of the accelerating extinction rates during the past centuries (see Frankel and Soulé, 1981; Myers, 1988; Wilson, 1988; McNeely et al., 1990).

The challenge then is to answer a fundamental question raised by the paradox: why is it that human utilization of and human impacts on biosphere resources must expand to such a scale as to pose a threat to the biological systems upon which humanity depends? Demand for solution of this paradox is heightened by human intelligence and technological capacity, which suggest solution may be feasible. Answers to the question raised by the paradox require consideration of the nature and possible roles of 1) human values, 2) human institutions, and 3) technology.

The relevance of the conflict to humanity

The existence of the paradox of humanity and biodiversity has been raised most recently within the context of the implications of and threats posed by technological society (e.g., Osborne, 1948; Carson, 1962; Ehrlich, 1968; Meadows et al., 1972; Eckholm, 1978; Lovejoy, 1980; Raven, 1980, 1987; Ehrlich & Ehrlich, 1981, 1990; Myers, 1985; McKibben, 1989). While the conflict between humanity and biodiversity may be widely recognized, the relevance of that conflict to humanity deserves consideration. The relevance of the paradox of humanity and biodiversity must be viewed as originating from human values that grant value to biodiversity. At the same time, however, the existence of the conflict speaks to the existence of a set of human values which, when operationalized in behavior through social, political, and economic processes, has led human activity to degradation of biodiversity. Within the categorization of value adopted here both utilitarian and nonutilitarian value can coexist and would provide a basis recognition of the conflict as relevant by humanity.

From an anthropocentric perspective, if the state of the conflict in fact poses a threat to humanity, then utilitarian values constitute a logical source for strong imperatives for action or change. Nonetheless, many have argued that the scientific evidence concerning the existence of such a threat is highly uncertain, given past experience that demonstrates human ingenuity and capacity to exploit technology to escape constraints that emerge from natural processes. This argument, however, ignores the uncertainty that must be accepted to characterize human ingenuity as well as the timing of its successful application and diffusion of its products. Science has often found limits in its capacity to produce solutions to alter biological, ecological, and biosphere processes. Certainly, science has not provided comprehensive solutions to the problems of interdisciplinerity; in fact, global environmental problems are interdisciplinary in nature. The rivet popper parable (where continued removal of rivets from an airplane is perceived as riskless given past experience) speaks to the wisdom of this logic (Ehrlich & Ehrlich, 1981).

Going beyond imperatives based on utility benefits of consumption of the biosphere's limited resources, biosphere services that generate transformative or transcendental value provide a further basis of an imperative for solution to the paradox of humanity and biodiversity. Satisfaction of biophilia must also be recognized as a utilitarian benefit from biodiversity (Wilson, 1984; Kim & Weaver, 1994). Technological society has abstracted human individuals from nature and biodiversity through urbanization, acceptance of the paradigm of human dominion, and acceptance of the feasibility and certainty of technological innovation as a means of satisfying all human needs (McKibben, 1989; Weiner, 1990). Coincident with this abstraction of humanity from nature and biodiversity, opportunity for satisfaction of humanity's biophilia has eroded to a point where aesthetic and socioeconomic activities have been called for by some to restore and satisfy it (see Tukey, 1994). Nature is relevant to the cultural inheritance and capacity of humanity, and as a source of spiritual enlightment, aesthetics, lessons, and knowledge (Kim & Weaver, 1994). This anthropocentric, utilitarian value of biodiversity is crucial for humanity (see Norton on Thoreau and Leopold in Norton, 1987; and Norton, 1994). Clearly, the aesthetic benefits of biodiversity might be accessible to some limited extent through geographically restricted preserves, or even zoos; however, the transcendental or transformative value of biodiversity would seem to call for the generalized scope of focus on biodiversity adopted in this chapter. While many reject the transcendentalist movement's interpretations of the transcendental value of biodiversity and nature, at a more general level, systems with which humanity interacts must be recognized as providing humanity with an opportunity to view and consider human behavior or relationships with its habitat. This opportunity for observation and learning must be granted some credible value if the experience leads to transcendence, transformation, or more simply, change in human values, cognition, or actions.

A moral imperative for action based on nonutilitarian values such as an ecological ethic or intrinsic value may also be envisaged (Norton, 1987; Katz, 1994). While the extent of the existence of motivating conditions for a moral imperative is difficult to assess, the extent of the crisis from the perspective of threats to human civilization can be assessed, as can the extent of threats to nonhuman elements of the biosphere.

The origins of the conflict

Human domination

Granting that changes in biodiversity and the biosphere are due to human life and activity, human activity must be related back to the motivating human values, recognizing that actions taken based on such values are also limited by the scope of human cognition. The shear growth in human population has resulted in substantial changes in the intensity of alteration and occupation of the landscape as well as in the intensity of its management to satisfy human needs (e.g., Holdren & Ehrlich, 1974; Myers, 1994). In both cases, these processes reflect the nature of human values as well as the scope of human cognition as manifested in human acceptance of dominion over the biosphere. Logically, rejection of the outcomes of past human activity must lead implicitly to rejection of the underlying human values (as in Katz, 1994, or Cairns, 1994).

Increasing aggregate human consumption demand has evolved as a result of three processes: 1) increased human population, 2) substantial aging of the population, and 3) aspirations for increased standards of living (Kim & Weaver, 1994). The latter process has been accelerated by the recent evolution of autocratic, centralized economies to decentralized, market-oriented economies in which individuals have been granted autonomy from the state in making their consumption decisions. However, in each case, these observations of underlying processes fail to cut to the roots of expanding human consumption and, instead, cite only the outcomes of human choice – symptoms of the conflict, rather than the origins and causes of the conflict, which must lie in human values and human cognition.

Human cognition and values

Human values and the limits of human cognition are revealed by the nature and extent of human consumption and utilization of elements of the biosphere. The demands for biodiversity and biosphere products and services have included demands for private goods and services available both directly or indirectly through transformation of limited biological resources. The scale of expansion of consumption of these products and services discussed earlier in this chapter attest to the presumption by humans of a right to limitless consumption. As already noted, human demands have also increased for the use of landscape and biosphere characteristics as disposal sinks for the products of production processes and economic activity that have no value for human consumption. This process further elaborates the nature of human values that must provide grounding to the evident presumption of human rights to such actions.

The extent of human knowledge of biological systems and cognition of impacts of human activity on those systems is extremely limited. For example, our knowledge of global biodiversity has been estimated as less than 15% of total biodiversity (Kim, 1993; Kim & Weaver, 1994). While our understanding of biological function, adaptation, and evolution for some species and ecological systems is measurable, knowledge of vast numbers of biological systems is lacking. This limited knowledge base testifies to our inability to control the impacts of human activity on biological systems, even where such human intent exists. The combination of human acceptance of dominion over biological systems and only a weak acceptance of stewardship responsibility are sufficient conditions for human actions that may negatively impact biodiversity. Repeatedly, humanity has proven itself capable of accepting the limits of its cognition (e.g., uncertain possible impacts of its actions on biodiversity and the biosphere) and to proceed, basing action solely on human values.

This phenomenon suggests a strong faith in the ability of humans to react to control impacts, even in the absence of initial knowledge that would be sufficient basis to plan and implement outcomes with certainty. For example, despite the known implications of CFC and CO_2 emissions, only limited actions have been taken to change human activities that generate these emissions. As another example, though on a more limited geographic scope, consider the impact of the draining and clearing of marshes in Israel. Estimates suggest that 26 plants became locally extinct, accounting for about 1% of the vascular flora in the area (Dafni & Agami, 1976; Vermeij, 1986). Acceptance of any rationale for this type of impact suggests an implicit willingness to accept unmeasurable uncertainty concerning its dynamic implications on the remaining biodiversity and, in parallel, implicitly asserts and exercises human dominion over biodiversity, even in the absence of knowledge of longer-term implications of the actions.

The limits of human cognition and its role in affecting the relationship between humanity and biodiversity are also implicitly defined by the nature and extent of human utilization of the biosphere. The agricultural and industrial revolutions have been key elements in the emergence of technological society (Kim and Weaver, in this volume) and resulted in the intensification in the use of biological resources for nearly a century before diminishing biological capacity and damage to biological systems was generally recognized. As Thoreau wrote in *Walden* [Thoreau, 1854 (1960)] human dominance over landscapes and biological systems was predominantly accepted and often the subject of marvel [e.g., Emerson's acceptance of the railroad's intrusion into the landscape (Norton, 1987; Marx, 1964)]. As the implications of human activity for biodiversity began to be recognized in the second half of the nineteenth century, the relevance of such implications were not generally appreciated and often disregarded. Clearly, these limits to human cognition remain an important constraint on the nature of human response to the paradox of humanity and biodiversity.

A dramatic testament to the acceptance of human dominion and the role of human cognition is the widespread acceptance of the paradigm that human activity should proceed until scientific evidence confirms biodiversity effects (cognition) and at the same time until consensus exists within human society that the observed effects are undesirable (human values). Acceptance of such a paradigm implicitly accepts the notion and value of human dominion. The role of myopic human cognition appears in several contexts both at an individual and collective or societal level. At an individual level, the capacity of humans to recognize the long-term impacts of their actions may be limited by the extent of scientific or otherwise credible data, however, other innate limits in human cognitive capacity may exist as well. The debate at the turn of 1960s over the existence of limits to growth illustrates the extent and nature of limits on human cognition at a societal level.

The role of human population pressure in the current process of biodiversity degradation (see, e.g., Ehrlich, 1988) might be viewed as a further illustration of the limited nature of individual, and therefore, collective, human cognition of the implications of human actions. Mainstream economists view the problems associated with population pressure as ones of congestion in the utilization of a public good, a problem that is commonly acknowledged to result from the failure of decentralized markets, and many other pricing mechanisms, to value limited capacities for the consumption of many public goods and services. Individuals pursuing their own hedonistic goals will not manage the allocation of such limited capacity to allow the greatest social good to be achieved. From this perspective, the source of the conflict between humanity and biodiversity could be argued to result from the failure of individuals to coalesce for the purpose of actions that jointly serve their society.

Human dominion

The notion of human dominion is jointly grounded in human values and human cognition, though it nevertheless deserves special attention. The validity of the notion of human dominion over biodiversity has been supported as feasible through past human experience. The acceptance of human dominion also has strong cultural bases in Western cultures, in general, and in the Judeo-Christian tradition, in particular, as transformed by the acceptance and pursuit of human potential for salvation during the Renaissance and Reformation (see White, 1967; Taylor, 1984; Berry, 1979; Callicott, 1986; Norton, 1987).

The revolution in agricultural domestification of plants exemplifies both human domination of nature and human acceptance of a right to dominance over nature. It broke natural feedback loops that otherwise posed limits to production (see Wilkes, 1994). As a further example, Dutch agriculture was considered a miracle of the twentieth century. For an over thousand years, the Dutch wrestled land from the sea and converted swamp and unstable river delta into prosperous farmlands. Because of its damage to the Dutch landscape and the excessive costs of managing the sinking lowlands, the Dutch government has initiated an ambitious master plan to restore nature by buying up tens of thousands of acres to be rehabilitated to its natural state (Simons, 1993).

Perceived human dominion over nature brought with it the opportunity for abstraction from nature and biological systems. The extent of human abstraction from nature is substantial. Urban populations survive with little consciousness or understanding of natural systems. Rural populations involved in modern agriculture may view biological systems as stores of resources while remaining relatively isolated from the biological systems themselves. Humanity has evolved to view technology and human intelligence as providing freedom from the constraints of nature, implying that humanity no longer has to adapt to nature, but instead has within its capacity the control of nature (McKibben, 1989; Weiner, 1990; Cairns, 1994). Going further, some authors view human abstraction from nature as having evolved to a point where humans, in our technological society, are unconscious of nature in a general sense (Cairns, 1994). At a further extreme, it can be argued that anthropocentric perspectives on the value of biodiversity are simple implications of both the abstraction of humanity from nature as well as of the depth of cultural acceptance of domination of humanity over nature (see, e.g., Norton, 1994).

The Crux of the paradox

The current state of dominance of humanity over biological systems could be argued to be evidence of dysfunction of humanity in the natural system. While dominance has emerged in many senses, human dominance has been exercised with only limited interest in or ability for control of the human impacts on biological systems. This situation suggests human willingness to act is based on human values with little consideration of the existence of any limits in human cognition. For example, extractive resource activities and expansion of human manipulation of landscapes through urbanization or agricultural development are stereotypically viewed as occurring with a myopic purpose that excludes any interest in immediate or longer-term ecological impacts. Nonetheless, it must be recognized that this view is held by individuals who apparently recognize, and value, the more general implications of these human activities. In many cases, these individuals have successfully coalesced to argue for central governmental regulatory action to influence the course of these activities. Thus, while it would seem safe to acknowledge the roles of human values and limited human cognition in the paradox of humanity and biodiversity, these same ingredients can also be cited as essential for the solution of the paradox. This logic suggests that it is the balance or equilibrium of social, political, and economic processes based on human cognition and human values that must be cited as lying at the root of the paradox.

Stepping past human values and cognition as determinants of human impacts on biodiversity, an important question is whether and under what conditions could collective human motivation be marshalled to manage those impacts even if adequate knowledge were to exist. Focusing on human cognition and values as crucial determinants of the paradox, it is useful to distinguish two cases that will be useful for the consideration of the possibilities for change in the relationship between humanity and biodiversity. In the first, suppose humans are universally myopic, in the second, suppose only some humans are myopic, and others are nonmyopic. In addition, suppose we add that individuals in case one are universally hedonistic, concerned only with fulfillment of personal demands and aspirations. In case two, we suppose some individuals are altruistic. From a Darwinian perspective, the orientation of case one would seem both natural and necessary for survival. However, no change in human activity could be expected given case one. Logically, change could occur for case one only if 1) a change in human cognitive capacity occurred, i.e., the myopia shifted to nonmyopia; or 2) individual hedonism shifted toward some form of altruism.

For case two, a heterogeneous society is envisioned with at least two types of human values and levels of cognitive skill: a) hedonistic and myopic and b) altruistic and nonmyopic. The case allows the question: under what conditions could the actions of a collection of type b individuals be expected to respond to recognition of the impacts of their collective or individual actions? Western economic thought supposes that individual economic actions can be orchestrated by competitive markets to utilize limited resources to achieve the greatest good for the greatest number (maximum social welfare). Nonetheless, can such an invisible-hand process be hoped for in the case of management of the biological impacts of human activity? Given that the answer to such a query is typically a resounding no (Weaver, 1994), what other social or institutional arrangements are both acceptable to human society and capable of generating improvements in the nature of human impacts on biodiversity and the biosphere?

To answer these questions, it is essential to begin by considering the origins of human institutions as lying squarely in the balance of human values and human cognition. Human institutions evolve in response to recognized failures (performance failures) in individual actions or behavior to generate acceptable outcomes (Mueller, 1993). This process of recognition relies upon the arbitration of heterogeneous levels of human cognition and human values. In addition to recognized performance failures, construction of new institutions by a coalition of individuals requires collective recognition of some value-based incentive. For example, such an incentive might be that collective action through rule making and enforcement to regulate individual activity could improve the human value generated by associated, new outcomes of human activity as influenced by the adopted institutions. Human values establish the relevance of any recognition of performance failure of individual action. Importantly, values held by individuals who might benefit from collective action also will determine their willingness to collude to take collective action. Since the values held can be expected to vary across individuals, typically not all individuals will perceive the coalition as beneficial. In such a case, evolution of institutional changes may be stifled (Olson, 1965).

The human impacts on biodiversity are viewed by economists as *externalities*, or peripheral impacts of economic or other human activity that are not valued and so do not feed back to the individuals generating them (Weaver, 1994a). This view is equally relevant and appropriate independently of the nature of the economic, social, or political system in which the human activity is conducted. In all systems, externalities result in individuals being impacted by actions of other individuals, implying the impacted individual could improve the personal value of outcomes experienced by controlling the externality. In many systems, individual values are discounted, implying no increase in system value would occur from managing extemalities to improve the welfare of the impacted individuals. In such cases, individual action to amend and manage externalities may be discouraged or prevented. However, where individual action is tolerated, it can be expected to pursue means of managing the extemality. One course for such solution is through institutional change motivated by shifts in the balance of human values and cognition.

Numerous social, political, and economic obstacles inhibit the coalescence of individuals and the taking of collective action to prevent or alter their actions. That is, existing institutions and the distribution of human values and cognitive powers among both individuals that resist and want change may strike a balance that makes change infeasible. In the case of human impacts on biodiversity, the externalities involved are only sometimes obvious and often institutional means for their management are not in place (Weaver, 1994). The absence of these institutional means can be inferred to imply an absence of either human cognition or values that might motivate institutional change given the costs of achieving the change and the expected outcomes of the changes. It is important to recognize, however, that it is both the nature and the distribution of human cognition and values across individuals in a society that determine whether change can occur. These issues will be considered in more depth in a forthcoming discussion.

Solution of the paradox of humanity, biodiversity, and landscapes and identification of actions which might be sufficient to alter the relationships among these elements of the biosphere must be grounded on a hypothesis concerning the roots and the origins of the conflict among the elements. Previous discussion has cited several possible anthropogenic origins: 1) the myopia of human cognition, 2) the nature of human values, or 3) the limited extent and feasibility of human collective actions.

The challenge for change

The imperative

Ultimately, the imperative for change must be motivated by widespread human recognition of a need to alter the dynamic relationship that has evolved between humanity and biodiversity. Consistent with the receding discussion, such recognition will ultimately rely on the balance of both current human values and cognition as well as the inertia of past human values and cognition. Given that the impacts of technological society on biodiversity are global in scale, human recognition of the imperative for change must be achieved at a global level that transcends cultures and their environmental contexts. At the same time, the nexus of humanity and biodiversity occurs at a micro level, and human recognition must evolve across a very heterogenous set of physical conditions, experiences, and initial positions of human values and cognition. Nonetheless, both pragmatic and ethical arguments have been raised to support and rationalize change in the human role in the biosphere.

From an instrumental perspective, preservation of biodiversity can be defended as desirable to ensure high standards of living. However, such instrumental arguments de- emphasize the nonutilitarian value of biodiversity and are insufficient bases to rationalize conserving biodiversity, since alternative means may be substituted for any instrumental use to achieve equivalent use value (Katz, 1994, Rolston, 1987). Along similar lines, others have noted that while humans benefit from natural beauty and even satisfaction of their biophilia, this exposure is certainly not necessary for the continuation of human life (Katz, 1994). From these perspectives, humans possess the capacity to withdraw from nature and become technologically and culturally abstracted from nature, implying that biodiversity and nonhuman nature is expendable (Hargrove, 1994). Whether this will ever be possible is certainly unclear, but its possibility would seem to suggest weakness in arguments for change based on the instrumental value of biodiversity or on its necessity for human survival.

From an ethical perspective, ecological ethicists have argued that rationalization of environmental policy or change in the human impacts on biodiversity cannot be logically based on anthropogenic value systems such as utilitarian or instrumental value of biodiversity to humanity (see, e.g., Hargrove, 1994; Katz, 1994). Instead, an ecological ethic is proposed that transcends human use value and establishes a moral value system (Katz, 1994). Katz argues that such moral value can only emerge as a result of the extension of the human sense of community to include all life in the biosphere. As both Norton (1994) and Katz (1994) have noted, both Thoreau's and Leopold's sense of community might also be viewed as a sufficient basis for motivation of such an ecological ethic. Logically, if technological society has resulted in an abstraction from nature (see, e.g., Kim & Weaver, 1994; Hargrove, 1994), then emergence of such a sense of community would require deep changes in the social, cultural, and economic processes that have resulted in the current extent of abstraction from nature.

Opportunities and constraints

Conservation strategies were cited throughout the volume as providing one recourse for changing the conflict between humanity and biodiversity. However, these strategies must be viewed as likely to be ineffective unless the basic conflicts outlined above, and postulated as lying at the root of the paradox, are resolved. In this sense, the past relationship between human activity and biodiversity is viewed by many as incompatible with sustained increases in human consumption. This incompatibility is implicit even in elements of past conservation strategies such as reserves, ranging from gene banks to biosphere reserves, which attempt to isolate biodiversity from humanity. This motivates, for some observers, the conclusion that a fundamentally different course for economic activity must be taken to allow and ensure a sustainable relationship between human activity and biodiversity. Consistent with the preceding discussion, such a shift in the course of human activity can only occur as a result of supporting changes in underlying human values and cognition.

This conclusion is consistent with the argument that performance failures of market based economies can be amended (see Weaver, 1994). While this possibility is accepted, implementation of a strategy to amend the performance failure ultimately depends upon the nature and distribution of human values and cognition. For example, the following bases for performance failure of market-based systems have been cited: 1) failure of markets to price biological resources, 2) impossibility of quantification of the benefits and costs associated with actions affecting biological resources, and 3) difficulty in appropriation of the benefits of stewardship of biological resources (e.g., McNeely et al., 1990, Chapter III). In fact, concerns 1) and 3) are a direct result of the economic nature of the impacts of human activity on biodiversity (Weaver, 1994) and concern 2) represents simply a constraint on the extent to which performance failures can be corrected.

More fundamentally then, human cognition and human values must be viewed as at the core of the challenge of resolving the paradox of humanity and biodiversity. For case 2 above, two types of human cognition and values were identified as useful in considering the potential for change in human activities: 1) myopic and hedonistic, and 2) nonmyopic and altruistic. These extremes might be viewed as the diagonal cells in a two-by-two table of possible types, e.g., myopic and altruistic might lie in one of the off-diagonal cells. The opportunity for change can now be assessed.

Any strategy to change the human impacts on biodiversity would face the challenge of changing individuals of type 1. While myopia might be changed through education and information processes, it is unclear what strategy could be used to end hedonism. In the absence of such opportunity, it is conceivable in a heterogeneous society involving some individuals of type 2 that political power might be secured by a coalition of the type 2 individuals and hedonist behavior might be regulated to achieve outcomes such as resolution of the paradox of humanity and biodiversity. In fact, so long as the society includes individuals with some level of nonmyopia and altruism, change through regulation of myopic and hedonistic behavior is conceivable. Going further, substantial change might be achieved through empowerment of the nonmyopic and altruistic individuals to relax the impacts of their activity on biodiversity.

Given these possibilities, the focus shifts from the need to change underlying human cognition or values to implementing shifts in power within society to achieve change. Important lessons with respect to these challenges are available from the economics of collective action and by analogy from recent literature concerned with transition of social, political, and economic systems (e.g., Olson, 1965; Przeworski, 1992). Within this literature, the notion of social change is considered from the perspective of conflicting stakeholders finding evolving equilibria as the social, political, and economic conditions change, shifting opportunities and possibilities for formation or adjustment of coalitions of individuals.

In the case of human impacts on biodiversity, these perspectives clarify the challenge of achieving change. At any point in time, previous and new conditions and outcomes establish stakeholders or individuals who personally benefit (hedonistically) from particular types of consumption of biodiversity. Similarly, stakeholders with altruistic motivation based on utilization and nonutilization values might benefit from particular actions. The equilibrium achieved results from the resolution of conflict among competing stakeholders within the institutional rules of conflict admitted by them. These rules of conflict may or may not be supported by a general consensus. In some cases, rules may be granted by empowered coalitions in order to perpetuate their own goals. To conclude, the distribution of stakes, the relative power of stakeholders, and the rules of conflict can be viewed as crucial constraints on the process of change.

The goal of change

To conclude this section, the solution of the paradox of humanity and biodiversity will require changes in the balance of human cognition and values. The goal of these changes must be viewed as sustainable relationships among biodiversity and humanity. Within the context of economics, this goal must be distinguished from economic growth, which carries with it no conditions for its implications for biodiversity. Instead, the goal must be viewed as achieving intensification of the use of limited resources, including those of the biosphere. That is, intensification must not increase the volume of use, but rather expand the productivity of those resources both in generating human consumption goods and services as well by reducing the burden that use imposes on biodiversity, e.g., on the biosphere through use of its limited capacity as a disposal sink. Others have distinguished this change in orientation as a shift from development to environmentally sensitive or sustainable development (South Commission, 1990; Haavelmo and Hansen, 1991).

Key to the achievement of this goal will be changing the balance of or equilibrium of human cognition and values. This can be achieved through at least three tactics: 1) the empowerment of individuals and coalitions with nonmyopic cognitive capacity and altruistic values, 2) presentation of information and knowledge of long-term external effects of human activity to myopic hedonists, possibly altering their orientation toward nonmyopic perspectives and altruistic values, and 3) alteration of the behavior of myopic hedonists through regulation with enforcement, institutional change, and shifts of incentives to reflect social net benefits of human activity. The goal of these tactics must be articulation and implementation of the conservation of the generalized notion of biodiversity introduced in the first section of this chapter.

Through these changes in the social, political, and economic equilibrium of human cognition and values, a variety of changes in human activity and behavior might be expected, e.g., reduced throughput consumption of biological resources, recognition and management of biological capital in the interest of a global view of society, or underestimation rather than overestimation of technological opportunities for reversing human impacts on biodiversity.

Options and alternative paths of change

Adoption of tactical options for effecting a change in the relationship between humanity and biodiversity will require a shift in the social, political, and economic consensus from that which supports and perpetuates the current nature of the conflict between humanity and biodiversity. Thoreau's *Walden* has been interpreted as recognizing that nature itself offers humanity a path away from the current paradoxical conflict with nature (Norton, 1994). In this interpretation, Thoreau views humans as having the capacity to adapt to challenge and imperatives for a new role of humanity in nature by finding paths to simple lives, to transcend what Thoreau views as a larval stage of exhaustive consumption to a freedom from addictive consumption. Within this context, Thoreau viewed nature as an instrument of learning, rediscovery of wonderment, and ultimately of transcendence over materialistic consumerism to a new worldview (see Norton, 1994).

Shifting the consensus

From a more conventional perspective, opportunity for society to move beyond its current equilibrium of conflict between humanity and biodiversity ultimately relies on the empowerment of new coalitions of individuals who support as a goal the achievement of a new equilibrium in the relationship between humanity and biodiversity. Going back to the two extreme types of human cognition and human values cited in the previous section, the prerequisite for change requires establishment of a new consensus that society will be better off if political, social, and economic power is transferred away from myopic hedonists toward individuals with broader perspectives (nonmyopic) and some element of altruism.

This shift in equilibrium may require tactics based on education to enhance the understanding and appreciation of myopic hedonists of the extent, magnitude, and utilitarian net costs of the current conflict between humanity and biodiversity. Alternatively, the shift may be facilitated through external intervention of global or other national agencies as in the case of Brazil (Binswanger, 1987; Repetto, 1988). If such efforts are successful, a new societal consensus may be established for empowerment of individuals with less myopic and more altruistic views. The Green movements of the past decades provide examples of this type of strategy for shifting societal consensus (Nerfin, 1987; Sen, 1981).

A crucial aspect of the challenge of changing the consensus is that it must occur at the local level, between national and subnational jurisdictions, as well as between global and national jurisdictions (Potter, 1990). Starting in the 1980's, a common element of natural resource management projects in many developing countries has been empowerment of local communities through decentralization (Klee, 1980; McNeely and Pitt, 1984; Marks, 1984; de Camino Velozo, 1987; Potter, 1990). At the same time, an extensive history already exists of national and international efforts to shift consensus. International actions to change the impact of humanity on biodiversity have a long history, e.g., the regulation of whaling in 1946, the Ramsar Convention on Wetlands (Unesco, 1972), the CoCoyoc Declaration (UNEP, 1981), the Bali Declaration (see McNeely and Miller, 1984), the Brundtland Report (WCED, 1987), and the recent convocation of the United Nations Conference on Environment and Development or Earth Summit (see Barton, 1992). McNeely et al. (1984) present a thorough review of this legislation.

Implementing change with a new consensus

Having established such a shift in power, a variety of tactics are available to effect change in human activity. To facilitate changes in the behavior of members of the new coalition (call it the Greens), new institutions may be created, and the Greens could be empowered to manage their impacts on biodiversity. A substantive challenge will be to alter the impact of nonGreen activity on biodiversity. This can be achieved in one of two ways: 1) alteration of the understanding and appreciation of nonGreens of the biodiversity impacts of their actions through presentation of information and 2) regulation with enforcement of nonGreen activity. The first tactic can involve any of a variety of educational and market-based approaches including altering market incentives and opportunities as well as institutional change from the local to global levels. Regulatory approaches might include what could be thought of as the old notion of biodiversity conservation including the limiting of access and use of biodiversity resources, *ex situ* collection, or restriction of specific types of human activity. As a safety first strategy (Bishop, 1982), regulation by enforcement allows for immediate redress of aspects of the conflict between humanity and biodiversity, which might otherwise lead to reversible outcomes. Nonetheless, adoption of this tactic will require establishment of the necessary societal consensus.

References

Alagh, Y. K. (1991). Sustainable development. From Concept to Action – Techniques for Planners. Paper prepared for the UNCED Secretariat.

Barton, John M. (1992). Biodiversity of Rio. *Bioscience*, November.

Berry, W. (1979). *Standing by the words: The biblical basis for ecological responsibility*. Sage Chapel Convocation, Cornell University, Ithaca, N.Y.

Binswanger, H. P. (1987). *Fiscal and Legal Incentives with Environmental Effects on the Brazilian Amazon*. World Bank Report ARU 69:1–48.

Bishop, R. (1982). Option value: An exposition and extension. *Land Economics*, **58**(Feb.), 1–15.

Brady, N. C. (1988). International development and the protection of biological diversity. In *Biodiversity*, pp. 409–418. Washington, D.C.: National Academy Press.

Briggs, J. C. (1991b). Global species diversity. *Journal of National History*, **25**, 1403–1406.

Cairns, J., Jr. (1988). Can the global loss of species be stopped? *Spec. Sci. Technol.*, **11**(3), 189–196.

Cairns, J., Jr. (1994). Technology and biodiversity conservation. In *Biodiversity and Landscapes*, ed. K. C. Kim & R. D. Weaver, pp. 327–338. New York: Cambridge University Press.

Callicott. J. B., (1986). On the intrinsic value of nonhuman species. In *The Preservation of Species*. ed. B. G. Norton, pp. 133–172. Princeton, N.J.: Princeton University Press.

Carson, R. (1962). *Silent Spring*. Boston: Houghton, Mifflin Co.

Clark, C. W. (1973). The economics of overexploitation. *Science*, **181**, 630–634.

Collar, N. J., & Andrew J. (1988). *Birds to Watch: the ICBP World Check-list of Threatened Birds*. ICBP Technical Publication No. 8. ICBP, Cambridge, UK.

Conservation International. (1990). *The Rain Forest Imperative*. Washington, D.C.: Conservation International.

Constanza, R. (1991). The ecological economics of sustainability: Investing in natural capital. In *Environmentally Sustainable Economic Development: Building on Brundtland*, ed. Goodland et al., pp. 83–90. Paris: UNESCO.

Constanza, R. (1992). Natural capital and sustainable development. *Conservation Biology*, **6**(1), 37–46.

Cutler, M. Rupert. (1990). The watchdog role of nongovernmental environmental organizations. In *Biodiversity and Landscapes*, ed. K. C. Kim & R. D. Weaver, pp. 371–380. New York: Cambridge University Press.

Dafini, A., & Agami, M. (1976). Extinct plants of Israel. *Biological Conservation*, **10**, 43–52.

Daly, H. E. (1990). Toward Some Operational Principles of Sustainable Development, *Ecological Economics*, **2**, 1–6.

Daly, H. E. (1991). From empty-word to full-world economics: Recognizing an historical turning point in economic development. In *Environmentally Sustainable Economic Development: Building on Brundtland,* ed. Goodland et al. Paris: UNESCO.

Daly, H. E., & Cobb, J. B. Jr. (1989). *For the Common Good: Redirecting the Economy Toward Community, the Environment, and a Sustainable Future.* Boston: Beacon Press.

Dasgupta, P. (1982). *The Control of Resources.* Cambridge, MA: Harvard University Press.

Davis, S., Stephen, D., Droop, J. M., Gregerson, P., Henson, L., Leon, C. J., Villa-Lobos, J. L., Synge, H., & Zantovska, J. (1986). *Plants in Danger: What do we Know?* Gland, Switzerland: IUCN.

Davis, S. H. (1988). *Indigenous Peoples, Environmental Protection and Sustainable Development.* IUCN Sustainable Development Occasional Paper 1, 1–26.

de Camino Velozo, R. (1987). Incentives for community involvement in conservation programs. *FAO Conservation Guide,* **12**, 1–59.

Dobzhansky, T. (1962). Mankind evolving. In *The Evolution of Human Species.* New Haven: Yale University Press.

Eckholm, E. (1978). Disappearing species: The social challenge. *Worldwatch Paper* 22.

Ehrlich, P. (1968). *The Population Bomb.* New York: Ballantine.

Ehrlich, P. (1986). *The Machinery of Nature.* New York: Simon and Schuster.

Ehrlich, P., & Ehrlich, A. (1981). *Extinction: The Causes and Consequence of the Disappearance of Species.* New York: Random House.

Ehrlich, P. R., & Ehrlich, A. H. (1990). *The Population Explosion.* New York: Simon and Schuster.

Ehrlich, P. R., & Mooney, H. A. (1983). Extinction, substitution, and ecosystem services. *Bio-Science,* **33**(4), 248–254.

Emerson, R. W. (1982) Nature. In *Selected Essays,* ed. L. Ziff. New York: Penguin Books.

FAO. (1981). *Tropical Forest Resources Assessment Project (GEMS): Tropical Africa, Tropical Asia, Tropical America* (4 Vols.). Rome: FAO/UNEP.

Frankel, O. H., & Hawkes, J. G. (eds). (1974). *Plant Genetic Resources for Today and Tomorrow.* London: Cambridge University Press.

Frankel, O. H., & Soulé M. E. (1981). *Conservation and Evolution.* New York: Cambridge University Press.

Goode, D. A. (1984). Conservation and value judgements. In *Planning and Ecology,* ed. Roberts, R. D., & T. M. Roberts. London: Chapman and Hall.

Goodson, J. (1988). *Conservation and Management of Tropical Forests and Biodiversity in Zaire.* Unpublished document, USAID, Zaire.

Goodland, R. (1991). The case that the world has reached limits: More precisely that current throughput growth in the global economy cannot be sustained. In *Environmentally Sustainable Economic Development: Building on Brundtland,* ed. Goodland et al., pp. 15-27. Paris: UNESCO.

Goldsmith, E. (1985). Is development the solution or the problem? *The Ecologist,* **15**(5&6), 210.

Gray, J., & Shear, W. (1992). Early life on land. *American Scientist,* **80**, 444–456.

Groombridge, B. (ed.). (1992). *Global Biodiversity, Status of the Earth's Living Resources* (compiled by World Conservation Monitoring Center). London: Chapman and Hall.

Haavelmo, T., & Hansen, S. (1991). On the strategy of trying to reduce economic inequality by expanding the scale of human activity. In *Environmentally Sus-*

tainable Economic Development: Building on Brundtland, ed. Goodland et al., pp. 41–49.

Hafernik, J. E. Jr. (1992). Threats to invertebrate biodiversity for conservation strategies. In *Conservation Biology. The Theory and Practice of Nature Conservation, Preservation and Management,* ed. P. L. Fiedler & S. K. Jain. New York: Chapman and Hall.

Hargrove, E. (1994). The Paradox of humanity: two views of biodiversity and landscapes. In *Biodiversity and Landscapes,* ed. K. C. Kim & R. D. Weaver, pp. 173–186. New York: Cambridge University Press.

Harris, L. D. (1984). The Fragmented Forest: Island Biogeographic Theory and the Preservation of Biotic Diversity. Chicago: University of Chicago Press.

Heilbronner, R. (1991). Lifting the silent depression. *The New York Review of Books,* **38**(17).

Holdgate, M. W. (1989). the implications of climatic change and rising sea level. In *Proceedings of the International Congress on Nature Management and Sustainable Development,* ed. W. Verwey. Amsterdam: International Organizing Services.

Holdren, J. P., & Ehrlich, P. R. (1974). Human population and the global environment. *Am. Science,* **62**, 282–292.

Hsu, K. J. (1982). Mass mortality and its environmental and evolutionary consequences, *Science,* **216**, 225.

Hughes, D. J. (1975). *Ecology in Ancient Civilizations.* Albuquerque: University of New Mexico Press.

Humphrey, S. R., & Smith, B. M. (1990). A balanced approach to conservation. *Conservation Biology,* **4**(4), 341–343.

Ingram, G. G. (1990). Management of biosphere reserves for the conservation and utilization of genetic resources: The Social Choices. *Impact,* **158**, 133–142.

IUCN/UNEP. (1968a). *Review of the Protected Areas System in the Afrotropical Realm.* Gland, Switzerland: IUCN.

IUCN/UNEP. (1986b). *Review of the Protected Areas System in the Indo-Malayan Realm.* Gland, Switzerland: IUCN.

Jayawardena, L. (1991). *A Global Environmental Compact for Sustainable Development: Resource Requirements and Mechanisms.* Helsinki: WIDER/UNU.

Katz, E. (1994). Biodiversity and ecological justice. In *Biodiversity and Landscapes,* ed. K. C. Kim & R. D. Weaver, pp. 61–74. New York: Cambridge University Press.

Kellert, S. W. (1986). Social and perceptual factors in the preservation of animal species. In *The Preservation of Species,* ed. B. G. Norton, pp. 50–78. Princeton, N.J.: Princeton University Press.

Kim, K. C., & Weaver, R. D. (1994). Biodiversity and humanity: paradox and challenge. In *Biodiversity and Landscapes,* ed. K. C. Kim & R. D. Weaver, pp. 3–27. New York: Cambridge University Press.

Kim, K. C. (1993). Biodiversity, conservation and inventory: why insects matter. In *Biodiversity and Conservation,* **2**, 191–214.

Klee, G. A. (ed.). (1980). *World Systems of Traditional Resource Management.* New York: Wiley.

Kothari, R. (1990). Environment, Technology and Ethics. In *Ethics of Environment and Development – Global Challenge, International Response,* ed. J. R. Engel and J. G. Engel. Tuscon: The University of Arizona Press.

Kristensen, T., & Paluden, J. P. (1988). *The Earth's Fragile Systems: Perspectives on Global Change.* Boulder, CO: Westview Press.

Lawton, H. W., & Wilke, P. J. (1979). *Ancient Agriculture in Semi-Arid Environments.* New York: Springer-Verlag.

Leader-Williams, N., & Albon, S. D. (1988). Allocation of Resources Conservation. *Nature,* **336**, 533–536.

Lovejoy, T. E. (1980). A Projection of Species Extinctions. In *The Global 2000 Report to the President,* prepared by the council on Environmental Quality, pp. 327–332. Washington, D.C.: U.S. Government Printing Office.

Lovejoy, T. E. (1986). Species leave the ark one by one. In *The Preservation of Species,* ed. B. G. Norton, pp. 50–78. Princeton, N.J.: Princeton University Press.

Lovejoy, T. E. (1988). Diverse Considerations. In *Biodiversity,* ed. B. G. Norton, pp. 421–427. Washington, D.C.: National Academy Press.

Mace, G. A., & Lande, R. (1991). Assessing extinction threats: Towards a reevaluation of IUCN threatened species categories. *Conservation Biology,* **5**, 148–157.

McKibben, B. (1989). *The End of Nature.* New York: Random House.

McNaughton, S. J. (1989). Ecosystem and conservation in the twenty-first century. In *Conservation in the Twenty-First Century,* ed. Western, D. and M. C. Pearl. New York: Oxford University Press.

McNeely, J. A. (1988). *Economics and Biological Diversity: Developing and Using Economic Incentives to Conserve Biological Resources.* Gland, Switzerland: IUCN.

McNeely, J. A., & Miller, K. R. (eds.). (1984). *National Parks, Conservation, and Development: The Role of Protected Areas in Sustaining Society.* Washington D.C.: Smithsonian Institution Press.

McNeely, J. A., & Pitt, D. (eds.). (1984). *Culture and Cosnervation: The Human Dimension in Environmental Planning.* London: Croom Helm.

McNeely, J. A., Miller, K. R., Reid, W. V., Mittermeier, R. A., & Werner, T. B. (1990). *Conserving the World's Biological Diversity.* Gland, Switzerland and Washington, DC. Inter. Union for Conserv. of Nature and Natural Resources, WRI, CI, WWF-US, the Word Bank.

McNeely, J. A., Miller, K. R., Reid, W. V., Mittermeir, R. A., & Werner, T. B. (1990). *Conserving the World's Biological Diversity,* Gland, Switzerland, and Washington DC.

Maler, K. G. (1974). *Environmental Economics: A Theoretical Inquiry.* Baltimore: Johns Hopkins University Press.

Marks, Stuart A. (1984). *The Imperial Lion: Human Dimensions of Wildlife Management in Central Africa.* Boulder, CO: Westview Press.

Marx, L. (1964). The Machine in the Garden. New York: Oxford University Press.

Meadows, D. H., Meadows, D. L., Randers, J., & Behrens, W. III. (1972). *The Limits to Growth.* New York: Universe Books.

Meadows, D. H., Meadows, D. L., & Randers, J. (1992). *Beyond The Limits Confronting Global Collapse, Envisioning a Sustainable Future.* Post Mills, Vermont: Chelsea Green Publishing Co.

Miller, J. R. (1981). Irreversible land use and the preservation of endangered species. *Journal Environmental Economics and Management,* **8**, 19–26.

Mitsch, W. (ed.). (1987). *Ecological Engineering.* New York: Wiley.

Mitsch, W., & Jorgensen, S. (eds.). (1989). *Ecological Engineering, an Introduction to Ecotechnology.* New York: Wiley.

Mittermeir, Russell A. (1988). Primate diversity and the Tropical Forest Case Studies from Brazil and Madagascar and the Importance of the Megadiversity Countries. In *Biodiversity.* Washington, D.C.: National Academy Press.

Mooney, H. A. (1988). Lessons from mediterranean-climate regions. In *Biodiversity,* pp. 157–165. Washington, D.C.: National Academy Press.

Morton, E. (1985). The Realities of Reintroducing species into the Wild: The Problem of Original Habitat Alteration. In *Animal Extinctions*: What Everyone Should Know, ed. R. J. Hoage. Washington D.C.: Smithsonian Institution Press.
Mueller, D. C. (1993). *Public Choice II*. New York: Cambridge University Press.
Myers, N. (1979a). Conserving our global stock. *Environment*, **21**, 25–33.
Myers, N. (1979b). *The Sinking Ark: A New Look at the Problem of Disappearing Species*. Oxford: Pergamon Press.
Myers, N. (1983). *A Wealth of Wild Species*. Boulder, CO: Westview Press.
Myers, N. (1985). *The Gaia Atlas of Planet Management*. New York: Doubleday.
Myers, N. (1986). *Tackling Mass Extinction of Species: A Great Creative Challenge. Albright Lecture*. University of California, Berkeley.
Myers, N. (1988). Environmental Degradation and Some Economic Consequences in the Philippines. *Environmental Conservation*, **15**, 205–214.
Myers, N. (1988c). Threatened biota: "Hotspots" in tropical forests. *Environmentalist* **8**(3), 1–20.
Myers, N. (1989a). *Deforestation Rates in Tropical Countries and Their Climatic Implications*. London: Friends of the Earth.
Myers, N. (1989b). Major extinction spasm: Predictable & inevitable? In *Conservation in the Twenty-first Century*, ed. D. Western & M. C. Pearl. New York: Oxford Univ. Press.
Myers, N. (1994). We do no want to become extinct: the question of human survival. In *Biodiversity and Landscapes*, ed. K. C. Kim & R. D. Weaver, pp. 133–150. New York: Cambridge University Press.
National Research Council (NRC). (1989). *Evaluation of Biodiversity Projects*. Washington, D.C.: National Academy Press.
Nerfin, M. (1987). Neither prince nor merchant: Citizen – an introduction to the Third System, *Development Dialogue*, 1.
Nicholson, W. (1990). *Intermediate Microeconomics and Its Application*. Fifth Edition. Homewood, IL: The Dryden Press.
Nordhaus, W., & Tobin, J. (1972). *Is Growth Obsolete?* National Bureau of Economic Research. New York: Columbia University Press.
Norgaard, R. B. (1984). Environmental economics: an evolutionary critiqe and a plea for pluralism. Division of Agricultural Sciences, University of California, Berkeley, CA. Working Paper 299, 1–24.
Norgaard, R. B. (1988). The rise of the global exchange economy and the loss of biological diversity. In *Biodiversity*, ed. E. O. Wilson, pp. 206–211. Washington, D.C.: National Academy Press.
Norton, B. (1986) *The Preservation of Species: The Value of Biological Diversity*. Princeton: Princeton University Press.
Norton, B. (1987). *Why Preserve Natural Variety?* Princeton: Princeton University Press.
Norton, B. (1988). Commodity, amenity, and morality: the limits of quantification in valuing biodiversity. In Wilson, E. O. (ed.), *Biodiversity*, ed. E. O. Wilson, pp. 200–205. Washington, D.C.: National Academy Press.
Norton, B. (1994). Thoreau and Leopold on science and values. In *Biodiversity and Landscapes*, ed. K. C. Kim & R. D. Weaver, pp. 31–46. New York: Cambridge University Press.
Norton, B. R. (1986). On the inherent dangers of undervaluing species. In *The Preservation of Species*, ed. B. G. Norton, pp. 110–137. Princeton, N.J.: Princeton University Press.
Odum, E. P. (1969). The Strategy of ecosystem development, *Science*, **164**, 266.
Odum, H. T. (1994). "Emergy" evaluation of biodiversity for ecological engineer-

ing. In *Biodiversity and Landscapes*, ed. K. C. Kim & R. D. Weaver, pp. 339–359. New York: Cambridge University Press.
OECD. (1991). *The State of Environment.* Paris: Organization for Economic Co-operation and Development.
Olson, M., Jr. (1965). *The Logic of Collective Action.* Cambridge, MA: Harvard University Press.
Osborn, F. (1948). *Our Plundered Planet.* New York: Little, Brown.
Page, T. (1979). Keeping Score: An Actuarial Approach to Zero-Infinity Dilemmas, Social Science Working paper no. 248 (Pasadena, Calif.: Division of Humanities and Social Sciences, California Institute of Technology).
Passmore, J. (1974). *Man's Responsibility.* New York: Charles Scribner's Sons.
Pearce, D. W., & Turner, R. K. (1989). Economics of Natural Resources and the Environment. Brighton, U.K.: Wheatsheaf.
Perrings, C. A. (1988). An optimanl path to extinction? Powerty and resource degradation in the open agrarian economy. *Journal of Development Economics.*
Peters, R. L., & Lovejoy, T. E. (eds.). (1992). *Global Warming and Biological Diversity.* New Haven: Yale Univ. Press.
Plotkin, Mark J. (1988). The Outlook for New Agricultural and Industrial Products From the Tropics. *Biodiversity,* 106–116.
Plucknett, D. L., Smith, N. J. H., Williams, J. T., & Anishetty, N. M. (1987) *Gene Bnaks and the World's Food.* Princeton, N.J.: Princeton University Press.
Potter, C. S. (1990). Policy and Program Implications of Incorporating Biodiversity Conservation as a component of Agriculture Development Projects. A paper prepared for the 1990 North American Regional Conference of the Society for International Development. "Development Strategies for the Nineties: North American Perspectives." October 19–20, 1990, East Lansing, Michigan. *Biodiversity and Agriculture* 1–15.
Prescott-Allen, R. (1986). *National Conservation Strategies and Biological Diversity.* Gland Switzerland: IUCN.
Prescott-Allen, R., & Prescott-Allen, C. (1982), *What's Wildlife Worth? Economic Contributions of Wild Plants and Animals to Developing Countries.* London: International Institute for Environment and Development (Earthscan).
Przeworski, A. (1991). *Democracy and the Market.* Cambridge, UK: Cambridge University Press.
Race, M. S. (1982). Competitive displacement and predation between introduced and native mud snails. *Oecologia,* **54**, 337–347.
Randall, A. (1986). Human preferences, economics, and the preservation of species. In *The Preservation of species,* ed. B. G. Norton, pp. 79–109. Princeton, N.J.: Princeton University Press.
Randall, A. (1987). *Resource Economics.* 2nd ed. New York: Wiley.
Raup, D. M. (1986). Biological extinction in earth history. *Science,* **231**, 1528–1533.
Raven, P. H. (Chairman). (1980). *Research Priorities in Tropical Biology.* Washington, D.C.: National Academy of Sciences.
Raven, P. H. (1987). *We're Killing Our World: The Global Ecosystem in Crisis.* Mac Arthur Foundation Occasional Paper, Chicago, Illinois.
Raven, P. H. (1988). Biological resources and global stability. In *Evolution and Coadaptation in Biotic Communities,* ed. S. Kawano, J. H. Connell & T. Hidaka, pp. 3–27.
Ravera, O. (ed.) (1979). *Biological Aspects of Freshwater Pollution.* New York: Pergamon Press.
Repetto, R. (1987). Economic incentives for sustainable production. *Annals of Regional Science,* **21** (3), 44–59.

Repetto, R. (1988). *The Forest for the Trees? Government Policies and the Misuse of Forest Resources.* Washington, D.C.: World Resources Institute.

Repetto, R., & Gills, M. (eds.). (1988). *Public Policies and the Misuse of Forest Resources.* Cambridge, UK: Cambridge University Press.

Robinson, M. H. (1988). Are There Alternatives to Destruction? In *Biodiversity,* pp. 355–360. Washington, D.C.: National Academy Press.

Robinson, G. R., Holt, R. D., Gaines, M. S., Hamburg, S. P., Johnson, M. L., Fitch, H. S., & Martinko, E. A. (1992). diverse and contracting effects of habitat fragmentation. *Science,* **257**, 524–526.

Rolston, H. III. (1987). *Environmental Ethics: Duties to and Values in the Natural World.* Philadelphia: Temple University Press.

Sachs, I., & Silk, D. (1990). *Food and Energy Strategies for Sustainable Development.* Tokyo: United Nations University Press.

Sagoff, M. (1974). On preserving the natural environment. *Yale Law Journal,* **81**, 205–267.

Sanders, W. T., & Webster, D. (1994). Preindustrial man and environmental degradation. In *Biodiversity and Landscapes,* ed. K. C. Kim & R. D. Weaver, pp. 77–104. New York: Cambridge University Press.

Saunders, D. A., Hobbs, R. J., & Marguler, C. R. (1991). Biological consequences of ecosystem fragmentation: a review. *Conservation Biology,* **5**, 18–32.

Savidge, J. A. (1987). Extinction of an island forest avifauna by an introduced snake. *Ecology,* **68**(3), 660–668.

Sayer, J. A., & Whitmore, T. C. (1991). Tropical moist forests: Destruction and species extinction. *Biology Conservation,* **55**, 199–213.

Schiechtl, H. (1980). *Bioengineering for Land Reclamation and Conservation.* Edmondton, Alberta: University of Alberta Press.

Schneider, S. H. (1990). Global warming: Causes, effects and implications. In *On Global Warming,* ed. J. Cairns, Jr., & P. F. Zweifel, pp. 33–51. Blacksburg, Virginia: Polytechnic Institute and State University.

Scott, J. M. (1990). Preserving Life on Earth: We Need a New Approach. *Journal of forestry,* **88**, 13–14.

Scott, J. M., Csuti, B., Jacobi, J. D., & Estes, J. E. (1987). Species richness: a geographic approach to protecting future biological diversity. *BioScience,* **37**, 782–788.

Sen, A. K. (1981). *Poverty and Famines: an Essay on Entitlement and Deprivation.* Oxford: Clarendon Press.

Silver, C. S., & DeFries, R. I. S. (1990). *One Earth One Future: Our Changing Global Environment.* Washington, D.C.: National Academy Press.

Simons, M. (1993). A Dutch reversal: Letting the sea back in. *New York Times,* Sunday, March 7, 1993, 1–12.

Soulé, M. E. (1988). Mind in the biosphere; mind of the biosphere. In *Biodiversity,* pp. 465–469. Washington, D.C.: National Academy Press.

Soulé, M. E., & Wilcox, B. A. (1980). *Conservation Biology: An Evolutionary–Ecological Approach.* Sunderland, MA: Sinauer Associates.

Soulé, M. E., Bolger, D. T., Alberts, A. C., Sauvajot, R., Wright, J., Sorice, M., & Hill, S. (1988). Reconstructed dynamics of rapid extinctions of chaparral-requiring birds in urban habitat islands. *Conservation Biology,* **2**, 75–92.

South Commission. (1990). *The Challenge to the South.* Oxford: Oxford University Press.

Stern, P. C., Young, O. R., & Druckman, D. (eds.) (1992). *Global Environmental Change: Understanding the Human Dimensions,* Washington, D.C.: National Academy Press.

Taylor, P. (1984). Are Humans Superior to Animals and Plants? *Environmental Ethics,* **6**, 149–160.

Terborgh, J., & Winter, B. (1980). Some causes of extinction. In *Conservation Biology: An Evolutionary–Ecologica Perspective,* ed. M. E. Soulé & B. A. Wilcox pp. 119–133. Sunderland, MA: Sinauer Associates.

Thoreau, H. D. (1960). *Walden*. New York: The New American Library. Originally published 1854.

Tinbergen, J., & Hueting, R. (1991). GNP and market prices: Wrong signals for sustainable economic success that mask environmental destruction. In *Environmentally Sustainable Economic Development: Building on Brundtland,* ed. Goodland et al., pp. 51–57. Paris: UNESCO.

Tiner, R. W., Jr. (1984). *Wetlands of the United States: Current Status and Recent Trends.* Newton Corner, Massachusetts: U.S. Fish and Wildlife Service, Habitat Resources.

Uhl, C., Verissimo, A., Banetto, P., Mattos, M. M., & Tarifa, R. (1994). Lessons from the aging Amazon frontier: opportunities for genuine development. In *Biodiversity and Landscapes,* ed. K. C. Kim & R. D. Weaver, pp. 287–303. New York: Cambridge University Press.

UNDP. (1991). *The Human Development Report.* New York: United Nations Development Program.

UNEP. (1981). *In Defence of the Earth. The Basic Texts on Environment, Founex, Stockholm, Cocoyoc.* Nairobi: United Nations Environment Program.

UNESCO. (1972). Convention Concerning the Protection of the World Cultural and Natural Heritage. Paris.

Vermeij, G. J. (1978). *Biogeography and Adaptation: Patterns of Marine Life.* Cambridge, MA: Harvard University Press.

Vermeij, G. J. (1986). The biology of human-caused extinction. In *The Preservation of Species: The Value of Biological Diversity,* ed. B. G. Norton, pp. 28–29. Princeton, NJ: Princeton Univ. Press.

Virginia Commission on the University in the 21st Century. (1989). The academic organization. In *The Case for Change,* Section VIII. Richmond, Virginia.

Vitousek, P. M., Ehrlich, P. R., Ehrlich, A. H., & Matson, P. A. (1986). Human Appropriation of the products of photosynthesis. *BioScience,* **36**, 368–373.

Warford, J. (1987b). Nature resource management and economic development. In *Conservation with Equity: Strategies for Sustainable Development,* ed. P. Jacobs & D. Munro, pp. 71–85.

WCED (World Commission on Environment and Development). (1987). *Our Common Future* (The Brundtland Report). Oxford: Oxford University Press.

Weaver, R. D. (1990). *The Economics of Plant Germplasm.* Washington, D.C.: National Research Council, National Academy of Sciences.

Weaver, R. D. (1994a). Economic valuation of biodiversity. In *Biodiversity and Landscapes,* ed. K. C. Kim & R. D. Weaver, pp. 255–269. New York: Cambridge University Press.

Weaver, R. D. (1994b). Market based economic development and biodiversity: an assessment of conflict. In *Biodiversity and Landscapes,* ed. K. C. Kim & R. D. Weaver, pp. 307–325. New York: Cambridge University Press.

Weiner, J. (1990). *The Next One Hundred Years: Shaping the Fate of Our Living Earth.* New York: Bantam Books.

Weiskel, T. C. (1989). The ecological lessons of the past: An anthropology of environmental decline. *The Ecologist,* **19**(3), 98–103.

Western, D. (1989). Population, resources, and environment in the twenty-first cen-

tury. In *Conservation for the Twenty-first Century,* ed. Western, D. and M. C. Pearl, pp. 11–25. New York: Oxford University Press.
White, J. (1967). The historical roots of our ecologic crisis. *Science,* **155**, 1203–1207.
White, L. (1973). The historical roots of our ecologic crisis, in *Western Man and Environmental Ethics,* ed. Ian Barbour, pp. 18–30. Reading, Mass.: Addison-Wesley Publishing Company.
Whitmore, T. C., Peralta, R. & Brown, K. (1985). Total species count in a Costa Rican tropical rain forest. *Journal Tropical Ecology,* **1**, 375–378.
Wilkes, G. (1977). The World's crop plant germplasm: an endangered resource. *Bull Atom Scientist,* **33**, 8–16.
Wilkes, G. (1994). Germplasm conservation and agriculture. In *Biodiversity and Landscapes,* ed. K. C. Kim & R. D. Weaver. New York: Cambridge University Press.
Wilson, E. O. (1980). Resolution for the eighties. *Harvard Magazine,* **70**, 20–25.
Wilson, E. O. (1984). *Biophilia.* Cambridge, Mass.: Harvard Univ. Press.
Wilson, E. O. (1985a). The biological diversity crisis. *Bioscience,* **35**(11), 700–706.
Wilson, E. O. (1985b). The biological diversity crisis: A challenge to science. *Issues in Science and Technology,* **11**(1), 22–29.
Wilson, E. O., & Peter, F. M. (eds.). (1988). *Biodiversity.* Washington, D.C.: National Academy Press.
Wilson, E. O. (1988c). the current state of biological diversity. In *biodiversity,* ed. E. O. Wilson and Francis M. Peter. pp. 3–18 Washington, D.C.: National Academy Press.
Wilson, E. O. (1988b). The diversity of life. In *Earth '88: changing Geographic Perspectives,* ed. H. J. De Blij, pp. 68–81. Washinton, D.C.: National Geographic Society.
Wilson, E. O. (ed.). (1988). *National Forum on Biodiversity.* Washington, D.C.: National Academy Press.
World Bank. (1984). *Toward Sustained Development in Sub-saharan Africa: A Joint Program of Action.* Washington, D.C.: World Bank.
World Bank. (1988). *The Challenge of Hunger in Africa: A Call to Action.* Washington, D.C.: World Bank.
World Bank. (1989). *Sub-saharan Africa: From Crisis to Sustainable Development.* Washington, D.C.: World Bank.
World Commission on Environment and Development. (1987). *Our Common Future.* Oxford, U.K.: Oxford University Press.
World Commission on Environment and Development (WCED). (1987). *Our Common Future.* Oxford, U.K.: Oxford University Press.

Index